YOUR STONE AGE BRAIN IN THE SCREEN AGE

ALSO BY RICHARD E. CYTOWIC

Synesthesia—MIT Press Essential Knowledge Series
Wednesday Is Indigo Blue (with David Eagleman)—Winner of the Montaigne Medal
The Man Who Tasted Shapes
Synesthesia: A Union of the Senses (2nd edition)
The Neurological Side of Neuropsychology
Nerve Block for Common Pain

Your Stone Age Brain in the Screen Age
Coping with Digital Distraction and Sensory Overload

Richard E. Cytowic, MD, MFA

The MIT Press
Cambridge, Massachusetts
London, England

© 2024 Richard E. Cytowic

All rights reserved. No part of this book may be used to train artificial intelligence systems or reproduced in any form by any electronic or mechanical means (including photocopying, recording, or information storage and retrieval) without permission in writing from the publisher.

The MIT Press would like to thank the anonymous peer reviewers who provided comments on drafts of this book. The generous work of academic experts is essential for establishing the authority and quality of our publications. We acknowledge with gratitude the contributions of these otherwise uncredited readers.

This book was set in ITC Stone Serif Std and Trajan Pro by New Best-set Typesetters Ltd. Printed and bound in the United States of America.

Library of Congress Cataloging-in-Publication Data

Names: Cytowic, Richard E., author.
Title: Your stone age brain in the screen age : coping with digital distraction and sensory overload / Richard E. Cytowic, M.D., MFA.
Description: Cambridge, Massachusetts : The MIT Press, [2024] | Includes bibliographical references and index.
Identifiers: LCCN 2023054556 (print) | LCCN 2023054557 (ebook) | ISBN 9780262049009 (hardcover) | ISBN 9780262379113 (epub) | ISBN 9780262379106 (pdf)
Subjects: LCSH: Social media addiction. | Digital electronics—Health aspects. | Brain. | Attention
Classification: LCC RC569.5.I54 C98 2024 (print) | LCC RC569.5.I54 (ebook) | DDC 616.85/84—dc23/eng/20240327
LC record available at https://lccn.loc.gov/2023054556
LC ebook record available at https://lccn.loc.gov/2023054557

10 9 8 7 6 5 4 3 2 1

For Stephen P. Gorman
In memoriam

And to the Virginia Center for the Creative Arts, where, as a ten-time Fellow over the past twenty years, I have worked amid the Appalachian foothills and been blessed with equal shares of camaraderie and silence.

Contents

Preface: Your Stone Age Brain, 3 BCE versus Today   ix

1   Engineered Addiction: Brain Drain and "Virtual" Autism   1
2   Selfies Kill More People Than Sharks   17
3   The Brain Energy Cost of Screen Distractions   27
4   The Brain Energy Cost of Multitasking   39
5   The Digital Difference: We Treat It Socially   51
6   Silence Is an Essential Nutrient   65
7   Your Brain Is a Hackable Change Detector   81
8   What Gets Caught in the Corner of Your Eye   95
9   Missing Critical Time Windows Degrades Empathy   105
10 How Blue Screen Light Wrecks Normal Sleep   123
11 Hooked in the Pursuit of Happiness   139
12 Pandora's Box: How Ambivalence Keeps Us Hooked   153
13 iPads in the Nursery or Not?   167
14 Human Contact Traded for a Googlized Mind   181
15 The Consequences of Forced Viewing   193
16 Does Heavy Viewing Induce Autism-Like Symptoms?   211
17 Social Learning: Kindergarten, Handwriting, and Dexterity   221
18 War Games: Is the Only Winning Move Not to Play?   235
19 Coda: Lessons from the Lockdown Years   245

Acknowledgments   257
Appendix: Keeping a Dream Diary   259
Notes   265
Index   321

Preface: Your Stone Age Brain, 3 BCE versus Today

Ages ago, on the Tanzanian coast, sunrise: Aurora creeps over the African hills. Life begins to stir as the sky glows pink, then slowly brightens to blue. Nocturnal animals hunker down and daytime creatures awaken, including the prehuman precursor of *Homo sapiens.*

Kushim is the clan leader, the first to venture beyond the cave while the others hold back. The air smells sweet. Birds flutter overhead and warble in the trees. Kushim tilts his head to check the thrum of insects pulsating in the background. He hears the normal morning soundscape that he has come to anticipate. Anything unexpected—a nearby growl, an acrid smell, a haze of smoke in the wind—will instantly put him on alert without his having to think about it, and will spur him to signal danger to the rest of the clan.

There appears to be no danger this morning. So Kushim grunts the understood signal for kinfolk to come out into the clean morning air. His small band numbers about three dozen. The coming daylight hours hold much for them to do, all except the infants strapped to their mothers' backs. Mothers need their hands free to hunt and forage and augment the work the men will do. As the sun warms their faces and drives off the damp chill of the cave, five of the women head down to the water's edge, where they will gather clams and oysters. If lucky, they may catch a few shrimps in the tidal pools. Each day it takes enormous cooperation to feed everyone, and harmony is essential if the group hopes to survive. From dawn to dusk life is active, physically demanding, and requires the group to expend many calories that everyone must help replenish.

Three young girls tag along behind the older women. Leaders Irg and Uma wade with them into a tidal pool, one not too deep, so that they can turn over rocks and see what creatures lurk beneath. Animals tend to emerge from their hiding places at slack tide. The girls must be quick if they hope to catch any. But first the women show them how to collect the plentiful sea urchins trapped along the bottom while avoiding a painful lesson from getting pricked or stepping clumsily on their sharp and sometimes poisonous spines.

Rocks close to the beach harbor abundant seaweeds and crustaceans. Periwinkles spend so much time out of the water that they are easiest to find. Blue mussels, snails, and sea lettuce are likewise easy to pick and deposit into the communal basket. If the girls are fortunate they will find a sea cucumber exposed. Rockweed, with its small air bladders at the tips of its fronds, is easy to spot, too. The salty seaweed can be chewed raw or used to wrap food into packets for roasting on hot stones around the fire, a newly discovered invention. Suddenly, something catches the corner of Irg's eye. "Ayiee!" she shouts to get the attention of a young one who has wandered off and is reaching for an outcropping that juts above the water line. Be careful of the razor-sharp barnacles: they can slice open your fingers, she mimes. An infection can be fatal, she warns, drawing an index finger across her throat.

Irg's orienting reflex has kicked in, one of the automatic circuits that, without conscious thought, effortlessly focus, shift, and sustain attention. Given the dangers lurking everywhere, an orienting reflex is essential to Stone Age survival. Irg had expected the youngster to be at her side still. Her peripheral vision, strongly connected to her emotional brain, swiftly registered the discrepancy, which goaded her to action. Often the orienting reflex is accompanied by a hormonal surge of adrenaline and norepinephrine that fuels the fight-or-flight response. Yet it is actually two pathways that throw the pre–*Homo sapiens* on alert. The first, quick pathway allows no time for deliberation, so that they jump at the stick they thought was a snake. Yet it is better to mistakenly jump than stand around deciding and succumb to a venomous bite. The second, slower pathway can override the quick one and allow time to consider a response (should I go right or left around that wildebeest I'm stalking?).

With infinite patience, Irg draws the girls to the center of the tide pool and shows them how striped shore crabs live under almost every rock. Up to thirty green crabs may cluster there, too, but within seconds they sprint to a new hiding place away from grasping hands. For her part, Uma points out telltale bubbles that drift up from the sandy bottom. These mark the places where scallops have burrowed in during low tide. The women will come back another day to dig for them with shell scoops they have reserved just for that purpose.

The girls eye their elders carefully. It is common for group members to watch what others do, not out of suspicion but because imitation learning strengthens social cohesion. The whole community has a hand in child-rearing, reinforcing good behavior and dissuading the bad each time they mete out praise or disapproval. For every member, winning approval and the reassurance that one belongs strongly motivates behavior. Ostracism from the band would mean certain death.

Just then, another shriek. A sea slug has been spotted. This mollusk, which has no shell, is large, the size of two fists, and swims lazily in the quiet water. The creature hasn't any defenses and stands little chance of getting away. Into the basket it goes, getting the morning catch off to a good start and promising a tasty midday meal. Now if only they are lucky enough to find a baby lobster or octopus hiding under one of the tide pool rocks!

* * *

The menfolk have already departed before daybreak to hunt. Daybreak is when animals are most likely on the move, and early humans have now advanced up the food chain from scavenger to apex predator. Hunting provides ten times the energy return of a diet composed of fruits and plants. Carnivores throughout the Animal Kingdom have high stomach acidity that protects them from pathogens in rotting meat. Human stomach acid is higher even than that of such scavengers as vultures, hyenas, and coyotes, an adaptation that lets them consume large animals over a period of days or even weeks.

Kushim and kin are skilled at hunting in groups using sophisticated, close-range techniques with which they target carefully selected gazelles, wildebeests, and pronghorn antelopes. Mature antelopes are a favorite prey. Hunters sit in trees waiting to ambush a herd passing below, then spear them point-blank. After a successful kill they gut the animal with stone knives, tie it to a pole using vines, then haul it back to camp. The rest of the morning they spend butchering the kill while keeping a watchful eye out for opportunistic predators. The orienting reflex assures their vigilance because it renders the brain exquisitely sensitive to the slightest change in conditions. The nervous system overall has evolved into one massive change detector because every novelty seizes its attention. The orienting reflex guarantees that sentries will turn toward and instantly judge whatever stimulus set it off, making them either freeze, flee, or attack. For now, things in camp are happily quiet.

Back from their excursion to the tide pools, the women stoke the fire and heat up the perimeter stones. They have become skilled at controlling fire not just for warmth but also to prepare food and fashion tools. Everyone lends a hand in replenishing the spears, knives, and stone tools on which the entire group depends. Unlike with chimpanzees, which live in aggressive, male-dominated societies with clear hierarchies, dominance does not shape social relations in early human collectives.[1] Each band, wherever it may have scattered, maximizes its survival because all members cooperate regardless of age or gender. Both genders are adept at creating symbolic art, too. Women dye and decorate marine shells while men build geometric sculptures from broken stalactites they haul from underground caves. Fostering harmony lets the collective function as a superorganism.

Besides time to make art, there is also time for games and play using objects at hand. Here is where the drive to compete becomes channeled in socially acceptable ways. Good-natured competition is rooted in emotion, and a fundamental basis of all emotion is comparison. Does someone have more than I do, or something desirable that I lack? This is the feeling that lies behind winning. Likewise, the fear of missing out or of being left behind rouses strong passions. And despite every effort of the group to maintain harmony, issues of hierarchy still threaten to arise. The group has conferred status on Kushim, Irg, and Uma because it acknowledges them as successful leaders. But no matter how egalitarian its intentions, wanting and pleasure remain forces rooted deep in the early human brain that can otherwise compel certain behaviors.

Unbeknownst to Irg and Kushim, many of their actions, such as snatching the young girl away from sharp barnacles or Kushim orchestrating other members in the hunt, are guided by molecules such as dopamine, a neurotransmitter that first evolved in much lower animals. In humans its functions are many because dopamine is the molecule that has a passport to every nook and cranny in the brain. Foremost in its dominion are reward and wanting, two basic instincts that lie behind survival and reproductive advantage. The impulse that pushes us to want is largely unconscious and nearly impossible to satiate, which is why as soon as we get something we've wanted, we typically want something else. A related class of neurotransmitters comprises the endorphins, the brain's natural opioids, like those released during the runner's high or while chasing a gazelle across the savannah. Compared to dopamine, the opioids' range is smaller, they are harder to activate, and the satisfaction they provide is shorter-lived.

Stone Age humans are highly sensitive not only to wanting and reward but also to reinforcement, nature's way of perpetuating desirable behaviors. Food, water, sex, and shelter are primary reinforcers because they satisfy strong biological desires. As the *Homo* species evolves, their secondary reinforcers have become more diverse and sophisticated, as in the case of child rearing or teaching the younger members to gather, hunt, maintain the fire, and fashion useful tools. Reinforcement is not a tangible thing but a relationship between a behavior and whatever propagates it. Reinforcement leads to habits good and bad, which then become established as part of larger cultural traditions.

The sun now stands directly overhead, marking noon. Time to rest and eat what the gatherers and hunters have secured, the group's first intake of calories since awakening in the cave. Settled routines like this make the contours of human life predictable, and familiar repetition instills a sense of calm. Yet paradoxically, the experience of daily life is one of continual change, sometimes slow, at other times sudden. Many established routines are dictated by the Sun. Special light receptors in the retina that adjust the brain clock are especially sensitive to short wavelengths. Short blue wavelengths best penetrate the ocean, where all life began and where the photosensor first evolved before making its way to land creatures.

Later on, as daylight fades, it will yet again be time to wind down and sleep. In the morning, when dawn once again approaches, one set of hormones will signal it is time to wake up. For now another set, one that includes melatonin, start to surge in the bloodstream to signal it is time to bed down for the night. During the first part of the night slow brain waves predominate and the most restorative phase of sleep takes place. These are the golden hours, when everything the clan, especially its youngsters, has learned during the day is consolidated and transferred from short-term to long-term memory. This "first sleep" lasts until about midnight. when members awaken to engage in activities like sex, games, storytelling, or stoking the fire. Then body temperature falls, triggering a "second sleep" that lasts until dawn. The pattern repeats itself naturally, season after season, without cease.

* * *

Three million years later: These basic survival mechanisms are now a door into modern brains that tech companies exploit for profit and competitive advantage. They hack our biology and hook us on their products because the brain hasn't changed since the Stone Age, let alone during the mere thirty-three years that the internet has been around. The story of why we are helplessly distractible began long ago, which is why the screen age feels like it has been with us forever. It certainly feels as though the digital devices that surround us day and night are part of an embedded iSelf, so much a part of us as to essentially constitute a planetwide hive mind like that of the Borg in *Star Trek*.

This is the predicament of the Stone Age brain in the screen age. The following pages explain Why you are so addicted to your screen devices, What you can do to push back against these forces, and How to go about it.

* * *

Your brain has been thoroughly conditioned by digital devices. Links, "likes," and "follow" buttons are so easy to click on that doing so has become a mindless reflex.

It is not your fault: companies employ reams of psychologists and behavioral scientists whose job it is to exploit the Stone Age brain's built-in vulnerabilities, especially its inability to ignore novelty and any change in prevailing conditions. Before the pandemic, one-third of the world's population spent the better part of a day fixated on a TV screen, a computer screen, or a phone or tablet screen—sometimes all three at once. Those numbers have since burgeoned. During the two years of pandemic-related lockdowns our closest relationships revolved around digital devices. We TikToked, Zoomed, swiped, and FaceTimed. On March 29, 2020, Tinder set a record of three billion swipes in a day. It then broke that record 130 more times by the end of the year, or once every two days.[2]

What can one say in defense of this riotous growth in digital engagement? Users do not appreciate how heavy an energy cost screen distractions exact on their Stone Age brain, which biology limits by the fixed amount of energy it has available. The cognitive load imposed by screen devices degrades attention, memory, and thinking, along with sleep, mood, and concentration. The screens we habitually gaze into compete with and substitute themselves for our otherwise natural drive to socialize. They confront us with highly unnatural gambits for attention that we are nearly helpless to resist.

To prove that the urge to connect is not merely a social custom, fMRI measurements that simultaneously record from two individuals lying in separate but linked scanners illustrate a basic biological impulse to connect with one another. Brains in close proximity literally synchronize, just as the ending of *The Matrix Resurrections* film illustrates. It is this kind of fundamental coupling that smart screens interfere with.[3]

The coronavirus pandemic showed how hard it is to wrest digital devices away from kids of all ages. What started as a servant to those working from home quickly became the master. People increasingly began to realize that social media had given birth to a force their creators neither understood nor could control. While smart devices do bestow benefits, they are still narcotizing agents. So easily do they hack our neurological defenses that the forces behind them don't even need to hide their agenda.

From the brain's perspective, I ask: How much energy does the deluge of texts, alerts, and push notifications exact from the limited stock we have? And why did nature give us a limited stock to begin with? What consequences follow from the nonchalance with which we shove screens in front of a child's developing central vision and willfully ignore how iPads mounted in bassinets, car seats, and potty trainers displace that child's natural inclination to socialize?

I use the term "Stone Age brain" because we have the exact same biological organ as our long ago ancestors. Brain circuits operate at dramatically lower speeds compared to their electronic counterparts. No amount of diet, exercise, Sudoku puzzles, meditation, or yoga can increase what we have to work with. We can only affect how we manage it, and how we do that determines its efficiency. We face the same challenge as our distant ancestors of how to marshal and apportion the energy needed for thinking, acting, feeling, imagining, anticipating, and most of all paying attention to what's going on around us.

The brain accounts for a mere 2 percent of the body's mass but consumes 20 percent of the daily calories we ingest. The adolescent brain consumes 50 percent and the infant brain 60 percent, which is why the young are disproportionately affected by heavy screen exposure. Adjusted for body mass and fat content, babies between the ages of nine and fifteen months expend 50 percent more energy in a day than adults do. This likely fuels their growing brain and immune system. In older individuals too, mental exertion has an energy cost. In one experiment a student burned 40 percent more energy during a math test and 30 percent more during a tough interview. It is hard to think of any other process that exacts energy consumption by anywhere near 40 percent.[4] At any age, the costliest things we can do in terms of energy expenditure are shift, focus, and sustain attention—a cycle that digital devices force us to endlessly repeat as if circling a drain.

The brain achieves its remarkable feats using only a few watts of energy—the equivalent of a dim light bulb. Most of it goes toward keeping up the physical structure by pumping sodium and potassium ions across membranes to maintain its electrical charge. Little remains left for thought, feeling, or action. Precisely because the brain is so efficient, its reserve margins are slim and eaten up by the demands made by constantly shifting the focus of attention. Think in terms of a budget whose currency is all the molecules that sustain our 86 billion neurons. As with financial budgets, we can run a deficit and go into the red. The brain must then terminate metabolic processes that are too expensive, resulting in mental fatigue, reduced focus, patchy memory, and errors.[5] Screens act like secondhand smoke, affecting anyone in the line of sight. Even the mere presence of a phone drains us because trying not to look at it sucks up energy, too.

One specter in today's screen-heavy environment is "virtual autism," the induction of autism-like behaviors in otherwise healthy individuals especially the young who spend many of their waking hours online. (I must stress this distinction between developmental autism and similar-looking autistic behavior caused by something else, and I discuss it at length later on.) Only recently have peer-reviewed studies begun to question the causal connection between the two as they tease out the similarities between developmental autism (now thirty times more common than it was in 1960) and the newly evident virtual kind. In both, social media compete with in-person engagement and interfere with the development of emotional circuits necessary to read other people. In both varieties of autism, affected individuals studiously avoid eye contact and bungle social interactions because they fail to grasp the meaning behind body language.

What matters more than having 86 billion brain cells is the ceaseless interweaving and rewiring of connections among them, a lifelong process known as plasticity, a word derived from the Greek *plastikos* (πλαστικός), meaning "capable of being molded." The brain absorbs experience from birth onward, plastically molding the organ's structure and function in response to its experience of the world. Our twenty-first-century mindset is itself trapped in the immediate present, in need of constant stimulation, and giving it too much early screen exposure looks to have terrible unforeseen effects. Ubiquitous screens promote sensation at the expense of thought because amped-up sensory pathways compete with the maturation of other pathways normally destined to support social relationships and emotional intelligence. Developmental autism never improves spontaneously during early childhood, but children with virtual autism do show dramatic improvements once digital screens are taken away.

Continual touchscreen use almost certainly reshapes the brain. Clever research records that subjects make up to 40,000 finger swipes a day, even while supposedly asleep(!). Merely swiping a screen rewrites the hand's representation in the brain's sensory cortex. The latter adapts and shrinks to become more efficient, meaning that our devices fiendishly habituate us to sensory overload. When more than 50 percent of first graders have smartphones, why are we not discussing these scenarios when an innocuous-looking swipe that physically reshapes the brain may also be permanently changing that individual's psychology and temperament?

We live in a paradox in which digital tech alleviates some social isolation even as it worsens it in other ways. We are technologically more connected than ever before yet bond on platforms engineered to make outrage and indignation infectious. They exploit the psychological principle that emotion, like yawning, is highly contagious. Raw emotion allows for neither nuance nor complexity, but it easily overwhelms critical thinking and leaves us swayed by propaganda and manipulation that is hard to recognize for what it is.

Two fashionable phrases in education circles are "critical thinking" and "connecting the dots." But surfing the internet encourages shallow gulps of the data stream, not critical thinking, while offloading factoids to external apps like Google leaves users with little common knowledge and thus few dots to connect. A mind capable of ascertaining connections thrives on quiet, not on endless texts and notifications. There is a reason the natural world is easy on the eyes and ears, yet we cut ourselves off from its restorative power, forgetting that the brain, the psyche, and the soul need rest and uninterrupted interludes rather than streaks, autoplay, and push notifications that turn friendships and achievements into ruthless competitions.

The Latin word *addictum* once described the length of time an indentured slave, or addict, had to serve their master. The word's root means "bound to." And are we not bound like slaves to the screens in front of us? If not, then why do so many people claim to be addicted?

1

ENGINEERED ADDICTION: BRAIN DRAIN AND "VIRTUAL" AUTISM

In one of my columns for *Psychology Today* I discussed different ways smart screens negatively affect the brain. To my surprise, anxious parents and grandparents from around the world wrote in to share stories of the detrimental effects that smartphones, tablets, and televisions seemed to be having on their families. A father from as far away as Iran wrote:

> We have a boy 13 month old. Unfortunately we letted him to watch TV and mobile app. Compared to other baby, he was different. He does not respond to his name. When I come home from work, he didn't pay attention to me especially while watching TV and apps. He had little eye contact and didn't hug me and his mother too.[1]

Concerned readers wrote that kids who once enjoyed sports, socializing with friends, and spending time outdoors were now glued to their devices for hours on end. As time spent on screens expanded, their ability to communicate seemed to wane. Once bright, loquacious youngsters now answered questions in grunts if they answered at all. They wouldn't look up, and acted irritated. Frustrated, even scared, by shrinking attention spans and the tantrums that ensued if they tried to take a device away, parents began to wonder whether the tools they once thought of as educational were instead turning their kids into zombified addicts.

I decided to look into it. A plethora of books talk about screen dependency, but few, if any, explore the topic from the brain's point of view—particularly the energy drain that screens enforce in the face of the fixed limits biology has placed on it. Why do I choose this perspective? Because the energy cost of any given process, whether biological or mechanical, is a fundamental aspect of engineering and thermodynamics. That's why biomedical engineering is called engineering and not simply mechanics.

It is why, out of so many effects that screens have on the human brain, energy consumption must be a paramount consideration. The existing literature doesn't address the question because hardly any of the authors have backgrounds in engineering or biochemistry, disciplines in which questioning energy costs are routine. The absence doesn't mean that the issue isn't worth discussing but rather that it is an unappreciated omission.

Almost everyone agrees that attention spans have collectively gone to hell. Worried parents are just one group that questions the influence of today's technology on the ability to reason clearly. Does the internet weaken memory by relieving us of the need to learn phone numbers, multiplication tables, and detailed facts, or does it enhance intelligence by placing millions of factoids at our fingertips? Do social apps bring us closer together or do they isolate us and turn what used to be two-way conversations into public performances ripe for outrage and moral grandstanding? Are screen-based devices addictive, and do they really induce autism-like symptoms, especially in young users? I will come back to this question.

In *Reader, Come Home,* literary scholar Maryanne Wolf blames screen addiction for the loss of deep reading ability.[2] Senator Ben Sasse, who holds a doctorate in history from Yale, says flat out that we are "addicted to distraction," while popular neuroscientist Daniel Levitin says that "multitasking creates a dopamine-addiction feedback loop, effectively rewarding the brain for losing focus and constantly searching for external stimulation."[3] We live a paradox in which digital tech alleviates social isolation in some ways even as it worsens it in others.

* * *

In addition to writing "The Fallible Mind" column for *Psychology Today*, I am a neurologist, a teaching professor at George Washington University School of Medicine and Health Sciences, and an author of neuropsychology textbooks and popular works. Until now I have largely been known as the person who restored synesthesia to mainstream science. Sharing a root with anesthesia, meaning "no sensation," *synesthesia* means "joined sensation." Four percent of the world's population are born with two or more senses hooked together so that otherwise normal individuals not only hear music or someone's voice, for example, but simultaneously see it, taste it, or feel it as a physical touch. Perceiving the days of the week, the alphabet, or numerals as colored is a common manifestation, as is tasting words or seeing calendar configurations hovering around the body in space. These extra perceptions aid recall, endowing synesthetic individuals with measurably superior, sometimes photographic memories (technically called "eidetic"). When I explored my first synesthetic subject in 1979, my neurology colleagues immediately asked where the lesion was on his CT scan.

"No, you don't get it," I told them. "He doesn't have a hole in his head, a missing piece. He has something extra."

They looked at me like I was crazy and warned me to drop the matter as "too weird, too New Age." If I persisted in pursuing it, they warned, "it will ruin your career."

I spent the next fifteen years countering naysayers who insisted that synesthesia was bogus and couldn't possibly be rooted in the brain. It is the nature of orthodoxy no matter what the profession to dismiss or explain away what it cannot or does not wish to understand. Time has proven synesthesia to be a perceptual trait, like having perfect pitch; you either have it or you don't, and you cannot learn it through practice. My persistence in trying to understand this offbeat but fascinating human trait eventually brought about a paradigm shift in the way science conceives the brain's configuration and how perception works.[4] For years now, young scientists have happily been writing papers, books, and PhD theses about this once forgotten trait.

In other words, it didn't ruin my career.

Science is full of exceptions to established thinking. Take leeches, the application of which was once considered standard medical practice but is now mocked.[5] Who is to say that today's standards might not be dismissed as tomorrow's leeches? We have embraced smartphones and tablets as near magical tools that promise unparalleled productivity, connectedness, and opportunities for learning. Digital devices are indeed fantastic—Uber coming to your door in a minute is magic; GPS is magic; the ability to reach across time and span enormous distances is magic.

But for all its wondrous utility, the smart screen is also a narcotizing agent. What if it has unintended, possibly harmful, side effects? Since 2016 the American Academy of Pediatrics (AAP) has recommended no screen time other than video chatting for children two years and under, and only one hour of exposure in the presence of a parent for children two to five years old, with the caveat that "less is better." The priority for young children should be face-to-face communication, physical activity, and sleep because their brains are changing in the most complicated ways. Yet given how rapidly screen intrusion is encroaching on otherwise normal development, the AAP is updating its guidelines, while the guidelines of the World Health Organization (WHO) stress no or limited screen time for children under five. Additionally, the NIH has launched a $300 million investigation into the cognitive development of screen-saturated youngsters from adolescence to young adult.[6] Pediatrician David Hill, who oversaw the AAP guideline revisions, says that the WHO is simply "applying the precautionary principle. If we don't know that screens are good and there is reason to believe it's bad, then why do it?"

The growing pool of readers who took the time to write forced me to question popular assumptions about screen media and its influences. I set aside synesthesia to explore whether screen media actually do have a detrimental effect on the brain, and if so, how. Two things immediately stood out. Screens of any sort act like secondhand smoke, affecting both the user and anyone within range, and screen exposure relatively easily induces "virtual autism," the emergence of autistic like behaviors in otherwise healthy individuals.[7] The father quoted above spoke to this and the apparent resolution of symptoms that followed screen removal:

> A psychologist told us we have to turn off TV and other screen and play with him. We now see the result. When I come home, he come to me and ask me to pick him up from ground. He hug wife and me. When we call his name he turn his face better than before, but still . . . we are worried for his future. Would you help us and advise us please?

An American woman contacted me about her thirty-month-old grandson, Parker, who wouldn't look anyone in the eye or respond to his name. He was under evaluation for autism spectrum disorder at the time when she wrote:

> He was exposed to almost constant children's "learning" programs on TV, tech toys that teach, and computer and phone games for most of his waking hours. We enrolled him in a day care two days a week for socialization. At the same time, I read about virtual autism, and we removed all electronic toys, phone play, and all children's TV from his environment.
>
> I believe the screen time and electronics really hurt his development. Now we play kitchen, coloring, puzzles, blocks, and other imagination and occupational games. Within 2 weeks, he was responding

FIGURE 1.1
Parker, age thirty months. *Left,* Before the ban on screens, he has a vacant stare and is unresponsive to his name being called. *Right,* Eleven weeks after screens were banned he smiles, looks at the camera, and "looks at the person talking to him," according to his grandmother. "He discovered his baby sister, too. He now looks at her and plays with her." Courtesy of Claire Thies.

most of the time to his name and looking at the person talking to him. I thought you'd be interested in our progress since removing "screens." It's 11 weeks and progressing every day (figure 1.1).

Finally, the head social worker for Chicago Public Schools wrote:

I work with children on a daily basis who are spending the vast majority of their waking hours exposed to media/screens. Parents are concerned about sleep disruptions, difficulty with emotional regulation and tantrums when access to technology is denied or removed. I read your article . . . and am working to inform parents on the connection between deficits in social functioning and the amount of time spent on screens.[8]

I'd like to say a few words about the critics, mostly academics, who think that "an unnecessarily negative view of screens" risks instilling "misplaced worries about digital technology."[9] They accuse anyone who advises caution of being one-sided and selective in the evidence they pick to support claims about the potential dangers of excessive screen exposure. Cautionary voices are guilty of scaremongering, they say, while categorically denying that virtual autism has anything to do with screens. Critics object that the rationale for restricting use isn't particularly grounded in solid evidence.

I heartily disagree. I am a clinician, meaning that by training I take a view centered on the care of the individual patient compared to the tut-tut of academics who mutter "there is no convincing evidence" while safely distancing themselves from having to deal with zombified kids and young adults. They don't have to deal with distraught parents or battle a two-year-old who shrieks and fights when you try to take away their iPad. The academics tell those parents, "there is no convincing evidence." To parents, however, the evidence is in front of their eyes. Yet naysayers continue to insist that the alarming behavior observed is due to something else, anything else, except the screens that young people refuse to relinquish.

How is it scaremongering to acknowledge parental alarm? It is parents and educators and addicted individuals themselves who sought my attention. None of them said "Give me more." Parents didn't write about how pleased they were that their offspring were spending the majority of their waking hours in front of a screen. On the contrary: I have heard many kids, even medical students, complain "I'm addicted to my phone." No one says "I need to use my phone, and my tablet, and my laptop a lot more than I do now." The benefits of digital technology are obvious; you don't need academic studies to convince anyone. But the downsides are also obvious if you are just willing to look.

There is some truth to the observation that once critical and in-depth studies of any new technology have been done, a revised point of view comes into shape that is different from the initial one. The iPhone has been with us since 2007, the iPad since 2010, but complaints of digital distractions are not diminishing. They are getting worse, as reflected in the feedback from users themselves, particularly students.[10] Yet naysaying tech enthusiasts still want everyone to wait for "more evidence." A call for more research sounds well meant, but it is calculated to silence dissent from the Panglossian view that screen technology is nothing to worry about. It dismisses the precautionary principle out of hand and misrepresents the concept of evidence-based medicine. Those who insist on "more research" want science to be perfect. But science has limitations in trying to deal with the complexities of the real world and so can never arrive at "definitive proof" for multifactorial behavior that spans multiple disciplines. If we wait for the evidence that critics insist on, we will be waiting forever.

* * *

From my neurological point of view, the online world is one of hyperstimulation, not a paradise of risk-free beneficence. Ray Bradbury's totalitarian classic, *Fahrenheit 451,* depicts a world destroyed by overstimulation. Its protagonist, Guy Montag, a "fireman" whose job it is to burn forbidden books, lives with his wife in a digitally stimulated hedonistic bubble that constitutes the totality of their home life. COVID-19 lockdowns felt a bit like this as Zoom fatigue set in and the shift to life-on-screen drove people stir-crazy. Overwhelmed parents struggled to work from home. School-less, camp-less, and vacation-less kids, already obsessed with their screens, now had unlimited time with them. They gorged on YouTube and TikTok, surfing late into the night and sleeping away the day.

Paradoxically, in the midst of this bustle, adults had an opportunity to feel what they had been missing, namely, a less hurried life and time to think. COVID-19 forced us to slow down. It gave us still periods in which to reflect, to think about who we are, what we want, and what matters. Alan Lightman's *In Praise of Wasting Time* champions this kind of unstructured time for its "replenishment of mind that comes from doing nothing in particular." The mind needs stretches of calm. Creativity thrives in it. In Italy, quarantined citizens sang from their balconies each evening, their serenade reaching across empty streets to lift their neighbor's spirits. Opera, Broadway, regional orchestras, and pop musicians ingeniously stitched together coordinated online performances from far-flung participants.

Illustrating the downside were adults accustomed to noisy streets, constant demands for their attention, and a hectic lifestyle without a minute to be wasted. The unstoppable tempo of being busy is a stimulant, like nonstop caffeine and equally pernicious, gnawing away just when you have a moment to yourself. Suddenly alone with their thoughts, these folks didn't like it. People who couldn't sit down in a restaurant without whipping out their phone or alight quietly anywhere for ten minutes suddenly had to confront silence and tolerate stillness—a new experience.

As a neurologist, I approach screen distractions from the perspective of energy consumption. It forces me to account for biological costs in terms of energy expenditure when engaging in a mental task such as committing something to memory and retrieving it later on. Other costly tasks include the executive weighing of options by the frontal lobes and arriving at a decision; switching back and forth from one frame of mind to another, as we do in multitasking; or juggling interruptions and competing demands for our attention. These led me to see that excessive use of digital screens is a bad deal when appraised from a basic energy perspective.

Let me be clear: I am neither a tech cheerleader nor a Luddite. Digital devices unquestionably make life easier. By serving as memory repositories, they put calendars, phone numbers, and contact information at my fingertips. Synchronization among devices is a godsend if I need something at home from my workplace or vice versa. Likewise, time zones are no longer a problem when communicating with far-flung colleagues and friends. An encyclopedia of facts sits at my fingertips (assuming I know what facts I even want to know). Amateur musicians turn to YouTube for lessons or lyrics; cooks, golfers, and weightlifters turn to the web for tips. GPS keeps me from getting lost, and map apps make it easy to explore and find my way, as well as discover eateries, gas stations, and local attractions. Screens have expanded written language even though that language may be immature, provocative, or just plain stupid. At the same time, the omnipresence of screens has left many exhausted, depressed, and lonely. "If you do not cultivate a capacity to think, imagine, and create," warns Adam Garfinkle, editor of the *American Interest*, "you therefore may not realize that anything more satisfying than a video game even exists."[11]

Michael Crichton, author of *The Andromeda Strain*, *Jurassic Park*, and *Westworld*, among other cautionary tales about the unintended effects of technology, held a medical degree from Harvard and had hands-on scientific experience in the lab. His techno-thrillers often portray scientific advancements gone awry, culminating in catastrophe. Dr. Crichton complained, however, that others frequently misunderstood his view of technology as "being out there, doing bad things to us people, like we're inside the circle of covered wagons and technology is out there firing arrows at us." Technological catastrophes exist not because technology is inherently bad, he said, but because "people didn't design [it] right."[12] And, perhaps, too, because we don't use it thoughtfully.

Like Dr. Crichton, I aim to put forth evidence for the neurological consequences of excessive screen exposure and the costs of screen distractions. Tech titans have gone to great lengths to commandeer the fixed slice of attention our Stone Age brains have to work with. We are up against brilliant software engineers armed with personal data gathered through relentless surveillance and determined to capture our attention for commercial ends. They are better at distracting us than we are at defending the inherent weaknesses of our biology.

Not only have tech titans done a great job in vying for our attention, but they have also excelled at convincing us of the advantages of their products. With everyone blathering about supposed benefits, few mainstream media bother even to consider the potential negatives of this increasingly prevalent element of daily life. I am not saying that digital technology is bad, but in light of how it dominates everyday life, I suggest we examine its downside. Think of how often a traffic light turns green but the driver in front of you is buried in their phone, forcing you to honk the horn. Or how often you have to wait for a machine at the gym while the user finishes texting. Or how, when the power goes out from a storm, you put down your devices, venture outside, and speak to your neighbors.

Imagine that, for a change, you have promised yourself a good night's sleep. You've eaten dinner, changed out of your work clothes, gathered your things for tomorrow, and brushed your teeth. Before turning in, however, you decide to relax by streaming a show on Netflix. Ignorant of your intention to catch up on sleep, Netflix's "next episode in 5 seconds" and "skip intro" features keep you anchored to the screen. Before you know it you have binge-watched half a dozen episodes because streaming companies need you to do so in order to make money. YouTube's "next up" and "auto play" algorithms likewise analyze your viewing history. With a sniper's precision they target you based on what they infer from the online choices you have made.

Companies effortlessly unscramble purportedly anonymous metadata to trace your entire online transaction history using just three pieces of information: two physical locations and a dollar amount spent.[13] Facebook and Google own some of the fastest machine learning supercomputers on

Earth, so this has become easy to do.[14] As far back as 2012 Uber could examine user patterns to identify intimate behaviors such as one-night stands.[15] Companies insist that the details they sweep up are anonymous, but they can track your movements to within a few yards. Even particular items we buy can predict our political beliefs, race, and education with frightening accuracy.[16] Privacy is nearly impossible now thanks to "data inference" technology.[17]

Tech giants know everything about you, including your mood, whether you're lonely, anxious, excited, or depressed. They know where you shop, where you go, the online sites you visit, how long you hover there, and much more. Their artificial intelligence (AI) learning machines build models based on everything they gather about you to predict what you are going to do next. The recent integration of ChatGPT, Bing, and similar chatbots into browsers and operating systems will give them even more data about you. Armed with that knowledge, they then try to sway you to do what they want you to do. It will be one or all of three things:

- Keep you engaged as long as possible.
- Invite friends to do what you are doing online, called growth hacking, which multiplies your influence and thus increases their revenue.
- Respond to advertising offered up as you endlessly scroll, with the end goal of all manipulations being always to maximize profits. Since 2006 Facebook has had a director of monetization who offers software to its advertisers to do exactly this.

Shoshana Zuboff, emerita Harvard Business School professor and author of *The Age of Surveillance Capitalism*, says "trading on human futures" earns trillions of dollars. By exploiting vulnerabilities in human psychology, tech giants "trigger real-world behavior and emotions without the user ever being aware. They are completely clueless."[18]

The industry's earliest backers and engineers are now among its fiercest critics, angry that the platforms refuse to confront their product's addictiveness and treat our minds as an extractable resource no different than if they were operating an open pit mine. These critics lament the deceptive attitude that social media are "free," citing the adage that if something is free, then you are most certainly the product on sale. That is true to a point, but Jaron Lanier, father of virtual reality, ominously warns what the product really is: "the gradual, slight, imperceptible change in your own behavior and perception. That *is* the product. It changes what you do, what you think, and who you are."

Apple, Google, and Microsoft control the operating systems we use, while Facebook, YouTube, Netflix, and millions of apps supply the content that painlessly but mercilessly hijacks our attention. At the height of Netflix's reputation, chairman and CEO Reed Hastings admitted that sleep was the company's top competitor. He wasn't concerned about Hulu, Amazon Prime, or HBO Max siphoning off profits. "Think about it, when you watch a show from Netflix and you get addicted to it, you stay up late at night. We're competing with sleep, on the margin . . . and we're winning!"[19] But binge-watching supplants hours normally devoted to fitness, socializing, and sleep, making it a health hazard. Binge-eating and binge-watching often go hand-in-hand with sedentary behaviors linked to heart disease, stroke, and type 2 diabetes.[20] Watching television does have positive aspects. The problem emerges with unbridled indulgence and its associated downside.

Netflix has data to back up its binge-inducing power: there is a 75 percent chance you will binge-watch an entire series once you have seen the first two episodes. "Binge-racing" fans try to devour an entire series within twenty-four hours of its release, behavior the company promotes for the "unique satisfaction" it brings, calling it a "sport . . . an achievement to be proud of and brag about."[21] The number of Netflix subscribers who put themselves through this marathon grew more than twentyfold between 2016 and 2019 to reach 9.4 million. The top countries that do it are Canada, the United States, Denmark, Finland, Norway, and Germany. Binge-racing "is a new status symbol," it claims. But perhaps the habit is prevalent because it activates that part of the brain responsible for rewards. Thirty-seven percent of Netflix users have binge-watched at work. Over 50 percent of all adult TV viewers (including college students) have stayed awake all night to watch a show's entire season.[22]

At George Washington University we give each new medical student an iPad and a personal Google drive loaded with textbooks, assignments, and resources (though most students still prefer hardbound physical books because they are more conducive to remembering the material they read). These future doctors belong to a generation that grew up with digital devices. Yet they too are acutely aware of the downsides of daily screen engagement and the effort required to cope with them.

It is both a blessing and a curse that smartphones and tablets rarely leave our sides. More than fifty years ago, science fiction writer Isaac Asimov predicted what life would look like in 2014. "The lucky few who can be involved in creative work of any sort will be the true elite of mankind, for they alone will do more than serve a machine."[23] These biomechatronic devices simultaneously enhance neurological faculties (supplementing memory) even as they diminish them (by reducing focus and fragmenting our attention). They present a dilemma: Should I delete my Instagram account because it wastes so much time or should I interrupt what I'm doing now and check my likes (a reinforced behavior that is linked to reward)? Screen devices let us do what we could never have done in the past even as they exact a cost in depleting a finite mental resource: the fixed amount of energy we have for thinking, memory, and focus. Even if we resist the temptation to check a screen, brain drain still occurs because it is simply there, ineluctably sucking up our attention and taking a mental toll.[24]

Evasion characterizes the tech giants. Apple, Google, and others have reluctantly admitted, but only under the glare of scrutiny, that their products are indeed addictive. Historically, Mark Zuckerberg's response for years has typically been to evade responsibility. He claimed to be helpless to prevent Facebook being overrun by Russian disinformation in the runup to the 2016 presidential election. Resorting to the passive voice—always a sign of blame shifting—he said, "For the ways my work was used to divide people . . . I ask for forgiveness and I will work to do better."[25] He continued to make empty promises, and *The Wall Street Journal*'s deep reporting on "The Facebook Files" seriously damaged his company's credibility, now called "Meta."

Apple boasts rosy academic outcomes it attributes to its technology. But it misleadingly extrapolates miniscule data from a single school to all of K–12 education.

In a seeming act of penance tech giants have thrown apps at the problem as if the disease were also the cure: more technology to remedy the damage they wrought in the first place. But tech-based distraction management has been a bust: users thwart software barriers just as alcoholics and addicts circumvent well-intentioned efforts to limit their access.[26]

Screen absorption is not merely counterproductive. It fundamentally alters cognitive development in young generations because the brain does not fully mature until age twenty-five or so. Screen exposure typically begins in infancy and has now surpassed traditional play and in-person social engagement as a child's single most frequent experience. The brain adapts exquisitely to every niche because from birth onward it is highly plastic, that Greek-derived word that means "moldable." Plasticity is why all brains, but especially young ones, change continually in response to whatever environment they find themselves in. Experience influences neurological development, alters the way genes express themselves in the brain, and sways the brain's long-term maturation. This is how cumulative screen exposure plastically changes brain structure and function. Tech companies know this as well and yet shun responsibility for any harm done.

Prolonged screen viewing is associated with reduced volume and delayed development in the microanatomy of brain regions such as the frontal lobes that govern impulse control, also a key aspect in addiction.[27] Not all individuals are equally predisposed to substance or behavioral addictions such as screen dependency. But addictions of all sorts do have a robust heritability of around 50 percent. Kids whose parents engage in high discretionary screen time are thus more likely to succumb to screen dependency themselves.[28] Because twenty-first-century technologies have transformed the social climate in which we live, the brain and mind are likewise undergoing unprecedented changes. Baroness Susan Greenfield, a neuroscientist and former director of Britain's Royal Institution, coined the phrase "mind change" to signify how computer games, the internet, and the spectacularly misnamed social media have changed the brain in ways both good and bad.[29]

I have cited some of the many benefits of smart devices. Among the negatives is stifling the sound of one's own thoughts. The mindset of individuals who grow up in this screen-saturated, constantly connected century will be characterized by:

- A reduced attention span, combined with a need for personal attention;
- Recklessness and a premium placed on sensation at the expense of sequential, reasoned thought;
- Increased susceptibility to addictions of all sorts;
- Poor to absent person-to-person skills, leading to . . .
- Isolation, indifference toward others, cruelty, and bullying;
- Virtual autism (i.e., autism-like behaviors induced by heavy screen exposure); and
- A shaky sense of identity

In later chapters I discuss positive intermittent reinforcement, a well-established psychological mechanism that perpetuates a particular behavior. Think how slot machines persuade players to keep feeding in money despite many near misses and outright losses. Software engineers use the exact same psychology of intermittent reinforcement to keep you tethered to a screen. Candy Crush was a wildly popular, highly addictive game once played on smartphones by more than 100 million daily users (other fads have since taken over, which is the nature of fads). Candy Crush is addictive because it doles out intermittent and unpredictable rewards such as arbitrary points, avatars, boosters, and blockers that are emotionally satisfying enough to keep users engaged. The satisfaction it induces is no different in kind from the fuzzy glow addicts feel while using.[30] The ability of recurrent screen stimulation to infiltrate ordinary perception is illustrated by one avid Candy Crush user who started seeing Candy figures from the game in his peripheral vision after suffering a stroke.[31] We do not yet know whether such brain changes are temporary or permanent, but the fact that screen games can alter the brain this way is unexpected and concerning.[32]

Bassinets, training potties, and car seats now come equipped with iPad holders. Either manufacturers have not thought through the consequences of blocking the developing central vision of a young child with unnatural mediated images or they don't care. In the West, a third of infants under age one currently play with smartphones and tablets. By age two almost all are interacting with digital devices despite warnings from pediatric experts that repetitive screen exposure, especially from fast-paced games and animations, can foster addictive behavior.

Why does this worry me? Because in 2020, diffusion tensor imaging (DTI) established a correlation between increased screen use in prekindergarten children and lower structural integrity at the microscopic level in their brain's white matter tracts that support language and developing literacy.[33] We need further study, but the immediate implication is that screen exposure is causing these changes during early stages of brain development.

It is but a short step from phone and game addiction to the consequences of social media that displace face-to-face interactions. Linguists tell us that words alone convey only 10 percent of an encounter's meaning, which is why we grasp context much better when engaged face-to-face than on the phone. Much meaning depends on vocal intonation, mutual eye contact, body language, and touch—none of which is available to online social networks. Without having the opportunity to rehearse social skills and nonverbal communication, how can one hope to read other people, let alone empathize with and understand them? An entire issue of the *Atlantic* explored why hook-up culture has supplanted dating and romance among young adults. "We hook up because we have no social skills. We have no social skills because we hook up," complained one coed. Another agrees: "We'd probably have a lot more sex if we didn't get home and turn on the TV and start scrolling through our phones."[34] Mobile technology, which Gen Z has never lived without, has acted like a security blanket but has also hindered Gen Zers from developing face-to-face skills and resiliency in the inevitable face of failure.[35]

Tech's cool factor dazzles millions. Many celebrate that the digital realm is "free." But as behavioral psychologist B. F. Skinner warned decades ago in *Beyond Freedom and Dignity*, "A system of slavery so well designed that it does not breed revolt is the real threat."[36] Addictive screen media now constitute a slave economy with users as the labor. Advertisers are the paying customer, social networks and media platforms are the store, and your eyeballs and brains are the commodity on offer. Repeatedly we are sidetracked and manipulated by a handful of companies that steer what three billion people—more than 30 percent of the planet's population—think and do every day. Even the most repressive theocratic and authoritarian regimes don't hold such sway.

A cautionary legend pertinent in this era of digital indulgence is Goethe's folk tale "The Sorcerer's Apprentice" (*Der Zauberlehrling*). Popularized in a later version by Disney's *Fantasia*, it tells of an ageing sorcerer who leaves his apprentice to finish his chores for the night. Weary of fetching pail after pail of water, the apprentice enchants a broom to carry them for him by using magic in which he is not yet fully trained. The apprentice panics when he does not know how to stop the enchanted broom—a tool he believed would lighten his workload—as it begins to flood the workshop with bucket after bucket. The apprentice hacks the broom to pieces only to see each piece grow into a separate new broom that fetches even more water at terrifying speed. Just when all seems lost, the old sorcerer returns to break the spell.

What will it take to become like the wise sorcerer who halts the chaos when automatic newsfeeds, "recommended" videos, and "people you may know" constantly vie for your attention? How can you prevent your private data from being repurposed and redirected to "personalized" advertising, conspiracy theories, biased news, rank misinformation, and an echo chamber of outrage and indignation aimed to narrow your point of view without you even realizing what's being done to you? Partisanship or ideology isn't the problem. The problem is the biological limits of attention and the fact that the digital world has already taxed it beyond the breaking point.

2

SELFIES KILL MORE PEOPLE THAN SHARKS

"Technoference" describes the way technology undermines one-on-one social interactions. Among anthropologists and evolutionary scientists, the "social brain hypothesis" proposes that the human neocortex grew in order to help people better navigate the complexities of increasingly larger social relationships.

People today have on average three to seven individuals with whom they are close. Half are not immediate relatives.[1] Yet social media can bestow five thousand "friends," ten thousand "followers," and make one an "influencer" to a hundred thousand or more. Busy cultivating the attention of fans, we have stopped paying attention to our surroundings. By 2014, the so-called year of the selfie, the U.S. Department of Transportation reported that 33,000 drivers were injured in all types of phone-related accidents. One year later, nineteen people died in selfie-related incidents, and scores more injured themselves. During the same period, shark attacks killed only eight people.

The United States averages about nineteen shark attacks annually and one fatality biannually, whereas lightning strikes along the costal United States kill more than thirty-seven people each year.[2] Heatstroke and dehydration claim lives every year in the Grand Canyon, yet a surprising number of "unwise" individuals fall to their deaths from the canyon rim while trying to snap spectacular selfies. Attention seeking seems a defining impulse of our time; we are snapping ourselves to an early grave.[3] The editors of the *Oxford English Dictionary* designated "selfie" Word of the Year in 2013. By mid-2018 the number of people killed while taking a selfie had reached two hundred—a worldwide increase of 952 percent compared to 2014. While it may seem insensitive to cite morbid, gruesome misfortunes, such mishaps are indicative of the situations ignited by careless, inattentive cellphone use.

A Google search for "selfie deaths" will turn up photos of amazingly foolhardy behavior, while YouTube features no shortage of videos with titles such as "bizarre selfie deaths" and "selfies taken moments before death."[4] Absorbed self-admirers have fallen from cliffs, crashed their vehicles, and electrocuted themselves atop train cars while posing for their followers. Three college students died while trying to snap themselves in front of an oncoming train. In one spectacular blowout, two Russians in the Ural Mountains posed with a hand grenade from which they had pulled the pin. Only their photograph survived as a memento. In Colorado a twenty-nine-year-old Cessna pilot crashed his plane while taking selfies with a flash, killing all the occupants. You can watch video footage of a Polish couple visiting Portugal fall from a cliff into the ocean after stepping over a safety barrier to take a selfie with their two children.[5] In British Columbia three daredevils died going over a waterfall while filming a stunt intended to garner clicks on YouTube's "High on Life" channel.[6]

A Russian graduate student fell from a bridge while attempting to capture herself in a dramatic pose, and then for good measure electrocuted herself when she grabbed on to some live wires. "Nobody wants to be outdone," says psychology professor Zlatan Krizan of Iowa State University. Social media goad us into outdoing others in order to showcase perfectly curated lives. The 2017 Academy Awards illustrates the ease with which this happens and, more important, why working memory *can only ever handle one task at a time.*

Brian Cullinan, an accountant with PricewaterhouseCoopers, was responsible for ensuring that envelopes with the winners' names were in the correct order and that the correct envelope was given to the right presenter. There were two sets, one for Cullinan and a duplicate set for his colleague, Martha Ruiz, on the opposite side of the stage, so that either of them could hand over the correct envelope depending on which side the presenters entered from.

Ruiz had just given the envelope for Best Actress, Emma Stone in *La La Land,* to presenter Leonardo DiCaprio. The next prize for Best Picture was the highlight of the evening. Unfortunately, Cullinan forgot to dispose of the Best Actress envelope and handed it to presenters Warren Beatty and Faye Dunaway, who mistakenly announced *La La Land* as Best Picture instead of the correctly vetted *Moonlight*. Viewers in the hundreds of millions watched chaos unfold live.[7]

Cullinan, a prominent Twitter user, posted a photo of Emma Stone holding her Oscar just after she exited offstage, the precise moment when he should have been discarding the Best Actress envelope and cueing up the next one. Cullinan's task may not seem difficult, but it isn't one that can be carried out automatically the way we might habitually drive the same route every day without needing to give it much thought. Keeping track of the envelopes drew on his working memory, and when he combined this task with another of composing a Tweet, he forced his working memory to execute both tasks simultaneously. And that is the heart of the problem: the brain cannot take on two tasks simultaneously when each of them requires it to use working memory. This kind of so-called "task switching" always costs us in terms of reaction time and the number of mistakes made.

Anything that requires your attention and thus by definition cannot be carried out automatically draws on the finite reserves of working memory. How much it consumes depends on the task's degree of complexity. For example, it takes more effort to straighten a messy house than one that is fairly tidy. Likewise, the cost of switching is lower when you know ahead of time what must be done (cue up the correct envelope) because it clears working memory and prepares you for the upcoming task. This is where outside interruptions can wreak havoc. You are more likely to be distracted in a shared office where coworkers can interrupt you than in a private workspace. Undistracted students wrote 62 percent more than a distracted control group, and their notes were more detailed.

When you are interrupted you don't have a chance to flush your working memory completely; a remnant of attention always remains behind, hooked onto the previous task. The larger the residue you hang on to, the higher the switching costs will be. Switching costs some people less time and energy than others, and so they are better at it. One factor that predicts how efficiently a brain can switch is a person's degree of multimedia use. An earlier study from Stanford University found that people who spend a lot of time on chat and social media have higher switching costs than those who turn to screens less often. You may conclude that heavy screen use is the cause of distractibility and poor attention spans, but it may also be that people who are more easily distracted are more inclined to use multiple media simultaneously.[8]

* * *

It is only natural to compare ourselves to others, and the comparison strongly motivates how we act. The psychological theory called social comparison, in which people measure their own worth by how they stack up against others, dates from 1954 and has long been considered a normal way by which individuals establish healthy self-esteem.[9] Yet in today's taunting atmosphere of "pics or it didn't happen," you must submit graphic documentation to prove you were at a particular place and engaged in a significantly postworthy event. Civil discourse and debate have become reduced to schoolyard "he said, she said" taunts. Smart technology has made us narrow-minded and self-centered, exemplified by the phenomenon of "sad fishing," in which adolescents post self-deprecating emotional confessions on social media apps in the hope that friends will stick up for them and boost their self-image.

Our Paleolithic emotions are just not equipped to resist the pull of modern technology. If you doubt that minds are altered by social media, then consider Facebook's like button. According to its lead inventor, software engineer Justin Rosenstein, it was originally invented to "spread positivity and love in the world." Instead, it threw many teens into deep depression. The reason? Want of enough likes. "Fake brittle popularity" is how the industry describes the precarious mindset of lonely kids that leads them to crave even more likes.[10] Steve Jobs, of all people, counseled, "Your time is limited, so don't waste it living someone else's life." Yet we do just that, shouting "Look what I'm doing!" while expending enormous energy to maintain an image meant to impress others. We may have evolved to care about what others think—but in the context of small social clans, not in the stew of torrential doses of approval and opprobrium that social media throw at users every day. That relentless feedback persists even while we sleep, waiting to confront us in the morning. We react to other people's perceptions instead of living our own life, and this atrophies our coping skills and reserves of resilience.

As standards for postworthy content escalate, so too do the extremes people go to in order to make an impression. The drive to appear fabulous has alarmed plastic surgeons: once, patients armed with photos asked for the nose, lips, or jawline of a particular celebrity; now they ask to look exactly like their doctored selfies. A paper in *Facial Plastic Surgery* from

Boston University School of Medicine cites a new form of body dysmorphic disorder (BDD), a mental illness on the obsessive-compulsive spectrum and more pathological than mere self-doubt or insecurity about one's looks. About 25 percent of individuals with BDD who are chronically unhappy with the way they look attempt suicide. "Snapchat dysmorphia" disorder refers to the unrealistic expectations that the free FaceTune app instills in people obsessed with their selfie image (an annual VIP subscription is also available). Instagram's and TikTok's "beautification filters" alter people's perception of beauty to an image that is both unobtainable and unrealistic.[11] "They will ask for Kim Kardashian's nose, even if their facial structure looks nothing like hers," says facial cosmetic surgeon Daria Hamrah in McLean, Virginia.[12] The latest annual survey (2022) conducted by the American Academy of Facial Plastic and Reconstructive Surgery (AAFPRS), the world's largest such group, reveals "skyrocketing demand" for facelifts, fillers, and buccal fat removal to achieve a slimmer facial appearance. Videoconferencing amplified by the COVID pandemic has taken "selfie awareness to the next level."[13]

Once available only to celebrities, Photoshop-like filters blur the line between reality and fantasy. They warp the confidence of twenty-year-olds, who insist they need a facelift. "I feel more like a psychologist than a surgeon," Dr. Hamrah says, then reflects that the more people use social media, the more they ruminate on perceived shortcomings. A phone camera held at arm's length produces a fisheye distortion that makes the nose and center of the face look 30 percent bigger than they actually are.[14] Men become obsessed, too. In Silicon Valley men in their thirties have convinced themselves that prejuvenation is essential to their career. "If you're over the age of thirty-five you're seen as over the hill," says Dr. Larry Fan, a popular San Francisco plastic surgeon.[15] Men who work in tech, where the whiz kid image is prized, suddenly must cope with the sort of beauty standards that have long plagued women. Careers, even one's social life, can disappear behind the crow's feet of a forty-year-old.[16] One young worker who spends $500 on Botox injections every three months calls it "an investment." His concern illustrates how tech has morphed from enabling invisible, behind-the-scenes wizardry to literally shaping highly public figures in charge of corporations where billions of company dollars are at stake. Contemporaries spring for neck lifts, eye lifts, liposuction, and CoolSculpting, which kills fat cells by freezing them. CoolSculpting gives men a slimmer, more fit appearance and requires little downtime.[17] In 2018, ProPublica reported that IBM, which earned $79 billion in revenue that year, terminated 20,000 older U.S. employees in an effort to "correct seniority mix" and build a younger workforce. Google similarly paid out $11 million that year to 230 applicants over age forty who had sued the company for a "systematic pattern of age discrimination."[18]

Communications professor Jesse Fox at Ohio State University studies obsessive selfie-takers. "They imagine, 'I'm getting the confirmation from other people that I'm awesome,'" but those same people score high on traits of narcissism and psychopathy. "You don't care about the tourist attraction you're destroying [or that] you're dangling off the side of the Eiffel tower."[19] Ignoring the risks to which you are driven to capture a perilous tableau, the anticipation of making an envy-inducing post sets the brain's dopamine loop in motion. Once it gets going it plays out to the end, the way the reels of a slot machine inevitably clunk into place. You reach for the anticipated reward as if your life depended on it. And often enough it does.

"Enough with the bison selfies at Yellowstone," pleads the National Park Service. The Disneyfication of nature has now gone on for more than three quarters of a century. Its Pollyannaish view has instilled a widespread belief that animals are cute and friendly. Tourists determined to post selfies of themselves with bison, moose, and bears blithely come within inches of two-thousand-pound animals that are "wild and unpredictable" and can run three times faster than a human can. Each year several tourists are gored, or worse.[20] At the Wildlife World Zoo in Arizona a woman was mauled after climbing over a barrier to get a selfie with a jaguar.[21]

The dangers of cellphone distraction extend even to walking—who hasn't seen absorbed pedestrians plow into lampposts and telephone poles? Over the past twenty years an estimated 76,000 ED visits related to distracted cellphone use were for injuries incurred when people fell, tripped, or walked into obstacles while absorbed in their devices.[22] So-called "phombies," a portmanteau word created from "phone" plus "zombie," endanger themselves, drivers, and fellow pedestrians. Phombies become so absorbed by their screens that they become oblivious.

Hospital data analyzed by Ohio State University found that injuries involving cell phones incurred while users were walking more than doubled between 2005 and 2010. By 2018 the National Safety Council was noting more than six thousand annual phone-related pedestrian deaths.

FIGURE 2.1
Left, Stairways at Utah Valley University marked for walking, running, and texting. *Right,* Lane warnings at National Geographic headquarters. Both warnings were routinely ignored. Courtesy of Utah Valley University.

These dangerously distracted perambulators have prompted countermeasures. Utah Valley University created separate stairway lanes so that slow-moving texters wouldn't trip up others (figure 2.1). But the effort failed miserably. Heads down and earbuds in, phombies were too absorbed to read the directions. "We're dealing now with an addiction . . . that is, frankly, all-consuming," says the head of the National Safety Council.[23]

Utah Valley University was not the first to experiment with dedicated walking lanes. The township of Montclair, New Jersey, tried it; the effort failed. Outside its headquarters in Washington, D.C., National Geographic painted lanes to divide pedestrians using cellphones from everyone else. Freshly stenciled commands said "No cellphones" on the left and "Cellphone users, walk in this lane at your own risk" on the right. A solid line divided the two. Few paid attention to the markings, and pedestrians absorbed in their mobile phones continued to collide with one another.[24] The most common reaction was for passers-by to stop and snap a picture of the sidewalk stencils while they stood blocking the lane.

After decades of declining traffic deaths thanks to safer automobiles and road designs, inattentive driving and walking have made highway fatalities rise. According to the Governors Highway Safety Association, pedestrian deaths increased by 5 percent during the pandemic despite a sharp decline in driving during that time. Only 22 percent of pedestrians struck as they stepped off the curb had even bothered to look for oncoming vehicles.[25]

A 2021 experiment showed that walking through a crowd is "like a dance we perform with those around us." It illustrated why merely a few distracted individuals can disrupt the dance and throw off the movements of more than fifty walkers in the vicinity. We normally rely on a variety of visual cues to anticipate where others in a crowd will go next. Intuitively, the crowd forms lanes: once an individual at the front finds a way through oncoming pedestrians, those in the rear fall in line, creating "ribbons of walkers" that glide past one another. The maneuver is effortless and nearly instantaneous. By contrast, distracted walkers at the head of a crowd dramatically slowed down the whole group. Rather than move smoothly with purpose, distracted walkers take sideways steps or weave in and out, which renders their movements hard to predict.[26]

* * *

Are we really that addicted to our smartphones? You don't have to be a neurologist to guess the answer. People take their phones with them to the toilet (yes, they do); clutch them while they sleep; check them repeatedly while they eat, wait for a stop light to change, and even while having sex. What if, instead of reflexively reaching for your phone, you cultivated what the Dutch call *"niksen,"* the art of doing nothing?[27] It can be excruciatingly challenging for Americans to sit still and do nothing because we have conditioned ourselves to be doing something all the time. But a moment's pause and an interlude of silence can replenish your equanimity. Try removing social media apps from your phone. Clear your home screen of distractions that goad you to engage. Better yet, turn it off altogether for a few hours. What is so crucial that you can't allow yourself a short interval of down time?

The digital world that promised to bring us closer together has instead brought face-to-face communication to a near standstill.[28] Illogically, we protest invasion of privacy and then voluntarily divulge reams of personal information to corporations that profit from it. Such contradictions are easy to account for because human nature is inherently divided: six decades of neuroscience research have repeatedly confirmed that the brain's right and left hemispheres differ. Each supports a separate set of skills, endowing the two hemispheres with similar but not identical concepts of self with respect to past, future, family, culture, and social history. What we normally regard as a singular, unified self is really the net interaction between two distinct modes of perception.[29] The biological arrangement helps explain why one part of us wants to put our phone aside while a different part, feeling anxious, feels compelled to engage with it.

It may help to consider the "alien hand" syndrome that affects individuals with a particular kind of brain injury, and particularly those who have undergone split-brain surgery. It beautifully unmasks the reality that within each of our skulls reside two conflicting minds. Split-brain patients are astonished when one hand acts as if it had a mind of its own. The right hand tries to lather a washcloth while the left one keeps putting the soap back, or one hand turns the pages of a book while the other keeps shutting it. "Why is 'it' doing that?" they demand. The noir humor in the film *Dr. Strangelove* depicts the alien hand phenomenon when Peter Sellers's wayward appendage tries to strangle him, gives disconcerting Nazi salutes, and steers his wheelchair astray. The alien hand, phombies, and Montclair citizens' disregard for their municipal street laws point to the strength of our own digital habits, which have become sufficiently ingrained that they seem like separate entities often acting against our best interests.

The existence of two minds in a single person is a counterintuitive reality difficult for most people to accept because they firmly feel themselves to be a singular "me." That unified feeling is an illusion and explains why it is our inner self with whom we are most often at odds.

3

THE BRAIN ENERGY COST OF SCREEN DISTRACTIONS

From an engineering perspective, the brain has fixed energy limits that dictate how much work it can handle at a given time. Feeling overloaded leads to stress. Stress leads to distraction. Distraction then leads to error.

The obvious solutions are either to staunch the incoming stream or alleviate the stress. Hans Selye, the Hungarian endocrinologist who developed the concept of stress, said that stress "is not what happens to you, but how you react to it."[1] The trait that allows us to handle stress successfully is *resilience*. Resilience is a welcome trait to have because all demands that pull you away from homeostasis (the biological tendency in all organisms to maintain a stable internal milieu) lead to stress. Screen distractions are a prime candidate for disturbing homeostatic equilibrium. Long before the advent of personal computers and the internet, Alvin Toffler popularized the term "information overload" in his 1970 bestseller, *Future Shock*. He promoted the bleak idea of eventual human dependence on technology. By 2011, before most people had smartphones, Americans took in five times as much information on a typical day as they had twenty-five years earlier. And now even today's digital natives complain how stressed their constantly present tech is making them.[2]

Visual overload is more likely a problem than auditory overload because today, eye-to-brain connections anatomically outnumber ear-to-brain connections by about a factor of three. Auditory perception mattered more to our earliest ancestors, but vision gradually took prominence. It could bring what-if scenarios to mind. Vision also prioritized simultaneous input over sequential ones, meaning that there is always a delay from the time sound waves hit your eardrums before the brain can understand what you are hearing. Vision's simultaneous input means that the only lag in grasping it is the one-tenth second it takes to travel from the retina to the primary visual cortex, V1. Smartphones easily win out over conventional telephones for anatomical, physiological, and evolutionary reasons. The limit to what I call digital screen input is how much the lens in each eye can transfer information to the retina, the lateral geniculate, and thence to V1, the primary visual cortex.

The modern quandary into which we have engineered ourselves hinges on flux, the flow of radiant energy that bombards our senses from far and near. For eons, the only flux human sense receptors had to transform into perception involved sights, sounds, and tastes from the natural world. From that time to the present we have been able to detect only the tiniest sliver of the total electromagnetic radiation that instruments tell us is objectively there. Cosmic particles, radio waves, and cellphone signals pass through us unnoticed because we lack the biological sensors to detect them. But we are sensitive, and highly so, to the manufactured flux that started in the twentieth century and lies on top of the natural background flux. Our self-created digital glut hits us incessantly, and we cannot help but notice and be distracted by it. Smartphone storage is measured in tens of gigabytes and the hard drive of a computer in terabytes (1,000 gigabytes), while data volumes are calculated in petabytes (1,000 terabytes), zettabytes (1,000,000,000,000 gigabytes), and beyond. Yet humans still have the same physical brain as our Stone Age ancestors. True, our physical biology is amazingly adaptive, and we inhabit every niche on the planet. But it cannot possibly keep up with the breathtaking speed at which modern technology, culture, and society are changing.

Attention spans figure prominently in debates about how much screen exposure we can handle, but no one considers the energy cost involved. A much-cited study conducted by Microsoft Research Canada claims that attention spans have dwindled to below eight seconds—less than that of a goldfish—and this supposedly explains why our ability to focus has gone to hell. But that study has shortcomings, and "attention span" is a colloquial term rather than a scientific one. After all, some people's Stone Age brains have the capacity to compose a symphony, monitor the data stream from a nuclear reactor or the space station, or work out heretofore unsolvable problems in mathematics. Individual differences exist in the capacity and ability to cope with stressful events. To give California its due, Gloria Mark at the University of California, Irvine, and her colleagues at Microsoft measured attention spans in everyday environments. In 2004, people averaged

150 seconds before switching from one screen to another. By 2012 that time had fallen to 47 seconds. Other studies have replicated these results. We are determined to be interrupted, says Mark, if not by others, then by ourselves. The drain on our switching is "like having a gas tank that leaks." She found that a simple chart or digital timer that prompts people to take periodic breaks helps a lot.[3]

* * *

Neuroscience distinguishes sustained attention, selective attention, and alternating attention. Sustained attention is the ability to focus on something for an extended period. Selective attention speaks to the aptitude for filtering out competing distractions to stick with the task at hand. Alternating attention is the capacity to switch from one task to another and back again to where you left off. In terms of the energy cost incurred by repeatedly shifting attention throughout the day, I fear we have hit the brain's Stone Age limit. Exceeding it results in foggy thinking, reduced focus, thought blocking, memory lapses, and feeling overwhelmed.

Humans have adapted to new technologies ever since the advent of fire and the invention of the wheel. From a simple stone flint or bristle broom to a titanium tennis racket or precision calipers, any tool quickly comes to feel like an extension of oneself. The same applies to smart devices. Two centuries ago when the first steam locomotives reached a blistering speed of thirty miles per hour, alarmists warned that the human body could not withstand such speeds. Since then ever-faster cars, communication methods, jet planes, and electronics have diffused into the culture and become absorbed into daily life. In earlier times fewer new technologies appeared per decade, fewer people were alive, and society was much less connected than it is today. By contrast the invention, proliferation, and evolution of digital technology have put the status quo in constant flux. Unlike analog counterparts such as a landline telephone or a turntable, smart devices repeatedly demand and command our attention. We have conditioned ourselves to respond to texts and incoming calls the moment they arrive. Admittedly, sometimes jobs and livelihoods do depend on an immediate response. Yet we pay a price in terms of energy cost incurred by constantly shifting and refocusing attention.

For years. the "searchlight" hypothesis of attention reigned, according to which brains somehow zero in on whatever stimuli interest them. But the searchlight metaphor is backward: the brain does not intensify the illumination on an item of interest but rather dims it on everything else. Filtering is more important, and ignoring impertinent details may be the better way to make sense of sensory deluge. "The brain seems to be wired to be periodically distractible," says Dr. Ian Fiebelkorn at Princeton's Neuroscience Institute.[4] Yet suppression, filtering, and distractibility all exact a price in terms of energy cost incurred.

Silicon Valley's latest answer to overload is "dopamine fasting," a kind of dieting meant to "reset the brain to be more effective" by depriving it of this important neurotransmitter.[5] But dopamine is fundamental to dozens of basic processes in both brain and body, and there is no plausible basis behind the claim that the brain can somehow be reset. And to what prior state would that be? Proponents never say. The best way to reset one's state of mind is to turn off your devices, put them away, and go outside for a walk. We seem to have forgotten that the brain, the psyche, and the soul need interludes of solitude and quiet. Purdue University president and former governor Mitch Daniels advocated just such an unplugging in his 2020 commencement address to "a lonely generation" besotted with social media.[6]

Ironically, computer chip engineers increasingly base their design architecture on circuits found in the brain's cortex. These are "neuromorphic" chips (*neuro*, denoting related to nerves or the nervous system, plus *morpho*, form or shape). In 2011, IBM's supercomputer Watson famously beat human contestants on *Jeopardy!* but it could do so only because Watson was a "brute force" machine able to evaluate thousands of variables and crunch data at lightning speed before a human contestant could even reach for the buzzer. Its approach was totally opposite to the one the brain actually uses. Compared to the mere 25 watts needed to run the brain, the room-sized Watson consumed enormous amounts of electricity and required 240,000 BTUs of cooling, equivalent to that provided by ten two-ton commercial air conditioners. Watson's neuromorphic successor, called True North, needs only a fraction of that power and physical space. A single neuromorphic chip smaller than a fingernail contains one million neurons and 256 million programmable synapses that can perform 46 billion synaptic operations per second—a supercomputer in the palm of your hand that puts Watson to shame. True North draws only 70 milliwatts of power compared to the 80,000,000 milliwatts (80 kW) needed to crank up Watson. Your smartphone battery could run the neuromorphic array nonstop for two weeks.[7]

The point is that the brain achieves its amazing feats using an amount of energy equivalent to that of a dim light bulb. Most of the energy goes toward maintaining its physical structure, largely by pumping sodium and potassium ions across cell membranes to maintain an electrical charge. Little is left over for thought, emotion, and action. Because the brain is so efficient, its margin of energy reserves is slim and easily eaten up by the power demands made by the constant shifting of attention that distractions oblige. Just 3 to 5 watts of extra demand amounts to 12 to 20 percent of capacity, an unsustainable imposition that leads to brain drain, whose manifestations start with inattention, thought blocking, and so-called brain fog.

Compare the astounding capabilities of modern technologies to the way they progressively tax the very wits that invented them. Surrounded by and immersed in our technological creations, our engagement with digital culture puts us in a very short feedback loop: we invent increasingly sophisticated digital devices, and they in turn shape the way we think and act. Only recently have researchers begun to probe this recursive dynamic in detail. Studies from the Institute of Neuroinformatics in Zurich, led by Dr. Arko Ghosh, confirm that "the brain is continuously shaped by touch-screen use." His group showed that merely swiping a phone screen remaps the representation of the hand area in the brain's sensory cortex. The effect is proportional to the motor effort made, making swipes a reliable indicator of causation rather than a mere correlation. In other words, the more you swipe the more you rewrite your brain circuitry, particularly circuits that underlie motivation and reward.[8] Established research tells us that the more habitual a movement becomes, the smaller the amount of motor cortex needed to represent it—meaning that over time, we unwittingly habituate ourselves to sensory overload.

One might suppose that repetitive hand movements made in other contexts, such as knitting or playing a musical instrument, similarly remap the hand's sensory representation in the brain. Because brains adapt when they learn to use new tools, similar motions should logically cause similar results. But the smartphone is unlike any tool we have ever created, and the remapping observed looks to be specific to swiping movements and not other kinds. Earlier I defined neuroplasticity as the shaping of neural circuits by contextual experience. What we already know about the plastic rewiring changes in the brains of musicians that result from practicing repetitive movements does not shed any light on smartphone users because the latter use the same phone for different purposes in different contexts. Dr. Ghosh confirmed to me that the way a smartphone changes the brain is not a matter of how much but rather the way one uses it. Compared to knitting, musical practice, or repeatedly using hand tools, we use the phone much more frequently throughout the day for a broad variety of purposes. What has received scant attention is the role that incessant swiping and the rapid shifts of attention necessitated by that repetitive sensory input might play in fostering virtual autism (autism-like behaviors that emerge in otherwise normal individuals who engage in heavy screen use). Neurologists have typically regarded autism spectrum disorder (ASD) as arising exclusively from aberrant white matter connections in the developing brain. But molecular biologist Lauren Orefice at Harvard Medical School has discovered that mutations related to ASD affect sensory neurons in the extremities—the first time that changes outside the brain have been documented in autistic behavior. Neurons that receive touch signals from the skin disrupt central brain development and cause such ASD-like behaviors as sensory overreaction, social impairment, and anxiety.

Tactile hypersensitivity is so characteristic of ASD that the latest, fifth edition of the *Diagnostic and Statistical Manual of Mental Disorders* (DSM-5) considers it diagnostic of the condition: an ordinarily innocuous touch such as a hug or a breeze on one's face can be disagreeable, even painful.[9] The brain does not develop in a vacuum, of course: sensory inputs help shape its development. But Orefice found that mutations in sensory neurons from the extremities are sufficient to alter specific brain circuits. Her findings suggest a natural treatment for virtual autism as well as developmental ASD: boost inhibitory signals with drugs that target GABAA (γ-aminobutyric acid type A) receptors in sensory neurons, or simply lessen outside stimulation by swiping less or taking screens away. Insofar as one in fifty-nine individuals in the United States is diagnosed with developmental ASD and untold more exhibit symptoms of virtual autism, this fresh thinking highlights the interplay between genes and environment and suggests a probabilistic model that predicts who will respond negatively to excessive screen swipes and who will not. If touch sensitivity is perturbed during development, reasons Orefice, it could affect a youngster's central "brain development and lead to large changes in behavior."

Fear and anxiety, both core components of physical addiction, are also at play in digital addictions. The Center for Internet and Technology

Addiction has a "smartphone compulsion test" to find out whether you are in fact addicted (see the web page at virtual-addiction.com/smartphone-compulsion-test). Today's impulsive need for constant connection feeds the fear of missing out, or FOMO. For some individuals, separation from their phones for as little as ten minutes triggers intense anxiety akin to an addict's craving for the next fix. One can easily measure personal anxiety levels by applying an electrode pair to the base of the thumb and reading the galvanic skin resistance (GSR). Strong emotions increase the electric (i.e., "galvanic") resistance of sweaty skin because the amygdala of the emotional brain shares a link with the circuit that joins sensory cortex to the nerve ganglia that innervate sweat glands. Psychologically, the fear of missing out is very much like the fear of abandonment: the phone has become a security blanket of sorts that, through frequent use and repetition, has morphed into being our companion and escape valve, our drug of choice. In evolutionary terms, abandonment equates strongly with poor survival, and so we associate our phone with our virtual tribe and community.

Severing this sense of belonging triggers angst. Our Stone Age brain evolved to reward social acceptance and make belonging feel enormously salient: each is associated with an increased chance of survival and procreation. Whatever we find emotionally salient is noticeable, prominent, piques our interest, creates an attitude and a frame of mind. Salience develops over time, and your brain's subjective point of view shades it to make it uniquely yours. This evolutionary framing explains why smartphones easily capture our attention to produce technoference. Evolutionary adaptation favored the formation of small social circles that were crucial to human survival 3 million years ago. Setting up and maintaining intimate relationships requires reciprocal self-disclosure, being responsive to others, and forming attachments. Dr. Joseph Sbarra, director of psychological training at the University of Arizona, argues that smartphones and social media have "hijacked the human need to self-disclose" to the point that we now engage ancient "intimacy processes with the farthest reaches of [our] social networks."[10] Digital devices enlist Stone Age adaptations in ways that don't benefit us but instead undermine being present, attentive, and responsive to family, friends, and acquaintances. A phone by its mere presence competes for our limited amount of attention, reduces the capacity of working memory, and can make it seem as if we are uninterested or lacking in empathy.

We do not yet know whether technoference is a function of how relationships degrade in response to diminished interpersonal attention or whether the numerous social roles raised by smartphones create new contexts in which people can feel rejected. Intimate and meaningful relationships can develop online, but, Sbarra points out, the anthropological literature on network size reminds us that "people cannot maintain intimacy with the entirety of their social network."

* * *

In the chapters ahead I explore the similarity between screen dependency and compulsive behaviors that likewise incur negative consequences, such as smoking, alcohol dependency, overeating, gambling, shopping, and pornography obsession.[11] I show why people commonly surf the web instead of doing their work even as they curse themselves for wasting hours online, or text while driving despite knowing that doing so could be fatal. Case in point: an experimental self-driving Uber car killed a pedestrian in Tempe, Arizona. The backup driver, Rafaela Vasquez, told the National Transportation Safety Board (NTSB) that she was observing road conditions and watching the self-driving system at the time of the crash. But an investigation by the Tempe Police Department revealed that she was streaming NBC's *The Voice*, a singing competition television show, in the minutes prior to the collision.

Uber fired Vasquez from its employ. Because she had used her phone in a setting she knew to be dangerous, she went on to face negligent homicide charges in 2020.[12] A search warrant let the NTSB look at records from the video streaming service Hulu, which showed Vasquez's account to be active and streaming the program when the crash happened.[13] Based on footage from cameras inside the vehicle, the Tempe police said that Vasquez was "distracted and looking down" during the twenty-two minutes before the fatal collision. While the system was not designed to recognize a pedestrian outside the boundaries of a crosswalk, the pedestrian's death "would not have occurred if Vasquez [had been] monitoring . . . conditions and was not distracted."[14]

Human psychology is nothing if not counterintuitive. Knowing that screen checking can be injurious or even fatal is not enough to stop people from doing it. Indeed, Ohio state senator Andrew Brenner—someone who should know better—participated in a government Zoom meeting while driving. He appeared, wearing a seatbelt, against a video background of his office during a meeting in which his colleagues were introducing a bill cracking down on distracted driving. He excused himself by claiming it was hands-free, but hands-free does not mean risk-free or that your attention is fully focused on the road where it should be. Clearly it was focused on the meeting. Conversing on Zoom while driving a car consumes attention and working memory, inescapably distracting you from the immediate surroundings.[15]

What happens in the minds of heavy screen checkers is qualitatively no different from what happens in the mind of any addict. Society does not yet disparage screen dependency and distraction the way it does drug addiction, although Vasquez's prosecution suggests we may be getting there. Like drug dealers, tech platforms have enormous financial incentives to push us toward greater use. They know there are only 1,440 minutes in a day and compete ruthlessly for every second of our attention. They prey on parental FOMO and doggedly steer conversations away from the risks of screen dependency and virtual autism and toward the latest apps, often touted as "educational," without offering any evidence that they actually are.

Like secondhand smoke, the invisibility and seeming innocuousness of digital screen intrusion render it more malign. Addicts of any kind in the early stages of dependency genuinely don't think of themselves as hooked even as friends, family, and coworkers see changes in their behavior. It takes a while for the realization to penetrate—if it ever does—and for defensive justifications to fall away. Users who insist that their habit isn't a problem usually put forth the following four general arguments, after each of which I give a rebuttal. Together, these rationalizations unwittingly reveal how dangerous screen addiction really is:

Rationalization 1: It's just a matter of willpower, unplugging isn't complicated, and people are free to choose.

Rebuttal: The forces that ensnare users are typically inapparent and unconscious. Diabetes drugs like Wegovy and Mounjaro that cause enormous weight loss should put paid to options that willpower can overcome biological forces.

Television and radio never drew on massive supercomputers to predict what broadcasters should show to keep users glued to their devices or steer billions of people's thoughts in a preferred direction the way tech corporations do now. Users usually are not consciously aware of the forces that create digital addicts. Having an unlimited source of entertainment that encourages binge-streaming is designed to instill addiction in the same way that alcohol and drugs instill euphoria. What needs to be addressed in the "it only takes willpower" argument are addictive personalities and the way that addiction complicates choice but does not eliminate it (discussed in chapter 5). The smartest people in the world do not understand how their own mind is vulnerable to manipulation because only a fraction of the population thinks about psychology and the way the mind works. This gap is why magic and illusions easily fool people.

Rationalization 2: It's not a drug, it's just a tool.

Rebuttal: This rationalization relies on the illusion of self-control and downplays the negative costs of digital distractions. There is a large literature on this, and the illusion of having self-restraint actually promotes impulsive behavior. Users say their devices are no different from conventional tools such as a pen, an appliance, or a car. The fallacy here is that neither earlier technologies (cars, radio transmitters, cinema, microwaves) nor the plethora of pre-internet gadgets affected brain circuity and personalities the way digital devices do. No former device has kept such a precise ledger of everything we have said, shared, clicked, and watched in order to influence our behavior.

Rationalization 3: We always adapt to new technology.

Rebuttal: Yes, we adapted to trains, electricity, telephones, automobiles, jets, and early computers. But what is different today is that screen addiction is driven by exponentially advancing technology. Since the 1960s computer processing power has increased a trillion-fold while over millennia our Stone Age brain hasn't evolved at all. No artifact in human history has evolved even a fraction as rapidly as modern handheld devices.

Rationalization 4: I can stop anytime.

Rebuttal: Have you visited virtual-addiction.com/smartphone-compulsion-test? "I can stop anytime" is the exact language used by individuals addicted to alcohol, gambling, recreational drugs, and pornography. To physicians everywhere, hearing this rationalization is close to diagnostic proof that the individual is addicted. To test yourself for possible screen dependency you only have to go five days without checking any of your social media feeds or messaging apps. If during cold turkey abstinence you find yourself feeling anxious, afraid, left out, or ruminating about what online "friends" are doing, then you may have a problem. Can you admit the possibility? Or is your first impulse a defensive, out-of-hand denial?

4

THE BRAIN ENERGY COST OF MULTITASKING

All creatures that have a backbone—the animal class known as vertebrates—share the same blueprint for the general anatomy of their brain. The human organ is no different in its fundamental architecture from the brain of the other 64,000 vertebrate species, which include soft and bony fishes, amphibians, reptiles, birds, and other mammals (figure 4.1). Vertebrates make up only 4 percent of all living creatures, yet their brains evolved to become the most complex, with human ones the most complex objects in the known universe. Yet ever since the vertebrate blueprint came into existence hundreds of millions of years ago, the brain's basic arrangement has not changed, while technology has advanced more than a thousand-fold.

Brains evolved in a much different way than man-made designs have. The first modern washing machine was a corrugated tin washboard (1797) that later morphed into a hand-turned rotary drum (1851). Both were more efficient than pounding clothes on riverside rocks or rubbing them with abrasive sand as the ancient Egyptians and Romans did. Laundry became less of a chore with the advent of a motor-driven agitator and a manually fed wringer that had two rollers, fittingly called a mangle (1908) because it sometimes mangled the operator's hands or hair. After World War II, fully automated machines appeared, and then energy-efficient models that offered a choice of cycles. Today, home laundry systems can cost thousands of dollars and include luxuries such as drying closets, steamers, and dry cleaning modules.

The history of the washing machine illustrates how human inventors discard the old to make room for the new. A similar principle—let's call it redesign—goes for computers, the most brainlike machines we have ever built. The word "computer" originally referred to a person, almost always a

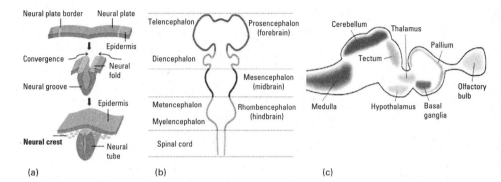

Figure 4.1

The nervous system of every vertebrate embryo starts out as a line of cells on its surface called the neural streak. (a) These then spread in two dimensions to form a sheet called the neural plate. If you were to take a sheet of paper and bring its long edges together you'd create a tube, which is what the neural sheet of the embryo does. (b) Under instructions from DNA that has copied itself for hundreds of millions of years, the neural tube segments into three parts, the hindbrain, the midbrain, and the forebrain (which becomes the cerebrum proper). A peripheral nervous system buds out below it to form the spinal cord, which will innervate muscles, skin, and organs. (c) These fundamental sections then divide further into structures that are common to all vertebrates. For example, the forebrain is well developed in four-legged vertebrates, whereas the midbrain dominates in fishes. In general, a central nervous system consisting of a single hollow tube that has a series of outpouchings, typically paired one to a side, is characteristic of all vertebrates. *Sources: a, c* Creative Commons; *b*, Eric Greif, Radiopaedia. [See also color plate 1.]

woman, who performed calculations with a slide rule, a mechanical calculator, and a lead pencil. During World War II, roomfuls of women computed tables of parabolic firing trajectories so that male gunners could accurately aim their artillery. In 1946 the University of Pennsylvania unveiled a massive redesign called ENIAC (Electronic Numerical Integrator and Computer), a thirty-ton, room-size analog computer powered by 18,000 vacuum tubes and operated by a team of six women. "It was just a monstrous thing," recalls Betty Snyder Holberton.[1] ENIAC took weeks to program a problem that entailed wiring it with patch cords (figure 4.2). What we know today as software didn't yet exist.

The analog vacuum tubes used in early machines burned hot and frequently blew out. In time, these tubes gave way to transistors, then to integrated circuits, central processors, math coprocessors, graphic coprocessors,

FIGURE 4.2
Marlyn Wescoff Meltzer (*left*) and Ruth Lichterman Teitelbaum were two of ENIAC's first programmers. *Source:* National Archives Education.

and finally the microprocessor circuits in our contemporary desktops, laptops, and smartphones. Today's cheapest prepaid burner phone is far more powerful than the 1960s Apollo rocket guidance computer, which weighed seventy pounds. In fact, a first-generation iPhone (2007) had several thousand times more computing power than all of NASA during the Apollo era. Put this enormous power in the context of a feedback loop, and it is easy to see how the machines at our fingertips could influence human behavior so profoundly.

First articulated in 1965, Moore's law predicted that the number of transistors that could fit on a central processor chip would double inexorably every eighteen months. Since then, transistors have steadily shrunk, allowing more components to fit in ever-smaller devices. The ones today approach the size of atomic particles, and we are reaching the physical limits of Moore's law because components can't shrink any further than the size of an atom. This obstacle has led engineers to turn instead to a new design of brain-inspired, or "neuromorphic," chips.[2]

Engineers typically replace old systems with something they consider more efficient: automakers introduce new models and more efficient engines, programmers develop new operating systems and apps. But evolution does not jettison the biological equivalent of river stones, washboards, transistors, and good enough chips from the past. Nor does it redesign new and improved versions of the vertebrate brain every millennium or so. Instead, it follows the principle of accretion: retaining whatever worked well enough for survival and cobbling new features on top of it as the brain adapts to changing environments. Consider the residual so-called junk DNA that every one of us carries from the distant past and that supposedly doesn't do anything (at least as far as we can tell at present). There is no reason some future evolutionary pressure won't arise to make use of what we now call useless junk, endowing us with wholly new skills.

Accretion is admittedly a less efficient way to build organisms compared to starting from scratch. But evolution doesn't deal in perfection; what matters is survival and procreation. Once a viable solution to a problem emerges and proves "good enough" to get the job done, it remains part of the brain's repertoire so long as an individual can pass on its DNA to offspring. The accretion principle holds for structure as well as function, which you can think of as the greatly simplified analogs of hardware and software.[3] The modern brain is an amalgamation of old and new systems, like an old New England add-on house that has seen one addition after another tacked on over the decades. The principle of accretion creates new pieces with fresh functions while leaving us with the same foundation as our Stone Age ancestors, a fact that has major consequences when facing today's onslaught of digital distractions.

Our large cerebral cortex sets us apart from other mammals. Credit typically goes to our prominent frontal lobes for the ability to pass on cultural knowledge from one generation to the next. The efficiency of formal education is huge compared to the glacial gene-by-gene mutation that physical evolution allows. But more important than size is the brain's pattern of connections, which blesses us with the mental flexibility to adapt. For an example we can look to the Australian spiny anteater, the echidna, an egg-laying monotreme, like the platypus, that evolved early in the mammalian family line. Its tremendously developed frontal cortex is far larger relative to its body size than that of our closer primate relatives. If modern humans had frontal lobes proportionately as large as the anteater's, we would need to carry our heads in front of us in a wheelbarrow.

The contradiction of a simple-minded mammal with disproportionately large frontal lobes suggests an evolutionary strategy gone down a dead end. Merely having extra analytical space for computations produces neither a more efficient brain nor one that is particularly smart, any more than adding additional memory or processing chips makes your laptop intelligent. What matters for both machine and animal is its algorithms (the rules it must follow), the physical layout of its components, and the wiring between them. Architecturally, the brain is structured to reduce wiring distances between components, thus optimizing conduction speed in the same way that ensuring short distances between elements on a printed circuit board speeds its performance. "Cytoarchitecture" (from the Greek *kytos*, meaning "container" or "cell") is the arrangement of cells within the six layers of the brain's cortex, whereas "topology" denotes the patterns of connections within a defined region of cortex (from the Greek *topos*, meaning "place"). People tend to think of nerve cells, or neurons, as identical in the way that skin cells or muscle cells all appear alike. But neurons come in a dozen varieties, each with unique properties. This is only one reason why the brain can perform many specialized functions.

My Verizon fiber-optic connection shoots data into my home at 8,589,934,592 bits per second, roughly 71,600,000 times the rate my gray matter can handle. At most, a nerve cell can discharge electrical spikes 1,000 times per second down its axon. But the fastest a signal can cross a synaptic gap to another neuron is about one millisecond. Considering both numbers, the brain can therefore carry out a maximum of 1,000 operations per second, 10 million times slower than an old laptop. In terms of resource consumption, listening to one person speak uses about 60 bits per second of brain power, half our allotted bandwidth.[4] Trying to follow two people speaking at once is well nigh impossible for the same reason that multitaskers fare poorly: attempting to handle two or more tasks simultaneously simply exceeds our cognitive bandwidth and overloads working memory. You may object, adducing as examples that people commonly read while using the cardio machine, sing while playing the piano or guitar, or drive while talking on the phone. These actions are largely automatic and repetitive and demand little monitoring compared to those that require watchful attention. When pushed to their limit, living cells become fatigued, making it harder for us to weigh conflicting demands and sort the important from the trivial. Today's deluge of stimulation asks our Stone Age brain to sort, categorize, and prioritize enormous streams of information that it never evolved to handle.

The term "multitasking" is a misnomer whether applied to machines or human brains. Despite marketing claims, your computer does not multitask, and neither does your brain. The latter simply cannot, whereas a computer's processor divvies up each clock cycle and apportions a slice of time—200 milliseconds, say—to each task. Round and round it goes until everything is done. The inherent inefficiency of having to split up processor time is why your computer bogs down the more you ask it to do. Brains respond the same way when multiple tasks vie for attention. We lack the energy to do two things at once effectively, let alone three or five. Try it, and you will do each task less well than if you had given each one your full attention and executed them sequentially. According to Cal Newport, professor of computer science at Georgetown University, "Even when you jump over to check the inbox and come right back, it can be just as damaging as multitasking. We're working with disruptive new technologies that only emerged in the late 90s." He says even minor switching from a current task to a different one is productive poison.[5]

Stanford University professor Clifford Nass presciently saw multitasking as particularly invidious. One of his most cited studies assumed that heavy multitaskers would excel at ignoring irrelevant information and at switching tasks, and accordingly would have superior recall. He was wrong on all counts:

> We were shocked. . . . Multitaskers are terrible at ignoring irrelevant information. They are terrible at keeping information in their head nicely and neatly organized. And they are terrible at switching from one task to another.[6]

Nass assumed that people would stop trying to multitask once shown the evidence of how bad they were at it. But his subjects were "totally unfazed," continuing to believe themselves excellent at multitasking and "able to do more and more and more."[7] If individuals in a controlled experiment are this oblivious and refuse to change when confronted with proof of their shoddy performance, then what hope do the rest of us have as we wade through the daily sea of digital distractions?

The only professions that require multitasking are simultaneous translator, air traffic controller, and motherhood—and the first two do it for only forty-five minutes before getting a break. Watching television while using another smart device is so common that over 60 percent of U.S. adults regularly engage in "media multitasking." Compared to controls, media multitaskers have more trouble maintaining attention and a propensity to forget; their anterior cingulate cortex (a brain structure involved in directing attention) is physically smaller than controls'. Another study found that the more minutes children engaged in screen multitasking at age eighteen months, the worse their preschool cognition and the more behavioral problems they exhibited at four and six years. The authors advise positive parenting and avoidance of media screen multitasking before the age of two.[8]

* * *

Do digital distractions differ from the analogue kind? One example that persuades me comes from a training session with George Washington University medical students in which we scrutinize an incident that happened at another well-known teaching hospital. During bedside rounds on the pediatric cancer ward, the whoosh of an incoming text distracted the resident physician, who was entering medication orders and updating the electronic chart. She was a digital native; the message was about a friend's upcoming party, nothing crucial in the context of the moment. But it momentarily seized her attention long enough to flush her working memory in the same way that Brian Cullinan's tweeting caused the Oscar snafu. The clinical team at the bedside was discussing changing the dose of an intravenous drug, and she failed to enter the change. By the time the omission was discovered, the four-year-old patient had developed kidney failure and gone into shock. In a different setting of 257 nurses and 3,308 pediatric intensive care patients, medication errors occurred when a text or phone call came in on a nurse's assigned institutional phone "in the 10 minutes leading up to a medication administration attempt."[9]

We have all experienced how electronic medical records steal time from mutual doctor-patient engagement—one example of the negative consequences of multitasking. Instead of looking, listening, and laying on hands, physicians now must type and check boxes on multiple screens to satisfy bureaucratic demands. A doctor may spend the entire appointment time facing a computer screen.[10] Logically, it should not matter whether a doctor takes handwritten or electronic notes. But it does matter because "hearing" is not the same as "listening." The former is a passive act of perceiving audible sounds whereas the latter is an active effort to understand another's perspective, what they are feeling and trying to communicate.

Electronic records demand so much of a physician's attention and working memory that attentive listening has become impossible. Forced to focus on the screen, they miss reading facial expressions and patient body language. Doctors have been handling interruptions for decades without making these kinds of errors. But screen-based distractions are different in kind, leading to more frequent errors during situations that demand attention. Something powerful takes hold of attention's so-called spotlight. Perhaps it is time to resurrect a phrase that all parents once knew: "Look at me when I'm talking to you."

At George Washington University a colleague and I teach small groups of medical students over their four years of study. We instruct them in clinical reasoning and professional development. We watch smart young adults transform into singular professionals who master huge amounts of factual knowledge as well as the know-how to apply it judiciously in the practice of medicine. Accomplishing a transition like this demands a high degree of sustained focus. It is the art of medicine we teach because the human being is more than the human body.[11] Increasingly, though, I witness inattention undermining our students, especially the undergraduates I encounter. As I waited for the elevator the other day, the doors opened and a dozen undergrads spilled out. All were staring down at their phones, oblivious to my presence even as they jostled and bumped into me. The screen lock on their attention had created a blind spot that made me invisible, a normal feature of perception called "inattentional blindness" or "change blindness." You can see a mind-blowing example of it by watching the "invisible gorilla" test on YouTube. The phenomenon is not a flaw or an optical illusion: the brain evolved to ignore whatever lies outside its immediate focus even when it stares us in the face. Some types of brain damage suspend patients in a perpetual state of inattentional blindness, a variety of agnosia (from the Greek meaning "not knowing"). In common terms, neurology calls this looking but not seeing. A growing proportion of the population seems to be drifting through life, looking at their screens but not seeing what is going on around them.

Because attention is like a sharp-edged spotlight, we can never know what we are missing. Anything outside its perimeter lies, by definition, within our mental blind spot. The undergraduates who mindlessly careened into me had already acquired habits that were actively undermining their ability to learn, think, and remember. Worse, their screen fixation made them oblivious to their self-inflicted handicap. How will these future leaders be able to focus, prioritize, delegate, meet deadlines, and achieve?

* * *

An enduring myth says we use only 10 percent of our brain. The other 90 percent presumably stands idly by to serve as spare capacity.[12] If the premise of untapped intellectual capacity were true, then the depictions of film characters ranging from Johnny Mnemonic to Lucy and Limitless would be documentaries rather than thrilling science fiction.

Yet two-thirds of the American public and half its science teachers (yes!) mistakenly believe the 10 percent myth, which perhaps underlies assumptions that one can multitask and overcome distractions by sheer force of will. Worse, more than 75 percent of American science teachers believe that enriching a child's environment—with Baby Einstein videos or iPads clamped to bassinets, car seats, and potty trainers—enhances intellect, despite a dearth of supportive evidence they can do any such thing. On the contrary, ample evidence explains why introducing tech impedes the natural development of a child who would otherwise have normal amounts of person-to-person interaction. It is true that growing up isolated and deprived of human contact drastically stunts brain development. But it does not logically follow that using tech to supplement a child's typical environment will boost cognitive development. Too much stimulation is equally detrimental as not having enough. Besides, it isn't stimulation per se that is crucial but the social context in which it occurs.

Individuals raised in isolation, such as Victor of Aveyron, the feral twelve-year-old discovered living in a cave in France around 1800, have helped us understand what happens to a human mind raised apart from social nourishment. Brought into society, Victor was accustomed to exposure and cold. He was comfortable being naked among the Enlightenment intellectuals who debated differences between humans and animals and tried to civilize him. But attempts to teach Victor language failed, and he never learned to speak.[13] Nearly two centuries later Genie, an American girl kept locked away with limited exposure to language and other people until she was rescued at the age of fourteen, confirmed that there is a critical window during which optimal language learning is possible. Once the window closes at around age seven, any language acquired is highly abnormal—as is thinking itself. After her rescue by medical specialists and social workers, Genie communicated only by pointing, gesture, and grunts. She was slow to respond, limited in comprehension, and inappropriate in affect. One moment she flew into a rage, the next she clutched total strangers, one of many behaviors in addition to masturbating and urinating in public that violated social norms. Lacking the most rudimentary social interaction during what should have been her formative years, Genie's problems were not surprising once she was returned to the world. After years of work and patient teaching—and frequent exposure to the outdoors—Genie started to express herself in words, phrases, and sometimes sentences. She never learned to read or write. What she did learn was to lie and manipulate, not the most commendable use of her newfound communication skills.[14]

By living in a rich social environment, by contrast, our neural networks self-calibrate, self-assemble, and adapt to stimulation, experience, and context. Yet energy consumption trumps all other factors. When we measure how the brain actually uses energy, the proposition that we have untapped reserves doesn't hold up. There is no sluice gate to open that will provide more juice to multitask or think genius league thoughts. To see why this is so, look at brain size and how it scales to the energy it must consume just to stay alive. During the past 2.5 million years the human brain grew proportionally much faster than the human body. Our central nervous system is nine times larger than expected for a mammal of our weight. The cortex constitutes 80 percent of the brain's volume, and its prodigious consumption of energy begat higher caloric meals and the invention of cooking. Cooking renders the calories in food more easily absorbed and allows the consumption of meat protein and carbohydrates that are otherwise indigestible in their raw forms.[15]

A rat's or a dog's brain consumes about 5 percent of the animal's total daily energy requirement. A monkey's brain consumes 10 percent. I noted that an adult human brain accounts for merely 2 percent of the body's mass yet consumes 20 percent of the calories we ingest, whereas a child's brain consumes 50 percent and an infant's 60 percent.[16] These numbers are larger than one would expect for their relative sizes because, in all vertebrates, brain size scales in proportion to body size. Big brains are calorie-expensive to maintain, let alone operate,[17] and the discrepancy means that energy demand is the limiting factor no matter what size a particular brain reaches.

It is also costly in terms of energy consumption to generate electrical spikes in a cell. This we know thanks to research undertaken during the administration of general anesthesia. As a person loses consciousness, their brain activity gradually shuts down until it reaches the "isoelectric condition," the point at which half the calories burned simply go toward housekeeping—the pumping of sodium and potassium ions across cell membranes to maintain the resting electrical charge that keeps the brain's physical structure intact.[18] This never-ending pumping means that the brain must be an energy hog. Compared to muscle cells or even highly metabolic cancer cells, the brain consumes an astounding 3.4×10^{21} molecules of adenosine triphosphate (ATP) per minute. ATP is the unit of fuel burned in the body's furnace. That is 34,000,000,000,000,000,000,000 units every minute, every day, for a lifetime.

Even if only a tiny percentage of neurons in a brain region were to fire simultaneously, the energy burden of generating spikes over the entire brain would still be unsustainable. Here is where the innate efficiency of evolution comes in. Letting just a small fraction of cells signal at a given time—known as "sparse coding"—uses the least amount of energy but carries the most information because a small number of signals have thousands of possible paths by which to spread themselves out over the brain.

A major drawback of the sparse coding scheme is that it costs a lot to maintain our 86 billion neurons. If some neurons never fire (meaning that if cells don't generate a current strong enough to travel down the axon and cross the synapse to the next neuron down the line), then they are superfluous, and evolution should have jettisoned them long ago. But it didn't. What evolution did discover by way of natural selection was the optimum proportion of cells a brain can keep active at any given instant. That number depends on the ratio between a resting neuron's housekeeping cost and the additional cost of sending a signal down its axon. For maximum efficiency, it turns out that between 1 and 16 percent of cells should be active at any given moment.[19] We do use 100 percent of our brain, just not all of it at the same instant.

Maintaining this balance is part of homeostasis, which we can think of as a budget whose currency consists of all the metabolic molecules that maintain our 86 billion neurons. As with a budget based on money, you can run a metabolic deficit and have your ledger go into the red. When this happens, the brain slashes processes that are too expensive, resulting in fatigue, boredom, clumsy errors, and foggy thinking.[20] The window of maximum efficiency noted above refers to an instantaneous snapshot of energy consumed by whatever neurons are firing at the time. The need to marshal resources in the most efficient manner also explains why most brain operations *must* be unconscious. Keeping ourselves alert and conscious, along with shifting, focusing, and sustaining attention, are the most energy-intensive things our brain can do. The high energy cost of cortical activity is why selective attention—focusing on one thing at a time—exists in the first place and why multitasking is an unaffordable fool's errand.

5

THE DIGITAL DIFFERENCE: WE TREAT IT SOCIALLY

Stark differences stand out when you compare today's audiences with those that grew up watching the *Ed Sullivan Show* (1948), *Lassie* (1954), or *Sesame Street* (1969) on network TV. Even during the burgeoning cable era of the '90s, televisions were fixed pieces of household furniture. You sat yourself in front of them at times dictated by the broadcasters to watch from a limited menu of available programs. Broadcasters forced you to align your personal viewing schedule with whatever time a favorite show came on. Today's digital devices offer an unlimited assortment of instantaneous handheld viewing options. Earlier when a program ended or it was time for bed we turned off fixed television sets, whereas digital platforms that broadcast around-the-clock disgorge an infinite stream of viewing material on command. They exist in multiple iterations as videos, games, and a hundred thousand apps. Content comes to us instead of our needing to seek it out.

More than anything else, readers of my *Psychology Today* column wanted to know how to fight digital interruptions and the constantly increasing pull on their attention. Inefficient, indecisive, and unable to prioritize, they felt they had become "humans doing" rather than "human beings."

Stanford's Clifford Nass was one of the first academics to see that we respond to digital technology socially, the same way that we interact with other people.[1] Users invest technology and social media with intent, importance, and meaning. Ample evidence blames heavy technology use for impeding the emotional maturation of social skills in young people.[2] A major drawback of engaging technology socially is, as Baroness Greenfield put it, that there is no one to look in the eye. A number of TED talks explore why FaceTime and Skype are less satisfying than genuine eye-to-eye contact: computer and smartphone cameras are not well positioned to capture the wealth of expression in human faces.

Stone Age brains evolved in a setting of social cooperation and two-way, back-and-forth engagement. When cues are salient and compelling, social interaction can connect people emotionally. Whether information arrives through sight, sound, touch, smell, or taste, the brain's emotional networks react hundreds of milliseconds faster than those for perception or reason. This may not sound like a lot until you remember that brains operate at drastically lower speeds than electronic circuits. The difference is significant, and precisely why initial responses must be automatic and not slowed down by detours for conscious consideration. It also bears reminding that the brain is predictive, never waiting passively for some stimulus to arrive but actively anticipating what might be coming next. Predictions can cross over between sensory modes of perception. The time lag is sufficient, for example, for the brain to see the articulation of a speaker's mouth before hearing what they say, and to use the visual information to predict what they are about to hear.[3]

Young brains are distractible because white matter tracts and executive areas take three or four decades to reach their full functional maturity.[4] The long-term memory that lies behind learning requires protein synthesis, of course, but an old neurological adage adds, "To learn is to myelinate," meaning that the brain must create not only new nerve cells (neurogenesis) and the connections among them (synaptogenesis, which peaks between six and nine months) but also the supporting glial cells that sprout in response to experience. The role of hands-on experience in myelination makes it a powerful force behind anatomical plasticity and mental flexibility. This is particularly relevant in prefrontal areas that are involved in judgment and discrimination. Besides supplying neurons with oxygen and nutrients and cleaning up debris from dead neurons, glial cells do another important thing. They wrap a fatty insulation called myelin around nerve fibers, which boosts the velocity at which impulses zoom down the neuron's axon (saltatory conduction along the nodes of Ranvier). Myelinated nerve fibers need less energy to conduct action potentials even though those action potentials travel at faster transmission speeds than along unmyelinated fibers.[5]

Originally dismissed as inert insulation, myelin turns out to be functionally quite active. Besides reshaping white matter tracts in response to experience in every mammal yet studied, it also affects the timing of information flow among nerve cells. Maturation occurs earlier in sensory systems than in those for motor control. Crucially, the bulk of myelination takes place early in life—reason enough to ask whether exposing infants, children, and adolescents to the energy drain that screens impose might not have deleterious and long-lasting consequences. Older teens will quickly pick up whatever new technology they need for learning or the workplace, but how do you reverse stunted brain development that is inflicted by too early, too frequent screen exposure in the susceptible young?

Developmental illnesses such as schizophrenia, mental retardation, and bipolar illness all feature prominent defects in myelination. Adverse early life experiences such as social deprivation particularly interfere with the brain's expected myelination. The correlation is disturbing in light of the nonchalance with which we shove screens in front of a child's developing central vision while simultaneously displacing face-to-face interaction with actual people.[6] The iPad is the worst babysitter ever precisely because it monopolizes a youngster's limited bandwidth of attention. A young brain must simultaneously resist external distractions and its natural inclination to chase novelty. But this is a losing battle. Once in school the same youngsters must then come to grips with the frustration of having to master difficult, unfamiliar coursework while pushing back at tech's countervailing scripts that promise quick and easy gratification.

Without question, smart devices are cool. They offer attractive benefits but nonetheless remain addictive. Years ago my spouse and I used simple TracFones for making the rare outgoing call. But while traveling once in Manhattan (which is laid out on a grid, for goodness' sake) we became frustrated trying to find an address. At the time, public pay phones and their attendant phonebooks had been ripped out. We resolved on the spot to get an iPhone, "if only for the maps."

We were late adopters; it took about a decade for the share of Americans with a smartphone to go from zero to 80 percent. I had fun playing with the new gadget that everyone had been talking about for years. And then during the first days of fiddling with it, I said, "No wonder these things are so addictive!" I was staring behavioral addiction in the face. Both behavioral and physical addictions activate identical brain circuits and produce the same feelings you get from repeated actions such as those involved in pathological gambling, compulsive sex, online gaming, overeating, anorexia, smoking, or obsessive screen engagement. A cardinal feature of both physical and behavioral addictions is the level of ambivalence in them. An addict wants to stop but by definition cannot. They may beseech others for help even as they deny they have a problem. They promise themselves (and families, courts, and employers) that they will quit—only to break one resolution after another. The diabetic intends to lose weight, knows he must, but then gobbles down two buy one, get one free—known as bogo—cartons of ice cream. The smoker proudly eliminates his stash of cigarettes in the car and at home only to bum smokes at the office. I have watched an addicted acquaintance jab a needle in his thigh while pleading, "Dear God, help me stop!"

Lapses like these lead outsiders to moralize and blame lack of willpower. But addiction has nothing to do with willpower, as the new class of GLP (glucagon-like peptide-1 agonist) drugs like Ozempic and Mounjaro have proved. Genetics plays a part. So does environment. Together the two render some people more susceptible than others. The brain disease model of addiction has gained popularity but is perhaps more prominent than it ought to be. Addiction is not only a biological phenomenon, it is also a set of complex interactions with the environment. Dopamine is not the central actor, either. Some addictive drugs, such as cocaine and amphetamines, feature large releases of dopamine, whereas others, such as nicotine and alcohol, do so very little or inconsistently. Nor, as I've said, is dopamine strictly related to pleasure as was thought earlier; it is much more important for wanting or anticipating a reward rather than for enjoying it. Dopamine also turns out to be important for other aspects of cognition, such as memory and motivation.

* * *

There are parallels between compulsive screen use and drug addiction. Addictive drugs affect the brain's reward system, while the exact mechanisms in the case of screen use are not yet clear. They probably target evolutionary behaviors that affect survival and result in pleasure, thus positively reinforcing the behavior. Through negative reinforcement, screen use can also relieve stress the same way that drug and alcohol use do. The two mechanisms of reward and relief may explain why some forms of screen use are more likely to be addictive than others, but they do not clarify why some users become addicted while others are enriched by the same exposure. In January 2022 the eleventh revision of the *International Classification of Diseases* (ICD-11) added online gaming and gambling disorders to its roster of addictions, perhaps indicating how this authority is trying to come to grips with this complicated field.

Matthias Brand at the University of Duisburg–Essen has looked at the neural pathways behind addiction in a fresh way and reduced them to two circuits he calls "feels better" and "must do" (figure 5.1).[7] The feels-better circuit operates during both the positive reinforcement phase of pleasure and reward and the negative reinforcement of relief from stress, dysphoric mood, and craving. The must-do circuit encompasses habitual and compulsive behaviors and seems to engage later on. Brand's "simplified model

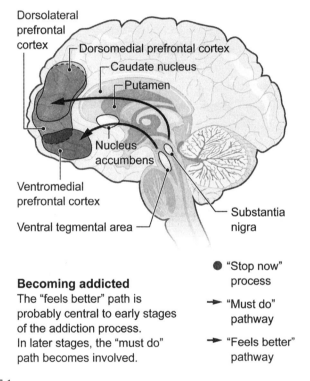

FIGURE 5.1

Neural pathways in human addiction illustrate the hypothesized main circuits in addictive behaviors. The "feels better" path includes both positive and negative reinforcements, along with relief from cravings. The "must do" pathway addresses compulsive behaviors and automatic, habitual responses to stimuli. "Stop now" concerns self-control, which the model suggests is out of balance with the two driving pathways. *Source:* Matthias Brand, "Can Internet Use Become Addictive?," *Science* 376 (2022): 798–799. Reproduced with permission. [See also color plate 2.]

is that individuals with online addictive behaviors have an imbalance between the driving paths and those underlying self-control." Functional connection imaging (tractography) and cognitive analysis support the similarities between substance addiction and the online kind.

While society isn't about to absolve distracted drivers of responsibility when they kill people and plead that they were addicted, one solution may be to establish environments that minimize the chance of addiction (by, e.g., surrendering devices to a phone box prior to classes, meetings, or athletic practice; parents inculcating good media habits and saying no to mandatory tech in schools; jamming WiFi and cellular signals). True, initial choice is involved, and while addiction complicates that choice—namely, the subsequent steps one must commit to in order to stop the addiction—it does not eradicate it. It also doesn't help that tech companies willfully exploit our psychological weak points to keep us engaged as long as possible. This is morally offensive, doubly so because corporations have made it the individual's burden to undo the addiction that they themselves have made so easy to succumb to.

Even when digital devices fill us with anxiety, many still can't keep away. This is the ambivalent tang many feel but don't know how to resolve. Berating yourself for time wasted online is counterproductive because shame and anger strongly predict relapsing into the very behavior you say you want to stop. Logically, addicts should be constantly high if their reward circuits were always switched on. But they aren't, despite a flood of pleasure neurotransmitters marinating their brains. What is more, as a physician I have never known an addict who *wanted* to be addicted. You keep checking your newsfeeds or influencer stats, but if doing so were truly pleasurable, would you still complain about being addicted to your phone?[1]

* * *

The flood of factoids that spew from your screen doesn't constitute knowledge, to say nothing of power, as often assumed in the phrase "knowledge is power." Knowledge by itself can never change behavior. If it could, then no one would smoke, scarf down junk food, procrastinate, or show up late. We'd all be thin, perfectly fit, and well rested. Numerous website-blocking apps fail to make us virtuous because users circumvent whatever restrictions they impose on themselves when in a temporarily resolute frame of mind. Longtime teachers tell me that no school firewall will keep determined kids out, and that unbeknownst to most parents, the youngest students are gorging themselves on pornography and sensational content: one second grader was sending around a video of an ISIS beheading. Teens smuggle burner phones to defy parental restrictions even as, according to the Pew Research Center, 64 percent of them freely admit to spending "too much time on their phones."[8]

Knowing how to contextually relate new facts to what you already know—connecting the dots, in common parlance—and acting on that knowledge is the real power. But fewer and fewer people know how to connect the two. By themselves, modern devices don't confer power because they only serve up factoids. What you do with them is the point. We are in the mess we're in because corporations have deftly commandeered our Stone Age brain, which can't help but glom on to novelty and change because that is what it evolved to do. Each time it redirects the focus away from one thing to another it throws your nervous system into high alert, the way a radar operator who shouts "Incoming!" alerts the rest of the squadron.

When push notifications, texts, and alerts have become irresistible—when they make you glance away for even a moment from what you are doing—then you are behaving like Pavlov's dogs. Aside from the classic psychological conditioning involved here, loss aversion is also at work. Loss aversion is an age-old script woven into the reptilian part of our brain, one of many unconscious forces that prod us to act in certain ways. These reflexive, unthinking responses were crucial to our ancestors' survival. Today the same automatic script still plays out even though many scripts have lost their usefulness or become counterproductive.

Dr. Clay Shirky, professor of media studies at New York University, has studied human computer interaction for nearly four decades. When he first began teaching, he was laissez faire in allowing students to use laptops, tablets, and smartphones during class. And then he banned them. What had changed? I asked. "The level of distraction in my classes seemed to grow," he said, "so I had to demand lids down." He was familiar with research showing that multitasking degrades the quality of work. But what surprised him was seeing how distractibility fanned out to affect others. Those not personally engaged with a device but within sight of a multitasker performed worse than students not similarly seated. Realizing that "the degradation of focus is social," Professor Shirky likens allowing social media in the classroom to bringing a boom box to a conversation: it will degrade the experience of everyone.[9]

The draw of a smart screen is so profound that its mere presence is enough to disrupt attention, memory, and relationships through what has been dubbed technoference. Something about simply having a mobile device nearby is aversive. The *Harvard Business Review* examined how "merely having" a smartphone nearby takes a toll on thinking. In two experiments, eight hundred people engaged in a task requiring arithmetic, attention, and memorization. Those whose left their phones in another room performed best, followed by those whose phones stayed in their pockets. Those who did worst were permitted to leave their phones in view. The latter's cognitive capacity fell to a level on par with those suffering chronic lack of sleep. Even resisting the urge to look when your phone dings or lights up undermines mental performance because it takes energy to resist the novel stimulus. Hearing your phone ring when you are otherwise occupied automatically raises your anxiety level. Missing the call or text makes your mind wander, which in turn further affects your concentration and efficiency.[10] Evidence mounts that the mere presence of smartphones disrupts fundamental cognition and the capacity of working memory.[11] Receiving a single text notification disrupts focus by prompting "task-irrelevant thoughts." A 2019 prospective study in *Pediatric Research* found that merely having a bedroom television during preschool years "does not bode well for long term cardio-metabolic wellness, mental health, and social relationships" when these factors were later reassessed at age twelve. Upon reaching that age, study participants were fatter, more emotionally distressed, physically aggressive, and less social.[12] My concern, therefore, would seem to extend beyond just screens of the smart kind.

Some contrary studies argue against the idea that lazy reliance on technology can lead to so-called "digital dementia," particularly a breakdown of memory retention. One study suggested that taking a picture causes a shift in attention toward the visual features of an experience and away from its auditory aspects, in this case paying attention to an actual museum exhibit. Participants could better recollect parts of the exhibit they had photographed than of parts that they hadn't. Subjects using a camera during their museum visit identified even nonphotographed aspects better than participants who hadn't brought a camera along.[13] In another study, psychologists at University College London found that taking a smartphone photo improves recall even for unsaved information. But this benefit comes at a cost: when the phones were taken away, subjects remembered low-value items but had forgotten the high-value ones they had entrusted to their devices. As the senior author says, "If a memory tool fails, we could be left with nothing but lower-importance information in our own memory."[14]

* * *

It would be lovely if we had technology like that depicted in the *Matrix* films that allowed us to jack in and download a chosen skill directly into our brain. Unfortunately, real-world learning takes time. A lot of time. Acquiring a skill likewise takes practice. So too does learning how to focus and sustain one's attention. We cannot speed up thinking the way self-appointed experts like Evelyn Wood once promised to teach speed reading and largely failed, because the demands of her method exceeded our set bandwidth for attention.[15] We cannot speed up personal experiences like intimacy. And what would it even mean for personal relationships to be more efficient? The Roman statesman Seneca said that "the love of bustle is not industry. It is only the restlessness of a haunted mind."

Such time-honored sentiments are inconvenient for champions of accelerated learning and early immersion who cannot tell the difference between enrichment and overload. Tech moguls won't let their own children near digital devices, and increasingly neither will neuroscientists. My colleague David Eagleman at Stanford University told me that he and his wife, also a neuroscientist, didn't allow their children to access "screens of any sort" until past the age of six. "The winners of tomorrow," he said, "will be those who can focus and sustain their attention for long periods of time."

Two common traits frequently mentioned in education circles as desirable are "critical thinking" and "connecting the dots," activities defined vaguely if at all, although book-length treatments such as Jonathan Haber's *Critical Thinking* does a good job explaining our "propensity to believe fake news, draw incorrect conclusions, and make decisions based on emotion rather than reason."[16] A capable mind thrives on quiet, not push notifications. It wants interludes of uninterrupted time rather than streaks, banners, badges, autoplay, and frantic look-at-me activity that turns friendships and achievements into competitions. Creative people don't necessarily know more facts; they can connect more dots because they see more dots than others do. They are attentive; their mind is open and receptive. Creative minds need space in which to wander without an agenda. Yet what tech columnist Christopher Mims calls "the distraction-industrial complex" does exactly the opposite by design. "Entire empires have been built on the push notification."[17]

Think for a moment about doodling: it flows out of an individual without conscious effort. Compared to the helter-skelter interruptions common to digital media, doodling is a great way to let your mind wander. A blank page, a napkin, even a scrap of paper can turn magically into a playing field on which the unconscious mind can draw. The *Oxford English Dictionary* defines a doodle as "an aimless scrawl made while a person's mind is more or less otherwise applied." This sounds a lot like Mihaly Csikszentmihalyi's concept of flow, a psychological state of immersion so complete that the ego falls away and time for a moment seems to slow. Colloquially, we describe this state as "being in the zone." When wholly immersed in a task it feels neither effortful nor much like an effortful task. No one sets out to doodle, but doodling shows that a person's brain can be simultaneously occupied in a conscious task such as talking on the phone while also solving creative problems or generating ideas intuitively. Whatever it draws on appears to be established early in life because an individual's doodles tend to be consistent and specific to a particular individual in the same way that one's signature is.

In *The World beyond Your Head* Matthew Crawford at the University of Virginia's Institute for Advanced Studies in Culture argues that technological overload is more symptom than cause. Paying attention becomes a cultural problem only when public surfaces become marketing billboards—from the miniscule to the grand to the inescapable. Ads appear on escalator handrails, airport security trays, even on credit card terminals during the few seconds it takes to authorize a purchase. Attention is no longer ours to direct at will. "Ours is now a highly mediated existence in which . . . we increasingly encounter the world through representations."[18] The disembodiment of perception—taking it literally from the natural world into an abstract digitized realm—and the growing scarcity of face-to-face engagement cause anomie and isolation in our supposedly always connected world. And what could be more mediated than the highly touted "metaverse" that is supposed to bring people together for virtual interactions and engagements such as concerts and hangouts?

Amid all the hype, no one seems to know what the metaverse is other than a revamped Second Life that you will spend in a holographic reality. Compared to picking up the phone, it sounds like a lot of work and not worth the bother. Perhaps metaverse is just a fancy word for old-fashioned cyberspace. Its enthusiasts promote it as a fundamental shift in how we interact with virtual reality, from sitting in front of handheld computers to being inside them. Though presented as a wider representation of digital space, it demands that we be always online, removed from physical reality as you and your avatar interact with others in a brave new digital environment. I can't imagine a more distracting environment or one antithetical to focusing one's attention.

Back in the real world, automobile manufacturers offer voice navigation and blind-spot assist to help keep you on the road and out of the ditch. Yet these features simultaneously take the driver's mind off the road by drawing it to ever larger dashboard screens and active side mirrors that beep at you. Computer chips in cars impose layers of electronic mediation that leaves the driver with less and less to do. By design, these technological innovations foster disengagement.

A small but growing body of neuroscience research is rediscovering what Marshall McLuhan, the iconic philosopher of media theory, surmised five decades ago:

> It is not only our material environment that is transformed by our machinery. We take our technology into the deepest recesses of our souls. Our view of reality, our structures of meaning, our sense of identity—all are touched and transformed by the technologies which we have allowed to mediate between ourselves and the world. We create machines in our own image and they, in turn, recreate us in theirs.[19]

McLuhan's famous adage, "The medium is the message," tells us not to scrutinize the information a delivery system conveys but the way in which that very system manipulates our perception. If we allow vested parties to choose for us and if we do not demand to know what they are doing to us, then we are passive, apathetic participants in the largest social science experiment in history, one being conducted without our consent. In *Understanding Media* McLuhan says:

> The "content" of a medium is like the juicy piece of meat carried by a burglar to distract the watchdog of the mind.[20]

The ultimate mediation today is the interposition of third parties between reality and oneself. The billions of smartphone photographs taken every day are but one example. If you are always taking pictures, how much do you honestly see what is in front of you? Chances are high that you are looking but not seeing when occupied in picture taking, focused on getting it right so that the image is good enough to post online. Not long ago when digital cameras were less ubiquitous, more complicated, and more expensive to own, a roll of film had a capacity of either twenty-four or thirty-six pictures. Taking a photograph took consideration. Today the cost of snapping a hundred pictures is negligible, and people publicly document the most mundane moments of their lives. Taking a large cache of pictures endows each of them with equal weight, and when everything is important, then nothing is important. I have watched in amazement people at the Chelsea and Philadelphia Flower Shows go through the motions of living through a viewfinder. Whether framing a selfie or trying to capture the grandeur of French châteaux, picture hounds never pause to take in the magnificent vistas before them. I too have sometimes tried to capture a spectacular vista, but the pictures on my smartphone pale in comparison to the scene I am looking at or the one registered in my memory.

Reading prose larger than a screenful at a time is an excellent way to hone your attention, especially when it is competently written and its progression is logical. "Linear" reading means following a sequential line of reasoning as a story unfolds. Following a path like that can be a restorative tonic for a mind degraded by interruptions and left to flit from one thing to another. In the same way that you exercise a muscle, reading strengthens your ability to focus and resist the intrusions that media companies have become deft at hurling your way. Linear reading builds emotional muscles, too. Individuals who habitually read fiction or narrative nonfiction have greater empathy than those who are not regular readers. Narrative requires you to impute motives and intentions, anticipate a character's actions, and predict upcoming turns of plot.[21] As Pulitzer Prize–winning novelist William Styron put it, "A great book should leave you with many experiences and slightly exhausted at the end. You live several lives while reading."[22] Multitasking, on the other hand, is mentally draining because it burns a lot of glucose, the brain's primary fuel.[23] Like muscle cells, living neurons can become metabolically fatigued. By contrast, Styron's pleasurable "exhaustion" from reading is the satisfying glow from fresh ideas that challenge and enliven your imagination.

Without a speck of irony, former Google chairman Eric Schmidt admitted that "the overwhelming rapidity of information . . . is in fact affecting cognition. It is affecting deeper thinking. I still believe that sitting down and reading a book is the best way to really learn something. And I worry that we are losing that."[24] Mark Zuckerberg's New Year's resolution once was "to read books."[25] The then Facebook chairman invited millions of his followers to read a new book with him every two weeks, boasting, "Books allow you to fully explore a topic and immerse yourself in a deeper way than most media today. I'm looking forward to shifting more of my media diet towards reading books." The moral contradiction between what he was advocating and what he was peddling evidently eluded him. The book club, called "A Year of Books," quickly went defunct.

In terms of moral vapidity, nothing quite equals the willful ignorance of accused fraudster Sam Bankman-Fried, who lost billions of his clients' money while living high on the hog. During an interview with Adam Fischer at Sequoia Capital he boasted, "I would never read a book."

> I don't want to say no book is ever worth reading, but I actually do believe something pretty close to that. I think, if you wrote a book you fucked up, and it should have been a six-paragraph blog post.[26]

In his belief that merely taking away instrumental facts is what matters, what escapes Mr. Bankman-Fried is that one reads for the pleasure of reading, not just to have read something. Similarly, rapper Kanye West (having now decided to call himself Ye) boasts of his disdain for reading, oblivious to how it indicates a much larger deficit of character.[27]

6

Silence Is an Essential Nutrient

The human body cannot produce all the nutrients it needs to function. A nutrient is therefore "essential" if it must come from outside the body—in most cases from the diet, with one exception. Micronutrients, which a person needs in small doses, include vitamins and minerals such as folic acid, B_{12}, thiamine, and iron. A deficiency here can lead to illnesses such as anemia, scurvy, pellagra, or rickets. Macronutrients must be ingested in larger amounts; these include water, protein, carbohydrates, and fats. Essential fatty acids are omega-3 and linoleic acids, without which we can't survive. Every cell in the body also needs protein to function properly. Of the twenty amino acids humans use to synthesize hormones and neurotransmitters, nine are indispensable and must come from an outside dietary source. The one exception is sunlight, which meets the requirements for an essential nutrient. Without it, the skin, our largest organ, could not synthesize vitamin D. More important, without the daily cycle of light and dark our circadian clocks could never reset and our sleep-wake cycles would lapse into free-fall chaos.

I consider silence another nutrient essential to health and well-being. It is remarkable how much people obsess over what foods they put in their bodies: organic, non-GMO, vegan, gluten-free, lactose-free, sugar-free, no artificial coloring, and so on. But why aren't they as picky about what they ingest through the senses? Mental garbage we take in is arguably more harmful than an occasional cheeseburger.

Some individuals fast for health reasons or as part of religious observance, while others fast to sharpen their awareness. What if it were possible to indulge in a sensory fast for a day or even an hour, free from texts, tweets, videos, emails, and other forms of digital junk food? We might call it a *caesura*, to borrow the term for a quiet pause in a musical composition or a line of poetry. In vocal scores, a caesura allows singers to catch their breath—an apt metaphor for the digital maelstrom we live in.

It has become impossible to find a quiet seat in a waiting room, a doctor's office, or an airport lounge without televisions and piped-in music assailing us. High-definition screens spew out visual noise, beckoning us to turn from whatever we are doing and LOOK HERE. You may believe, as the Borg of *Star Trek* insist, that "resistance is futile," but sometimes you must shut out the world in order to open your mind. The Borg are a collective race, a partly organic, partly cybernetic hive mind whose singular purpose is to subdue species, expropriate their technology, and thereby "raise the quality of life" of the civilizations they forcibly "assimilate" (It sounds a bit like Mark Zuckerberg's insistence on making us all "more connected").[1] Borg units are never aware of themselves as individuals, only as part of a collective conscious. Each unit hears the constant drone of the hive and cannot disengage from it. The hive mind is the ultimate in crowd sourcing: always "on," always connected, always engaged in real time. The Borg represent the modern fear that technology may overtake humanity, and my *Star Trek* example is a representation of the way screen media infect us right now. Universities such as Duke, Georgetown, and Syracuse teach *Star Trek*–related courses in genetics, evolution, philosophy, and social media. As far as I know, none addresses the challenge of being able to disconnect.[2] In *A World without Email*, Georgetown professor Cal Newport calls the hive mind a workflow centered on unstructured and unscheduled digital messages. He suggests regular office hours as a solution and advises students to obsess over quality rather than quantity.[3]

Silence has become the ultimate luxury. After *Washington Post* columnist Richard Cohen cut his cable TV, he reminisced about a time when "an ordinary American could close the door and keep the world at bay." Now, "only the offline can be serene. . . . The world comes elbowing in every time you go online."[4] If you want to escape the commotion you must pay to join exclusive airport clubs or VIP sections. Or at least shell out for a pair of noise-canceling headphones. It is silence rather than constant badgering that gives a person room to think. Our Stone Age brain evolved over eons hearing only the sounds of nature and often, for long stretches, nothing else at all. You have only to walk in the woods, stand on a beach or in the shade of a tree, to appreciate how restorative natural quiet can be.

From a psychological point of view, silence is more than the absence of sound. It is a singular mental space. But listening to your inner self is impossible when newsfeeds, notifications, and the infinite scroll constantly address you and hold you hostage. Rather than fearing silence as emptiness—which many people dread—you can shift perspective to see it as an essential nutrient brimming with possibility. There is a reason why yoga and meditation aim to distance body and mind from the everyday buzz: a caesura is good for your neurological health and mental well-being. One meditation aid is a kōan. A famous kōan goes, "Two hands brought together make a sound. What is the sound of one hand clapping?" The more your intellect tries to solve it, the farther it pushes an answer away. Sōtō Zen, the Serene Reflection School of Buddhism, says, "Just sitting is the essence of pure Zen." Just sitting with eyes open, trying neither to think nor not to think (closing the eyes invites daydreaming), is the natural kind of silence that constitutes an essential nutrient.

States of absence and nonoccurrence are physical as well as mental states. Yet we are more likely to notice when something happens rather than take note of an event that does not happen, as in the case of Sherlock Holmes's dog that didn't bark. The dog didn't wake up the household because it was familiar with the murderer, its owner; from the dog's perspective nothing had changed.[5] When *Gestalt* psychologist Kurt Koffka wondered why we "normally see things and not the holes between them," he was asking why we associate objects and people with one another rather than the space that surrounds them. This might seem an odd thought. But artists attune themselves to seeing meaningful negative spaces, pauses, and gaps. Cultural differences likewise condition what we see and hear or else fail to discern. Consider a row of columns in front of a building or a stand of trees set in a line. Where the Western mind sees only empty space between them, the Japanese mind apprehends both the physical elements and the interval, or *ma*, that separates them. The word "ma" translates as "gap," "pause," or "space between two parts." Violinist Isaac Stern called this space "an emptiness full of possibilities . . . the silence between the notes which makes the music."[6]

Listening for sound gaps can lead to more restorative forms of silence. Unfortunately, Western minds see quiet as a negative hole, whereas the Japanese have five words to describe an aesthetic of full, voluptuous quiet. *Sabi* ("loneliness") is the beauty of a solitary object such as the lone ancient

FIGURE 6.1
The top of the *wabi-sabi* ideogram represents quiet, the bottom solitude. Together, they mean "tranquility."

pine clinging to a mountainside, its branches molded by the wind. *Wabi* ("poverty") is the quality of things as they are, the poignancy of the simple and commonplace. *Shibui* ("bitterness") conjures the taste of green tea, an object reduced to its essence. *Awarē* ("pity") speaks to fragility and the transience of life. And *yugen* ("hidden" or "obscure") is the reality behind appearances, like the snowy heron hidden by bright moonlight or the vague object at the bottom of a pool. These nuances open a constellation of quiet. In the *wabi-sabi* ideogram (figure 6.1) the top character represents quiet, the bottom solitude. Together, the two mean "tranquility."

American architect Leonard Koren calls *wabi-sabi* the capacity to find beauty in imperfection; in that which is simple, slow, and uncluttered; in the natural cycle of growth and decline.[7] The concept's appeal, he suggests, lies in its ability to instill physical calm and counter the "pervasive digitalization of reality." Fear of missing out has become our default state, he says. "Today being an American requires being tethered to digital gadgets," the "de facto national policy. Is it an absurd way of life? I think so."[8] Japanese cinema often calls on the five concepts of quiet to emphasize the simplicity of everyday life. In films such as *Tokyo Story* or *Spirited Away*, scenes of pouring tea or walking through a forest can be as crucial to a story's essence as a lightsaber duel is in a *Star Wars* film. The books of Yasunari Kawabata, especially *Snow Country*, are likewise good places for American minds to learn how to see elements that otherwise seem foreign or simply not there.

Can the digital world be said to have *wabi-sabi* traits, or is making oneself quiet wholly an inside job? The Stone Age brain has always known periods of mental stillness, which is not the same as physical quiet or solitude. The latter has two dimensions: staying centered amid commotion and being physically alone. People often mistake solitude for loneliness, but the distinction is crucial. Solitude has; loneliness wants.

* * *

The physical isolation spurred by the coronavirus pandemic led to a big uptick in complaints of loneliness. Perhaps some confused solitude with loneliness, as my example above showed. I suspect that many newly isolated individuals had not previously had much chance to explore the pleasures of solitude given the baseline of near constant stimulation that reigned before the pandemic. When you are unpracticed at taking a pause, being suddenly alone with your thoughts can indeed feel frightening.

Yet everyone can spare thirty seconds. Moments of stillness snatched here and there quickly add up, particularly for a mind that already knows how to savor interludes of calm. I snatch moments of it while sitting at a red light, waiting in line, or buttoning my shirt in the morning. I can find quiet while walking up the stairs, folding laundry, or brushing my teeth. When addicted to your phone you render yourself open to nomophobia, a word that elides "no mobile" and "phobia." It is the fear of not having immediate access to your mobile device, and its anxiety is associated with elevated levels of the stress hormones cortisol, epinephrine, and norepinephrine.[9]

In *The Revenge of Analog: Real Things and Why They Matter*, journalist David Sax argues that devotion to digital devices sacrifices "first-hand experiences we have given up in the name of progress." Some people feel the loss instinctively: Sax points to the resurgence of fountain pens, vinyl LPs, photographs taken on film, and Moleskine notebooks for jotting down sketches and ideas. Ancillary aspects of vinyl records are their sensuousness: the "physical browsing of album spines," the "careful examination of the cover art," the "diligent needle drop and that one-second pause between its contact . . . and the first scratchy waves of sound." Analog aficionados speak of the "finishability" of reaching the end of a book, or the reading experience in terms of connecting with the material in a way they cannot when reading on a tablet. Having only one page visible in the window of an endless scroll forfeits a sense of place. What makes one tool superior to another has nothing to do with how new or cool it is, says Sax. "What matters is how it enlarges or diminishes us."[10] Oxford's Charles Spence has written eloquently on the multisensory experience of handling and reading physical books.[11]

Part of the renewed interest in analog objects is the sense of nostalgia they confer, a reasonable sentiment in response to the loneliness wrought by products that tech titans promised would make us "more connected." Yet look how common it is for people to whip out their phones at the first possibility of face-to-face engagement. Likewise, you can read the visible anxiety in people who must keep the television on constantly for what they call "background noise." You see it in the need for occupation with mindless activity and the rapid saturation of mobile games like Pokémon GO. It took radio thirty-eight years to reach an audience of 50 million, twenty years for the telephone to do the same, three years for the iPad, and just three days for the Pokémon app.[12] The latest fads have swarmed over us even faster. This onslaught has left many people no longer able to tolerate silence, mental quiet, or being alone with their thoughts. Instead, they soothe themselves with attention-sapping apps. Like a security blanket it may assuage anxiety, but at the long-term cost of peace of mind.

Evolutionary psychologists' conjecture that the ability to disengage from immediate surroundings is why early humans advanced beyond impulsive, knee-jerk behavior. Our ancestors developed imagination and the ability to consider hypotheticals removed from present concerns. But today, contemplating one's thoughts feels strange. Deliberately "thinking for pleasure is cognitively demanding" and requires "mental resources that people are either unwilling or unable to devote to the task," says University of Virginia psychologist Timothy Wilson.[13] When in *De Brevitate Vitae* ("On the shortness of life"), the Stoic philosopher Seneca says "life is long if you know how to use it," he is referring to the existential shock that comes on realizing that much of one's life has been squandered on unproductive activities. Even scarier is the prospect of getting to know yourself better.

Silence: The Power of Quiet in a World Full of Noise, by Thích Nhất Hạnh, the Vietnamese monk and peace activist, acknowledges that the central "goallessness" of Buddhism often baffles Westerners. Zen teaches that whatever you are chasing already resides in you. "To practice solitude is to practice being in a singular moment," he says, "not caught in the past or carried away by the future, and most of all not carried away by the pressure of the crowd." You can work among people, go about your routine, and still enjoy the pleasures of silence and solitude "while being aware of every feeling and perception that's happening in yourself."[14] That is the inside job.

If you are feeling lonely and long for connection, you might imagine a drug that induces a sense of intimacy without your having to put in the effort to make and retain friends. Aldous Huxley's *Brave New World* featured just such a pill, called Soma, that washed an individual's troubles away. "There is always Soma, delicious Soma," a character says, "Half a gramme for a half-holiday, a gramme for a week-end, two grammes for a trip to the gorgeous East, three for a dark eternity on the moon." But what if a drug for connecting you to others and to nature were something as simple as taking a walk in the woods? There is.

Sinrin-yoku became popular in Japan during the 1980s as a respite from that country's intense work ethic and cases of people literally working themselves to death. It translates as "taking in the forest atmosphere" or "forest bathing." The government spent millions promoting it, and the Tokyo-based Forest Therapy Society certifies "therapeutic" forests and urban wooded paths. The concept is new in America, but the Association of Nature and Forest Therapy certifies forest-bathing chaperons nationwide.[15]

Its therapeutic benefit may derive from inhaling phytochemicals: aromatic compounds given off by trees and plants (from the Greek phytón, meaning "plant"). In the laboratory, phytochemicals stimulate immune system killer cells, a type of white blood cell that secretes enzymes that can kill bacterial infections and some kinds of tumors.[16] Early pharmacopeia manuals in England and colonial America were heavily plant-based. Further back, the Egyptians used willow bark, from which we later derived aspirin, to treat pain and fever. Science today confirms that numerous airborne molecules that we cannot consciously detect influence behavior and mood. In principle, the idea of inhaling therapeutic plant vapors is no different from taking them by mouth or injection.[17]

The idea of taking a pause also comes into play because brief timeouts help reinforce long-term learning and productivity. You can learn more and access that learning faster when coming out of downtime than if you had never taken a break. Indulge in something mindless, gaze out a window, find a comfortable chair, and simply breathe. Rest is not idleness, as scans of the brain's default network show. You may feel that you are relaxing, but your brain never stops working in the background, or in meaningful ways.[18]

An objective of taking restorative walks is to slow down and stop neurotically busying yourself with a bevy of concerns. "What did Hillary [Clinton] do after she lost" the 2016 election? asks the Association of Nature and Forest Therapy. "She went walking in the forest" (their answer could well have been metaphorical).[19] We did not evolve to live in urban metropolises cut off from the sensuous embrace of nature. Humans can adapt to the harshest of environments as well as the most luxurious ones, but we cannot ignore our connection to the Earth and the planet's rotation without suffering consequences.

Exposure to forests and leafy landscapes has been proven many times over to change mood and perspective, to make people measurably less anxious and less prone to ego-centered worry. Ruminating, in the sense of chewing things over the way a ruminant animal like a cow chews its cud, is strongly correlated with depression and by nature is antisocial.[20] The acclaimed science fiction writer Ursula K. Le Guin reminisced about the solitary walks she took during World War II after her brothers had deployed overseas. "The summers in the Valley [Berkeley, California] became lonely ones, just me and my parents in the old house. There was no TV then; we turned on the radio once a day to get the war news. Those summers of solitude and silence, a teenager wandering the hills on my own, no company, 'nothing to do,' were very important to me. I think I started making my soul then."[21] Two centuries earlier, Romantic-era poets advocated a therapeutic back-to-nature regimen. Germans in Goethe's era called it *Naturliebe* ("love of nature"), while America had proponents in Henry David Thoreau's *Walden* and biologist E. O. Wilson's *Biophilia*. Thoreau and his fellow transcendentalists considered that communing with nature provided better benefits than traditional religious practice.

Studies in Japan, England, and the United States recently showed that fifteen minutes spent in the woods lowers blood pressure and concentrations of the stress hormone cortisol. After forty-five minutes, cognitive performance measurably improves. Green city parks work just as well as remote, wooded mountains. As little as five hours a month spent amid urban greenery has measurable benefits on mood and alertness, and the effect lasts for hours. Kids with ADHD who regularly play in parks exhibit milder symptoms compared to those kept indoors. Observed benefits scale up to the population level as well: In Britain, improved access to parks and green spaces has reduced the use of mental health services. If the average American spends 90 percent of the time indoors, then greater exposure to the forests or anywhere outdoors—without bringing your phone along, or at least turning it off to forestall notifications—would seem to promise benefits.[22]

Famous walkers shed insight as to walking's benefits. During her habitual, meandering walks Virginia Woolf honed her ability to portray consciousness and the character of thought. In one of her last novels, *The Waves*, she refracts six separate consciousnesses into one mind, the biographer named Bernard. Woolf uses this unifying character to explore different aspects of consciousness. In her later biographical essay, "A Sketch of the Past," Woolf said that *To the Lighthouse* burst forth while she was walking, "in a great, apparently involuntary, rush. . . . Blowing bubbles out of a pipe gives the feeling of the rapid crowd of ideas and scenes which blew out of my mind. . . . What blew the bubbles? . . . I have no notion."[23] With respect to "what blew the bubbles" is the axiomatic observation that artists, when inspired, often don't feel like the creator of their own work, the very thing that others ascribe uniquely to them.

Carving out mental space and freeing the mind of deliberate thought is a proven incubator of creative insight. In his essay "Walking," Thoreau explained, walking is nothing like exercise and "absolutely free from all otherworldly engagements." Nietzsche, too, walked so that he could think. In *Twilight of the Idols* he wrote, "All truly great thoughts are conceived while walking." Essayist Roger Rosenblatt in "The Boy Detective" explores the *flâneur* (a stroller who saunters and observes), a walker without purpose. "Wandering feet reflect the wandering mind, going wherever the associations of thought lead." We learn that Wordsworth was a *flâneur* of the countryside, the American Frank O'Hara a *flâneur* of the city. Walt Whitman was that elegant type of stroller, a boulevardier, whereas Montaigne believed in tracking random thoughts, engaging them as if in a *flânerie* of the soul. Rosenblatt develops a lovely concept: each of us has "two souls" that we take on our private walks, one "for the senses, one for the intellect." The two never meet, yet live connected "parallel lives . . . and side by side move into infinity."

*   *   *

Jonas Braasch, a musicologist at Rensselaer Polytechnic Institute, studies the psychology of soundscapes. He discovered that office workers who listened to a burbling stream performed better on written tests and felt more upbeat compared to associates who took the same tests while exposed to a background of white noise. "They were more patient and avoided more errors," he says, concluding that natural sounds have a restorative effect on our thinking.[24] Judging by the popularity of smartphone apps that let office workers listen to natural soundscapes, they (and their Stone Age brains) appear to yearn for the kind of natural world in which humans spent most of our history. What improves performance even better is an atmosphere of silence, a caesura from the thrum of modern life. While background music may boost your mood, it also disrupts reading comprehension and memory, especially if the music has lyrics. I can work while listening to an opera, for example, but not one in German, a language I understand. A few words are enough to snag my attention and break my concentration.[25]

Besides being easy on the ears, the natural world is also easy on the eyes. Dr. Arnold Wilkins, emeritus professor at the University of Essex, studies the metabolic effects on the brain of screen exposure and viewing artificial, mediated images. Untoward effects include migraines, seizures, fatigue, visual discomfort, and excessive oxygenation of the visual cortex. The biggest culprit is flicker, the nearly imperceptible fluctuation in brightness when either the number of frames per second is too small or the refresh rate is too low to produce the illusion of persistent vision. In sensitive individuals, flicker can induce a seizure. Watching television, reading, or driving past equally spaced telephone poles or a stand of trees that produce light-and-dark flicker are classic triggers of this kind of "reflex epilepsy." When TVs used old-fashioned cathode ray tubes, 50 percent of children who had such seizures had them only while watching television.[26]

Today's LED screens are much brighter, which makes the problem of flicker more noticeable and persistent (the iPhone has one of the few screens that does not flicker). Aware of this issue, Netflix posts a warning before some programs that says, "Some scenes have a strobing effect that may affect sensitive viewers." Increasingly, ambient lighting also comes from LEDs, yet another source of visual and mental strain, and not just for sensitive individuals. In a later chapter on disrupted sleep, I outline how dominant short-wave light emitted by LED screens disrupts the rhythm of melatonin secretion and the architecture of the brain's physiological sleep rhythms. Despite warnings not to do so, people still snuggle up to their devices before bedtime while they multitask or watch a video on a second LED screen. Many adolescents and young adults clutch their iPhones while asleep, and research shows that they still finger and swipe the screen as they do.[27]

I continue to stress that attention is a finite resource, and that the brain consumes much energy taking in mediated screen-based images. This includes movies and television images that shift rapidly and don't allow viewers enough time to make out details before jump-cutting to another scene. "We have designed an environment that is antithetical to what your visual system evolved in," Wilkins says. The spatial structure of natural images that we evolved to process is "scale invariant," meaning that no matter how much you enlarge them, they contain the same amount of detail. The brain processes invariant images efficiently using a relatively small number of neurons. Unnatural images, by contrast, vary with scale. The magnitude of variance determines how uncomfortable and energy-draining artificial images are to comprehend.[28]

Artificial patterns of lines that we see every day—stairways, lighting grids in grocery and department stores, corrugated and reticulated surfaces, the right angles of buildings—induce eye strain and are measurably uncomfortable to look at. One striped pattern we look at constantly is text. Isolating a line of text by masking the printed lines above and below it speeds up reading, whereas small type emphasizes the stripe and slows reading down. Children have traditionally learned to read from books that feature large type. Yet, Wilkins cautions, "modern text is getting too small for children too early in life."[29]

Artificial images induce eye strain and demand abnormally high oxygen uptake in the visual cortex because analyzing them requires more metabolic energy than decoding natural images such as trees, clouds, mountains, or bodies of water. Wilkins's research finds that the visual discomfort is a protective response that dampens oxygen consumption, which would otherwise diminish brain energy reserves. "In nature we don't get images with large color differences," he says, aside from ephemera such as autumn leaves, flowers, and sunsets. This is quite the opposite of Disneyfied and high-definition screen worlds. "We have known for a long time that nature is restorative," says Wilkins. "It's nice to go for a walk in the woods or on the beach. It makes you feel better. Part of the reason is that you're not looking at stripes all the time" (figure 6.2).[30]

Research like his contributes to the relatively new practice of ecotherapy, the prescribing of nature outings instead of pills. Increasingly, doctors around the country are writing prescriptions for trips to the park to assuage anxiety, depression, and chronic illnesses such as diabetes and asthma. Pediatrician Robert Zarr helped establish the nonprofit ParkRx.org, which curates hundreds of green spaces, from grassy triangles where streets intersect to large stately parks, all in the service of helping assuage various ailments. A 2023 study of nearly 62 million U.S. Medicare beneficiaries concluded that spending time in natural environments reduces the risk of Alzheimer's disease, related dementias, and Parkinson's disease.[31]

Designers of grand nineteenth-century urban parks such as Frederick Law Olmsted believed that green spaces provided emotional relief for workers living in rapidly industrializing cities. Hence the Boston Public Garden (1837), New York's Central Park (1857), San Francisco's Golden Gate Park (1871), Atlanta's Piedmont Park (1887), and Washington's Dupont Circle (1889). Dr. Zarr says, "We write prescriptions for all kinds of medicines. Now we're seeing nature and parks not just as a place to recreate but literally as a place to heal yourself." Prescriptions indicate which park to go to, on what days, and for how long because patients are more likely to accept concrete instructions over vague directives such as "exercise more" or "join a gym." The developed world elsewhere is seeing a rise in "social prescriptions" in which doctors prescribe pastimes from tango classes to ukulele lessons instead of pills.[32]

You probably can see a pattern here. Profit-making corporations sell hardware and software that hijack the brain's wanting network, ostensibly to make us feel more connected and happier but which often leave us feeling more isolated and anxious instead. Drug companies then sell us medications to deaden the anxiety of modern life. Increasingly, however, the most effective cure appears to be a return to person-to-person socializing and more time spent in the natural settings in which we originally evolved.

In the UK, some 20 percent of the population admit that they are often or always lonely, especially the elderly. One town in the west of England addressed the issue by erecting "Happy to Chat" benches in local parks (figure 6.3). "Sit here if you don't mind someone stopping to say hello," the

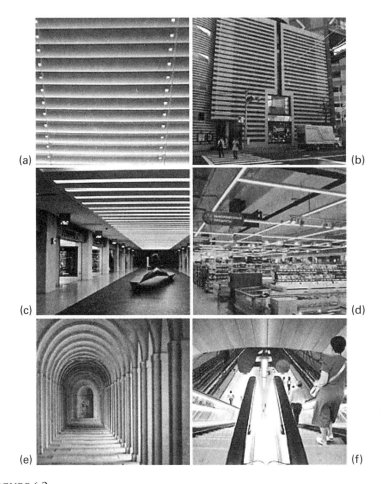

FIGURE 6.2
Uncomfortable stripes seen in everyday illumination. (a) Venetian blinds, Rupert Kittinger-Sereinig. (b) Wataki museum, Tokyo. (c) Commercial hallway, Harry Burgess. (d) Minsk department store, IdeaLight. (e) Arched walkway, Tiago Cardoso. (f) Underground escalator, Ana Ventura. *Sources:* Pixabay, Creative Commons.

Figure 6.3
A chat bench in the market square in Epsom, Surrey. *Source:* Epson & Ewell Borough Council. Reproduced with permission.

sign says, breaking down the barrier between strangers sitting side by side but each of them unsure about striking up a chat. The icebreaker has taken off in other countries. Situated in green parks, the benches demonstrate how silence, rather than being something to dread, can open up opportunities for conversation and connecting with others.[33]

If this book has a running theme, it is the conflict of Stone Age brains in the screen age and how to protect ourselves from being hijacked and biohacked by powerful forces that have anything but our best interests at heart.

Normally we consider noise pollution an urban problem, but increasingly it is encroaching on suburban and rural regions. Dr. Rachel Buxton and colleagues in the Department of Fish, Wildlife, and Conservation Biology at Colorado State University recently published a study on anthropogenic (i.e., man-made) noise that "reduces the capacity to perceive natural sounds."[34] "Noise pollution is pervasive in U.S. protected areas," she says, "even in national parks." So-called protected areas (dedicated to conservation of biological diversity, natural recreation, and cultural use) cover more than 13 percent of the world's land mass.[35] Dr. Buxton's team collected millions of hours of sound recordings from 492 federal, state, and municipal parks and wilderness areas. They found that human shouts and conversation, ringing cellphones, the rumble of traffic, the din of commercial development and resource extraction, and air traffic noise from planes, helicopters, and drones are drowning out natural quiet. The level of noise instigated by humans was more than double the ambient sound levels in 63 percent of protected areas. In 21 percent the measured increase was more than tenfold. To compensate and make themselves heard, city-dwelling birds such as sparrows and starlings have raised the pitch of their calls.[36]

In Grand Canyon National Park researchers have tallied an average of 150 overflights a day during peak tourist season. Sound levels reach 76 decibels, exceeding the 55-decibel limit known to raise stress hormone levels, blood pressure, and the risk of heart disease. Prolonged exposure to noise above 85 decibels leads to high-frequency deafness (sound decibels are measured logarithmically rather than linearly, so that a sound perceived to be twice as loud as baseline exerts one hundred times more sound pressure on your eardrums). Woodland sounds, by comparison, rarely surpass 40 decibels, equivalent to a babbling stream, whereas sustained city sounds regularly exceed 65 decibels.

The *New Yorker* has predicted that noise pollution will be the next public health crisis, while Bianca Bosker in the *Atlantic* explores the phenomenon in "Why Everything Is Getting Louder," and The Right to Quiet Society at quiet.org promotes awareness of the dangers that noise pollution poses to physical, emotional, and spiritual well-being.[37] Both Bosker and The Right to Quiet Society point out that the body does not adapt well to noise, and large studies confirm that a din kept up for days, months, or years raises the risk of high blood pressure, heart attacks, stroke, diabetes, dementia, and depression. In the wild, the concentration of stress hormones in elk and wolf feces spikes during snowmobile season and then returns to normal when the deafening machines retreat. The Environmental Protection Agency (EPA) last measured the volume of ambient national park noise in 1981.[38] Today a growing source of anthropogenic noise turns out to be data centers that house thousands of internet servers. The relentless din, which can travel for miles, comes from weighty steel chillers affixed to the sides of the buildings. They siphon off the enormous heat the servers generate as they crunch, store, and process data around the clock. The human ear can hear sounds up to about twenty-five miles away, depending on the weather and the environment.[39]

In a remarkable book, *Silence in the Age of Noise,* the Norwegian adventurer Erling Kagge reflects on his search for silence.[40] Kagge was the first person to complete the grueling Three Poles Challenge on foot: the North Pole, the South Pole, and the summit of Mount Everest. He calls them "extreme journeys to the ends of the earth." Which paths lead to silence? "Certainly trips into the wild," he says. "Leave your electronics at home, take off in one direction until there's nothing around you. Be alone for three days. Don't talk to anyone. Gradually you will discover other sides of yourself." After prescribing disconnection and a healing isolation from work, email, texts, and personal entanglements, the fabled adventurer ends with this thought: "The silence I have in mind may be found wherever you are . . . and is without cost." Is his remedy harsh? Perhaps. But perhaps we have become so sidetracked, unfocused, and frayed that only strong medicine can bring us to our senses.

We can gather the sundry ideas discussed here under the umbrella term "sensory overload." When the body and our Stone Age brain need respite from the busyness of daily life, silence can be a balm for the incessant noise and scurrying that surround us. Both mentally and physically, silence is an essential nutrient.

7

Your Brain Is a Hackable Change Detector

Modern humans have been around for about 200,000 years. For the vast majority of that time they did little more than eat, procreate, and survive. Climate extremes settled down about 12,000 years ago, after the Pleistocene ice age, which began 2.6 million years earlier. Agriculture appeared in the Middle East around 9,500 years ago. Living in proximity to crops and animals induced genetic mutations that allowed some ancestors to digest the lactose in goat, camel, and cow's milk and to metabolize animal fats. These mutations led to greater height at maturity and resistance to endemic diseases such as tuberculosis and leprosy.[1] Yet everyday life did not change much during the Holocene except in the phases of the Moon, the seasons, and the constellations overhead.

Even before that slow stretch of time the human brain had developed into a change detector distracted by anything out of the ordinary. Deviations from prevailing circumstances threw it into alert mode. With alacrity it responded to a sudden sound from behind, a fleeting shadow glimpsed in the corner of the eye, a rustle of movement, an unaccustomed taste or unfamiliar smell. Not just any twig snap would trip an alarm. Only one that was unexpected or out of context made our forebears freeze and then either fight or flee. Enough progenitors were sufficiently attuned to survival to reproduce and beget modern people like you and me. To be clear: there is no central "change detection" sensor in the brain. Organs such as the cochlea, the inner ear, the visual system, and various fields of touch all contribute to our ability to register any change in prevailing conditions.

To understand why novel stimuli automatically draw attention to themselves—a biological mechanism called "the orienting reflex"—I spoke to Harvard psychologist Steven Pinker, author of *Rationality*. Dr. Pinker addressed my query in terms of "the inherent value of information." He explained that information is necessary to take adaptive actions: you cannot react to threats, food availability, weather, relatives, friends, allies, enemies, or anything else unless you know they are there. "Something has arrived on the scene, or has changed its ability to impinge on you, which requires a recalibration of your current responses and expectations, which are set to the world as it was before the change, and which may have just become obsolete." More intense stimuli (louder, brighter, faster) indicate "more potent causal powers—something or someone that can hurt you or help you more." I put the same question to Gary Marcus, professor of psychology and neural science at New York University and the author of *Kluge: The Haphazard Evolution of the Human Mind*. He said that we learn more from novel stimuli than from those that are already familiar. Novel stimuli "inherently tend to carry more risk, too, and we need to be vigilant for them. Novelty is not unrelated to anomaly, and it's important to detect potential anomaly."[2]

In addition to novelty and intensity, context matters in determining what distracts or rivets your attention. Imagine yourself at a party where the music is loud and many people are talking at the same time. It takes effort to follow the person with whom you are trying to speak. But if someone across the room mentions your name, your attention may home in on it like a guided missile. Or consider a young mother who sleeps through sirens and honking horns outside her window yet awakens instantly if her newborn makes the slightest cough. Contextually important, salient stimuli such as these cut through our filters; the more important a stimulus is to our emotional needs, the more it takes precedence over everything else. In chapter 3 I explained the energy cost associated with salient objects. The question now is, how did digital devices acquire their status of being so highly salient? Software designers cannot engineer salience into their designs, although they can and do exploit our cognitive weak points. Rather, users themselves highly invest digital devices with social and emotional significance.[3]

Modern brains have the same change detectors as those of our Stone Age ancestors. Whether lolling in a woodsy hammock or dashing through an airport with a cellphone plastered to our ear, our change detector remains on perpetual alert. Even while we sleep it monitors surrounding conditions (the mother awakens on high alert if her infant makes the slightest sound).

This relentless vigilance, hardwired into the brain, ineluctably consumes a fixed portion of energy resources that we cannot increase but only replenish through quiet and rest. Think of the sheer number of items that vie for your attention in today's hyperstimulating world compared to the comparatively stable circumstances of your grandparents' era. Only one or two generations ago life was slower, less insistent, and imposed change less frequently than it does today.

I discussed nuances of context with Harry Whitaker, founding editor of the journals *Brain and Language* and *Brain and Cognition*. He framed the issue this way: captured attention depends not just on the novelty of a stimulus, or its intensity, or the fact that it carries new information but also on the expectation of what will fit into a plausible future that hasn't yet happened. There are contexts in which intense but predicted stimuli won't put you on high alert, and contexts in which almost imperceptible ones will. For example, consider that (1) you are in a room at home and hear a door open, quietly and slowly, and seconds later you hear it quietly shut, or that (2) you are in a room and hear a door open abruptly and seconds later slam shut. Scenario 2 is obviously more intense than 1. If the context is nighttime, you are in bed alone, and your partner is due to come home soon, you will alert to the loud sound more than the quiet one; if it's the same context but your partner is by your side and your teenage son is not home yet, you will alert to the quiet sound more than the loud one. This scenario can be developed in various ways. The point is that the brain establishes the context it is currently in (call it "present context") and continually predicts what aspects are likely to change, what aspects are of interest, and what new stimulus is predicted to accompany the aspects-of-interest that you expect (e.g., an animal you are hunting comes into view, someone you are expecting comes home—or maybe it's a burglar). To sum up, consider:

Present context, which is constantly being updated

Prediction—of what stimuli are expected in the next iteration of the pattern-context

Assessment—of whether the new stimulus was predicted or not

Response (if required or appropriate), with or without an emotional alert

Being forced to switch attention repeatedly incurs a high calorie cost on top of the background drain required to maintain round-the-clock change sensitivity. From a pocketbook perspective we should switch attention only for a very good reason; otherwise we are squandering a finite resource. While we are unlikely to starve from the calorie deficit, interruptions and flitting from task to task before finishing any of them still saps us enormously. Emotion, too, figures heavily in the way our change detector works because within a given neural network, emotion acts as what software engineers call an "interrupt function."

Early computer programmers designed the escape key to interrupt whatever program was currently running. Recall the way emotion is rooted in homeostasis, the propensity of all life forms to maintain a stable internal milieu. While all lower creatures have an orienting reflex, mammals, including humans, additionally have an interrupt function that nudges us back to a homeostatic steady state. The vertebrate brains of reptiles, amphibians, and bony fish do not possess such a function. Its apparent purpose is to make us stop what we are doing and switch to doing something else—perhaps because that something else is more meaningful and beneficial, such as securing a meal or a mate, or retreating to a safe harbor. Digital distractions hijack salient task switching to change the focus on novel matters such as likes and retweets that in the larger scheme are unimportant.

To appreciate how useful an interrupt function is, imagine yourself in a *Groundhog Day*-type situation, stuck in a repeating loop. Neurology is nothing if not an assemblage of wonderfully odd syndromes, and so it is no surprise that there does exist a phenomenon called *perseveration* that entails a failure to curb ongoing action and nudge the affected individual into doing something else. To an outside observer, perseveration looks as if the person were stuck—verbally repeating words, phrases, or nonsense sounds. Motor perseverations entail repetitive actions such as finger tapping, lip smacking, or opening and closing a drawer. Perseveration ignores the usual signals to stop when the reason for doing something ends or no longer serves a purpose. Social rules likewise dictate when to stop a behavior that is about to become inappropriate or embarrassing. Perseveration by brain-damaged individuals illustrates starkly the individual's inability to switch from one action to another. Asked to draw a square, they do so. Asked then to draw a circle, they draw a square. Asked to draw a cross, they again draw a square. Perseveration is curious in that it involves a fixation of attention while simultaneously showing how too much attention can be a drawback.

Frontal lobe damage is the most common cause of perseveration, although individuals with developmental autism, in which white matter connections among cortical areas are underdeveloped, often perseverate, too. Rocking, spinning, and obsessive counting are common. Being "set stuck" in a loop, such individuals repeat words or gestures after the stimulus that prompted them has long since ceased. The brain has optimal levels of focus and distractibility, with the two in steady tension. To be biologically useful, our change detector needs to alert us when conditions alter. Yet keeping vigilant keeps us from getting much accomplished. Our change detector therefore needs to have strong filters that separate meaningful changes from less meaningful ones, and a capacity to pick out relevant stimuli from the torrent of lesser distractions that bombard us every second. Paranoid schizophrenics live in an involuntary state of sustained hypervigilance. If you ask, they will tell you it is exhausting.

During the lengthy march of evolution, the early mammalian brain had to cope with the bombardment of energy flux coming at it from all directions—natural as well as man-made—by assigning salience to different aspects of it. Salience is the degree of relevance any organism attaches to something. Mothers attach salience to a crying baby because they are responsible for its survival. To a hungry frog, a fly is highly salient; to us, not so much. A related appraisal is valence, the amount of positive or negative value we assign to a person, place, or thing. Figure 7.1 illustrates the difficulty of culling relevant stimuli out of the objective torrent we face.

"Flux," we recall, is the flowing field of electromagnetic energy that surrounds all of us. Flux arises from terrestrial sources as well as those coming from space, most notably the Sun, which of course provides the energy for all life. What should give us pause is that our sense organs are able to detect only a minuscule slice of this incoming radiation energy. The figure illustrates the span of objective reality measured by instruments and the middle sliver of visible light to which we are sensitive—a mere ten-trillionth of the energy flux that in reality hits us every moment. Radio waves, cellphone conversations, X-rays, and cosmic particles pass through us unnoticed because we lack the biological sensors to know they are there. The flux of solar neutrinos that pass through each square centimeter of Earth is about 65 billion particles per second. Yet we don't feel a thing as they rip through us and shoot out the other side of the planet. We are insensate to all radiation beyond the ultraviolet and infrared wavelengths, including the flux

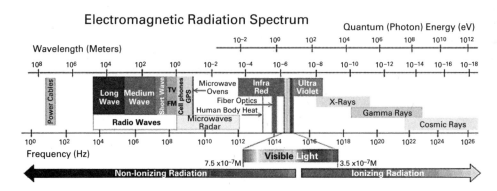

FIGURE 7.1
The visible light to which humans are sensitive is less than a ten-trillionth slice of the universe's energy spectrum, which covers a billionfold span. We simply lack the biological sensors to access other parts of the electromagnetic spectrum, and so our "reality," or *Umwelt,* consists only of what we can perceive. Brain-machine interfaces such as cochlear and retinal implants, as well as sensory-substitution devices, can change and enlarge this. *Source:* Richard E. Cytowic, *Synesthesia* (Cambridge, MA: MIT Press, 2018). Reproduced with permission. [See also color plate 3.]

of raw satellite, radar, and microwave signals, as well as the magnetic fields created by power lines, appliances, and MRI scanners (when you lie in a 1.5 Tesla scanner you are subject to a magnetic field 30,000 times stronger than the Earth's natural one, yet you don't feel a thing). Within the narrow band to which we are sensitive lies the plethora of screen-based media that we have created ourselves.

The bubble in which we live is our *Umwelt,* a nineteenth-century German term that describes the physical envelope we take to be reality. It implies not just "the surrounding world" but the lived-in one as experienced by an organism within it. Sonar defines the *Umwelt* of bats, porpoises, and whales; the frigid, high-pressure realm devoid of light constitutes the *Umwelt* of deep-sea creatures; the ultraviolet and polarized light invisible to us belongs to the *Umwelt* of insects and birds. The latter use it to navigate and hunt (rodent urine also leaves UV-visible traces that predators can see), whereas reptiles navigate their *Umwelt* via infrared energy that their tongues sense as heat or a chemical smell.

Every creature assumes that its *Umwelt* is the entirety of objective reality. After all, why would it imagine there is anything beyond what it already knows? A dog's *Umwelt* is largely centered on smell. Out on a walk, pausing to sniff every few feet, your dog might look at you and wonder how you could not know that earlier in the morning the neighbor's dog peed at this exact telephone pole. You can't know because you don't have anywhere near the acute sense of smell that your dog does. Yet you are blithely unaware of a sensory gap where smell ought to be any more than a congenitally blind person experiences an empty hole where sight ought to be. The latter's *Umwelt* doesn't factor in vision, so the concept is not just inapt but meaningless and incomprehensible. An *Umwelt* captures the idea of limited knowledge, a circumscribed reality beyond which an individual is necessarily ignorant.

Software and hardware designers deliberately hack our change detectors in an effort to manipulate our *Umwelt*.[4] "Tech companies fight cold-bloodedly for your eyeballs," says Gabe Zicherman, CEO of Gamification and cofounder of Dopamine, Inc. "I make a living by making things more addictive than they otherwise aren't. Broadly, it's called gamification because it incorporates game elements such as point scoring, rules, competition with others, and rewards. Companies succeed or fail by how much engagement they capture. . . . Every minute you take out of some company's pocket represents an enormous amount of dollars lost." While all applications of tech to workflows can have a benefit, such as faster service, self-service, and lower cost of operation, Zicherman says, "it's clear that each of these efforts simply increases our screen dependence."[5]

The infinite scroll is alluring but also seductive. Ramsay Brown, a neuroscientist who explores ways to counter phone addiction, says, "Your kid is not weak-willed because he can't get off his phone. . . . Your kid's brain is being engineered to get him to stay on his phone."[6] The reason users of all ages won't look up from their devices is that tech companies have been quietly using "persuasive design" and manipulative psychological techniques to hook them. There is hard science behind our evident tech dependency, and while software designers (euphemistically called "user experience researchers") don't broadcast the fact, its existence is no secret. Stanford University has operated its famous Persuasive Tech Lab since 1998, and American University has offered master's degrees in "persuasive play research" since 2012. Other institutions have hopped aboard the lucrative bandwagon, such as Drexel, Rensselaer Polytechnic, New York University, Carnegie Mellon, MIT, and the University of Utah.[7] Meanwhile, in the summer of 2018, sixty prominent psychologists petitioned the American Psychological Association (APA) to address the misuse of persuasive design for monetary gain.

> We are writing to call attention to the unethical practice of psychologists using hidden manipulation techniques to hook children on social media and video games. These techniques—employed without children's or their parents' knowledge or consent—increase kids' overuse of digital devices, resulting in risks to their health and well-being.[8]

The petition failed to change anything. Today the tech industry continues to prey on developmental vulnerabilities so it can sell children digital products even as the petitioners point out that an adolescent girl's typical need for peer acceptance makes it easy for social media to ensnare her. Similarly, a boy's "evolutionary need to rack up competencies" so he can feel good about himself makes him an inviting target for the intermittent-reward strategy used by all video games. Using psychological strategies is a perennial and inherent part of advertising, cautions Dr. Richard Freed, one of the petition signers and the author of *Wired Child: Reclaiming Childhood in a Digital Age*. Tech company "psychologists are helping to make products that are so stimulating and so good that they are better than real life. Overuse of tech has taken over their childhood."[9]

In October 2021, former Facebook (now Meta) employee Frances Haugen leaked tens of thousands of company documents to the Securities and Exchange Commission and the *Wall Street Journal*, which came to be known as "the Facebook Files."[10] They revealed that for three years, Mark Zuckerberg's company had been conducting internal research into how its products, especially Instagram, were harming a large proportion of young girls. "We make body issues worse for one in three teenage girls," the reports say. The company promoted posts intended to provoke "angry" reactions from users, its internal analysis showing that such posts led to five times more engagement than posts that merely received likes. It also pushed algorithms promoting stories to young users about anorexia and those containing self-harm photos. Haugen's whistleblowing gave an unprecedented look into how much Meta executives weigh the importance of their bottom line over common decency.

APA chief executive Arthur Evans Jr. addressed this outsize influence by saying, "The impact of technology and psychology's role in its development is becoming a major focus of APA's work." The APA's *Stress in America* report found that attachment to and constant use of social media apps is associated with high stress levels, and that many parents fret about the effect this has on their offspring's physical and mental health. It further noted a dispiriting study by the National Bureau of Economic Research, conducted before the COVID-19 pandemic, that found that many young men were choosing to play video games rather than look for a job.[11]

Physiological calculations for detecting salience begin in the limbic system, sometimes called the emotional brain (although that formulation is understood as overly simplistic). The limbic system roughly consists of tissues on the inside rim (from the Latin *limbus*) where the two hemispheres come together and meet the brainstem. Its network reaches from the top of the head down to the brainstem and into the spinal cord. I opted not to include an illustration because no stand-alone image can convey its scope or its complexity of connections. My preferred neuroanatomy atlas requires twenty-four separate illustrations and forty-three pages of text to detail its ramifications. The point is that two independent evolutionary trends in mammalian brains have been an expansion of the neocortex and the development of limbic structures. A given species tends to be high in one at the expense of the other. Monkeys, our close relatives, have substantial neocortical development but comparatively little limbic advancement. Rabbits show the opposite trend of robust limbic elaboration but poor neocortical sophistication. Humans are the exception in having robust development in both dimensions; the two structures burgeoned in tandem as reason and emotion co-evolved. Ironically, having multidimensional cognitive abilities is what makes us highly susceptible to screen-age distractions.

The physical architecture of cortical tissue provides analytical space and contains our model of reality, whereas the limbic brain determines what we do with the analysis once the cortex carries it out. The number of nerve fibers in the limbic system is greater than those in all other nerve tracts. It contains five times as many fibers as the optic nerves that ferry 85 percent of all sensory inputs.[12] Every input from external sense organs and internal viscera must feed into emotional circuits before looping back to the cortex for further analysis. The altered signals then feed yet again into the limbic system for a determination of whether the highly transformed, multisensory input is salient or not. If so, we might act. If not, we will be blind to it, just as we fail to register the flux of gamma, ultraviolet, and radio waves that pass through us without cease.

Professor Clay Shirky, NYU's vice provost for educational technologies, compared incidental screen distractions to secondhand smoke that wafts inexorably outward. Its diffusion explains our compulsion to look at digital advertisements on billboards, street kiosks, and even the checkout terminal when we swipe our credit card. Such are the distractions of the screen age.

* * *

The brain has separate fast and slow response circuits that enable us to judge circumstances in one of two ways. Each makes its own kinds of error because biological structures built by accretion can never be perfect in a constantly changing world; they have only to be good enough to get the job done. A fast, schematic route can set off false alarms so that we recoil at the stick we thought was a snake (it triggers our instinct for freeze, flight, or fight). While it is usually better to panic and dash than to pause and deliberate whether a shape that startled us is in fact dangerous, the slower, contextual path sometimes fails to terminate a fear reaction initiated by the fast route. It then becomes stuck, making us apprehensive about things that are not inherently scary, such as doorknobs, lampshades, or wall clocks. It is much too simplistic to label the fast route emotional and the more deliberate one rational because no thought is untainted by emotion, nor is emotion ever totally divorced from thought. Even problems in pure mathematics involve emotional salience, an intuition of whether one is on the right track or not, and a Eureka! sensation when one hits on a solution.[13]

Connections between cortex and limbic components are reciprocal yet asymmetric: the cortex receives more inputs from the emotional limbic brain than the limbic system receives from the more rationally acting cortex. Again, I simplify, but the significance of this arrangement is that emotion exerts a constant and sometimes decisive force.[14] We can see an example of emotion's strong influence in epilepsy. Affected individuals try to control excitement because they know emotional stress might trigger a seizure. Despite their best efforts, they will still have seizures because they have a neurological disorder and attempts at self-control can do only so much. Nonetheless their failure leads to self-recrimination, frustration, social withdrawal, and soon more seizures. Some individuals who have thirty to forty seizures a day come to research units dedicated to understanding epilepsy. They are taken off all medications and placed in a setting in which injury is unlikely. Continuous video monitoring and EEG telemetry collect data to determine exactly what kind of seizures they have. Feeling tremendous relief at being in the hands of perceived healers, epileptic individuals often stop having seizures temporarily. Of course, their relief cannot cure the underlying malady, but it does demonstrate the enormous influence that emotion can have in a particular contextual setting.

Emotional valence strongly influences how we respond to digital devices. Back in 1966 Joseph Weizenbaum at MIT showed how easily human thinking can be swayed when we interact with programs that simulate natural language as we do with today's smart devices and ChatGPT. In *Computer Power and Human Reason,* Weizenbaum explained how he composed a demonstration program called ELIZA "with which one could 'converse' in English." It parodied a Rogerian psychotherapist who listens without judgment and acknowledges the individual's experience without moving the conversation in another direction. ELIZA's users would type a message. The program would analyze it, compose a response, and spit it out on a teletype terminal (printers had yet to be invented). Simply parroting the "patient's statements encouraged them to keep talking, as this transcript demonstrates:

> My boyfriend made me come here. He says I'm depressed most of the time.
>
> I AM SORRY TO HEAR YOU ARE DEPRESSED.
>
> It's true. I am unhappy.
>
> YOU THINK COMING HERE WILL HELP YOU NOT BE UNHAPPY?
>
> I need some help, that much seems certain.
>
> WHAT WOULD IT MEAN TO YOU IF YOU GOT SOME HELP?
>
> Perhaps I could learn to get along with my mother.
>
> TELL ME MORE ABOUT YOUR FAMILY.
>
> My mother takes care of me.
>
> WHO ELSE IN YOUR FAMILY TAKES CARE OF YOU?
>
> My father.
>
> YOUR FATHER.
>
> You are like my father in some ways.
>
> WHAT RESEMBLANCE DO YOU SEE?
>
> You are not very aggressive, but I think you don't want me to notice that.
>
> WHAT MAKES YOU THINK I AM NOT VERY AGGRESSIVE?

You don't argue with me.
WHY DO YOU THINK I DON'T ARGUE WITH YOU?
You are afraid of me.
DOES IT PLEASE YOU TO BELIEVE I AM AFRAID OF YOU?
My father is afraid of everybody.
WHAT ELSE COMES TO MIND WHEN YOU THINK OF YOUR FATHER?
Bullies.
DOES THAT HAVE ANYTHING TO DO WITH THE FACT YOUR BOYFRIEND MADE YOU COME HERE?

On the MIT campus, ELIZA became famous for its superficial emulation of a psychiatrist. What struck Weizenbaum was how eager people were to "talk" to it and confide their "most private feelings." Weizenbaum grew "alarmed" at how easily people "became emotionally involved with the device and how unequivocally they anthropomorphized it," projecting their own meaning onto what the machine parroted back.[15] Even his secretary, who had watched him laboring on the machine for months, started conversing with it as if it were a real person—a situation dramatized decades later in the film *Her*. The film's hero falls in love with a female-voiced operating system designed to simulate a companion. That he is starved for attention is made clear by the way he heeds her opinions, follows her advice, and responds without complaint to her every interruption. He confides his most intimate secrets, illustrating our deep psychic need to feel that we are being listened to and understood. As I teach medical students at George Washington University, what patients want most is to feel that they are being heard and, even more, that they are understood.

After a few sessions with ELIZA, Weizenbaum's secretary asked him to leave the room so she could talk to it "in private." Instead, he shut it down, stunned that "extremely short exposures to a relatively simple computer program could induce powerful delusional thinking in quite normal people." Since then, psychological reactions to machine simulacra have not changed, as Rosalind Picard of MIT's Media Lab later showed in her groundbreaking *Affective Computing*, in which she recounts her attempts to get machines to first recognize user emotions, and then respond to them with synthetic emotions of their own. As Weizenbaum saw it, the danger of regarding a machine as a person lies in our ceding authority to it. Numerous authors such as Alan Moore (*V for Vendetta*) have illustrated the extremes to which false intimacy with AI can lead us. Its emotional tug is enormous, which is what makes our brain so hackable.

*   *   *

Passive learners often fail to acquire a knack for sorting out the meaningful from the irrelevant. They don't learn a skill by watching as well as those who do it themselves. The need for embodied or body-centered action to forge one's way in the world and understand it was demonstrated decades ago by Held and Hein's landmark "gondola kitten" experiment.[16] One kitten was free to roam about while another kitten from the same litter hung passively suspended in a gondola contraption that moved in parallel with the self-directed kitten. The passive kitten had its world presented as a fait accompli in the same way that curated screen images delivered to our waiting eyeballs are. Follow-on experiments confirmed that the passive kitten had learned nothing about its environment despite having seen everything the active, free-moving kitten had. Once set free it remained as functionally blind as it had been at birth, the critical window for developing vision having closed.

In an updated twist, an American child's Chinese-speaking nanny was videotaped so that a second American child the same age could see and hear exactly what the first one saw and heard. The first child's physical proximity let it take in vocal tone, gesture, eye contact, and a two-way emotional reading that neither nanny nor child was consciously aware of. The first child soaked up the language and began to speak it while the second child didn't learn a word of Chinese. Professor Patricia Kuhl at the University of Washington's Institute for Learning and Brain Sciences emphasizes how much language acquisition depends on social context and interaction as well as the "parental style" of speech used to address children. "Exposure to a new language in a live social interaction situation induces remarkable learning in 9-month-old infants, but no learning when the exact same language material is presented to infants by a disembodied source."[17]

Positive intermittent reinforcement is the prime culprit that makes us unable to disengage from screens. "Reinforcement" is the psychological term for a reward that prods us to repeat whatever behavior led to it. Predictable feedback fails to reinforce behavior: the fact that the light comes on when you open the refrigerator, for example, does not compel you to open it repeatedly. You know what is going to happen and therefore don't give it much thought until the bulb burns out and the shelves are suddenly dark. But when the brain's change detector alerts us that something novel has happened, we get a jolt of dopamine from cells that reside mainly in a small slice of brainstem called the substantia nigra. Despite their relatively minuscule number—400,000 out of a total of 86 billion neurons—these dopamine cells project their branches to every niche of the cortex. Dopamine is the traveler with access to every peak and valley in the brain. It is also the neurotransmitter involved in such familiar addictions as alcohol, cocaine, nicotine, narcotics, stimulants, depressants, gambling, food, and sex.

Slot machines and lotteries provide classic examples of positive intermittent reinforcement. Dopamine levels surge when we anticipate a reward, and the dose released by anticipation is greater than the amount released when we actually win an object of our desire. Gamification experts concede that the faster a product delivers an occasional payoff, the stronger it will hold a user's attention.

8

WHAT GETS CAUGHT IN THE CORNER OF YOUR EYE

The eye may look like a separate peripheral organ, but it is a proper part of the brain. The retina, made up of ten different cell layers, is a sublime piece of engineering, an intricate product of evolution. The entire visual system—from the retina at the back of the eyeball to the cortex at the back of the head, and from there to relays still elsewhere—is the linchpin that explains why our attention is readily hacked and we are so easily distracted.

Vision accounts for 85 percent of all inputs to the brain. Some two dozen cortical areas deal with the many different aspects of vision. Although conventionally we have five senses (in fact, more), they are not equal in terms of evolutionary importance or in their proportion of inputs. Visual alerts arrive hundreds of milliseconds faster than those coming in from other senses. They are also directional, whereas sound and smell signals come from all directions, putting us in a sphere of sensation for them but a tunnel when it comes to vision. Smell tells us little about its location aside from an object's identity. To orient to a smell, we would need an olfactory brain capable of following gradients the way ants and sharks do.

From an evolutionary standpoint of survival, sensing a changed situation is more important than analyzing one that remains static. A snapped twig is a change in situation that would have alerted our ancient ancestors just as it does us. But sounds don't localize nearly as well as sharp foveal vision does. For our ancestors to survive the perils around them, sensitivity to visual change mattered more than alerts from other senses. What mattered still more was the ability to detect any change quickly. The speed at which vision does this made it surpass all other senses.

Modern sound alerts include the whoosh of push notifications and the ding of text messages. But, just as for our Stone Age ancestors, the changes that alert us most powerfully and quickly are visual ones. As the secondhand smoke analogy made clear, movement caught in the periphery is impossible not to notice. Even the effort it takes to try and ignore it exacts a cost by stealing energy from the fixed bandwidth we have to work with. The exquisite sensitivity to change that helped humans survive in the past is, ironically, a liability in the present. Threats that were once a matter of life or death—being eaten alive, falling off a cliff, being poisoned by a venomous bite—have given way to existential ones today, self-inflicted psychological ones such as neurotic worry over one's likability or enragement over use of the wrong pronouns. Unfortunately, our Stone Age brain overestimates and misperceives digital interruptions as a matter of life or death. Tech companies know this and exploit it.

Software engineers hack our Stone Age biology and then go to great lengths to monopolize our eyeballs. Social media companies present themselves as benevolent, yet then invade our privacy and sell the profiles they built from our personal details to advertisers and still more other corporations. They have an agenda they don't bother to hide. It behooves us to understand that more is at stake than mere vulnerability to screen distractions.

Unlike a camera that indiscriminately records everything in its field of view frame by frame, the retina is selective in what it passes on to the rest of the brain. Stable, relatively homogeneous images such as a landscape don't produce much of a signal, whereas scenes that feature high contrasts and movement generate robust signals—one reason Zoom meetings and the flow of images that fly by as we swipe throughout the day can be draining. Brain-inspired or "neuromorphic" image sensors use the same energy-efficient trick that the human nervous system does: by prioritizing the dynamic, changing parts of a scene and ignoring whatever stays the same, machine sensors based on the sparse representations of only changes capture images more efficiently.[1] Currently, sensors that detect changing events are used in self-driving vehicles, machine vision, navigation, and safety features in your car. Motion cues are a particularly rich source of information for machine navigation.

Compared to a still camera, the retina is sensitive to changes both in space and over time. Think of how a flashing light grabs your attention more than a steady one does. Waiting at a stoplight, your attention may shift to adjusting the radio, sending a text, or consulting a map. Yet as soon as the light changes you notice it. Change seizes your focus, whereas sameness doesn't, which is why screens are such effective distractions. The retina also has an astounding ability to alter its sensitivity so that we can see in either sunlight or starlight, a ten billion-fold difference in intensity. This is why any flickering change in screen brightness helps capture your attention even when you try to ignore it. Humans can detect the flash of a single photon, the elementary particle that carries the smallest quantum of light.[2]

Light levels are measured in candelas per square meter. Relative levels of luminous intensity that we typically encounter are starlight (0.001), moonlight (0.1), indoor lighting (100), and sunlight (10,000). The intensity of computer monitors exceeds 100 candelas, with most shining in the 250–500 range, making them a common source of visual strain. The luminance value of an iPhone is quite high, especially when held close (intensity increases with the inverse square of the distance; a phone held one foot away is four times as bright, not twice, as one held two feet away). High-intensity visible light is a known cause of retinal damage.[3] The terms that refer to brightness and intensity are admittedly confusing. One is measured by the brain, the other by instruments. Brightness is the term we use, while intensity is the term we mean. Brightness is the sensation by which we distinguish differences in luminance. It describes the experience of a phenomenon (luminance), rather than the phenomenon itself.

Screen distractions begin in the most peripheral part of the visual field, represented in the lingual gyrus of the temporal lobe and rich in connections to the emotional brain (figure 8.1). This limbic connection is the reason we startle at objects caught in the corner of our eye. Things detected too briefly to register fully nonetheless seize our attention, ramp up emotion, and put us on alert. The orienting reflex kicks in too, yet another reason why screens distract us and sap energy. Their mere presence forces the brain to burn extra fuel.[4]

All mammals have an orienting reflex present from birth, meaning that they automatically turn toward novel stimuli. It's our evolutionary change detector at work, as demonstrated when I first held my newborn grandnephew, Max. His father had turned on the TV briefly to check the weather, whereupon Max turned away from me to the screen. The change detector in Max's young brain (triggered more strongly by sound than by vision at his age) was doing exactly what it was supposed to do, and what it did in our ancestors hundreds of thousands of years ago.[5] He wasn't in danger of

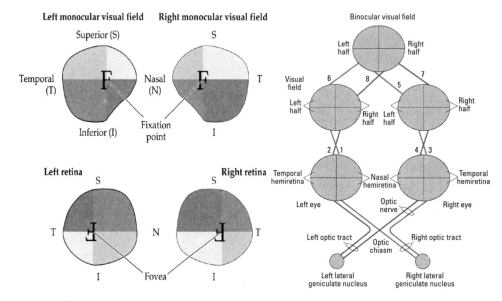

FIGURE 8.1
Left, The four quadrants of the visual fields (top) and their representation on the retina (bottom). The eye's lens makes the image inverted and backward. *Right,* The second row illustrates the outer quadrants, where most distractions catch our eye and seize our attention. The upper quadrants connect most strongly with the emotional brain. [See also color plate 4.]

being eaten by a lion, yet the moment illustrated the allure of technological distractions. This is why I argue that kids younger than three should not be exposed to screens or kindred digital devices when they would normally be socializing with actual people. The orienting reflex is encoded at the deepest levels of the brainstem and so fundamental to how brains operate that we find it nearly impossible not to look at TV monitors in airports or waiting rooms.

Corporations do not bother to specifically exploit the orienting reflex because they don't need to. Our own biology makes it all too easy to capture and hold our attention. They can instead manipulate us in more nuanced ways, as I discuss in chapter 6. All the major social media corporations employ teams of psychologists and behavioral scientists, including those versed in behavioral economics whose job it is to steer us away from person-to-person interactions and toward a manufactured reality of their own design that they now call "the multiverse." The illusion of having choice is a central pillar of consumer capitalism, and tech titans go to a lot of effort to conceal the fact that they have already made many choices for us.

If youngsters are ever to become emotionally intelligent, they need to experience auditory, visual, and emotional social interactions. Children learn by being spoken to and distinguishing their parents' tone and vocal inflections. The ability to read emotion is an innate skill all babies have. They can read gestures, body language, and most of all facial expressions. The face is so central to nonverbal communication that we use more than forty muscles to generate its infinite nuances.

Babies see where their caretakers are looking and follow their gaze. If the caretaker is glued to an electronic device, then that is where the baby will focus, too. Sadly, face coverings imposed during the pandemic have interfered with infants' neurodevelopment, to say nothing of playground and daycare center closures, which curtailed play opportunities with other children. An alarming longitudinal study by the Resonance Consortium and the Advanced Baby Imaging Lab at Brown University Medical School reports, "We find that children born during the pandemic have significantly reduced verbal, motor, and overall cognitive performance compared to children born pre-pandemic. Moreover, we find that males and children in lower socioeconomic families have been most affected."[6]

Children learn from person-to-person interaction, a physical type of engagement that an iPad can never approximate. Digital personalities and screen animations speak *at* a viewer, not *with* them, a distinction that Lamb Chop puppeteer Shari Lewis doggedly made in wanting "doers, not viewers" for her audience. For years she battled obtuse TV executives who couldn't grasp that children best soak up their environment when they are physically as well as emotionally engaged in it.[7]

What is it that makes digital distractions feel important and meaningful? Part of the answer is that peripheral vision projects to the limbic brain via the lingual gyrus. This outermost part of the temporal lobe is concerned with portent and meaning. A stroke in this location or an unstable nerve cluster that ignites a seizure here produces a neurological state called "the feeling of a presence." In technical terms, the mediobasal amygdaloid-hippocampal complex is associated with meaningfulness, an individual's sense of self, and the self's relation to space and time.[8] Relevant to screen distractions, and said in plainer English, individuals who experience the phenomenon speak of "just missing" objects or people in the corner of their eye while simultaneously feeling an emotional jolt. Stimulating this region electrically (during routine brain surgery, say) produces exactly these kinds of perceptions, along with a sense of portentousness.[9]

The frequent peripheral location of screens is a main reason they distract and drain energy. Instead of keeping screens out of their sightlines, many parents unwittingly plop their children in front of screens, teachers encourage it, and corporations push the practice, despite knowing that it is not in any youngster's best interest. Those who question whether there might be a downside are attacked as Luddites and technophobes. And so well-meaning but ill-informed parents have foisted devices like Fisher-Price's Apptivity Seat and the iPotty onto still developing infants, saturating them with bright screen media for hours a day. The device's inventors may have meant well, but pediatricians know that the fovea is not fully developed until age four (educators should know this, too), which is why dominating the visual field with an iPad during this critical window for developing visual acuity is ill-advised. The fovea is where 20/20 vision eventually develops. For infants who lack the strength even to lift their heads or turn away, the practice is appalling. Whether the practice constitutes child abuse is eminently debatable.

Peripheral distractions are no less of an issue for adults. Imagine meeting a friend for coffee and placing your phone face up on the table. A text or call comes in. The screen lights up, signaling a change in conditions. It whooshes, vibrates, or rings—another change. Both events ineluctably seize your attention; your limbic brain and that of your companion cannot help but react to the stimulus. The automatic response costs both of you physical and mental energy. And the more distractions your brain is forced to register and decide about, the more drained you feel. The real question is, what do you want to use your fixed allotment of energy for? The decision is yours. You could put the phone face-down, although the sound would still distract you. You could turn it off altogether, but that might trigger FOMO. Or you might fume that tech companies have forced you to decide which interruption is worth your attention: attending to your friend or to the ever-intruding digital world that demands your undivided attention NOW.

The film *Minority Report* first showed, presciently, Tom Cruise bombarded with personalized holographic advertisements that populated his peripheral vision as he sprinted through a futuristic concourse. Personalized ads are by definition meaningful, many times more salient than passive billboards plastered on bus stops, subway walls, or billboards. The manipulation depicted in the fictional film has now become reality. And it has grown more insistent. The failed Google Glass placed its alerts in the viewer's upper visual quadrant—exactly where they most distracted users and were impossible to ignore. Glass was brilliant technology but horrible psychology. It is not widely known that during its initial trials, Google Glass induced fatigue, eye strain, and visual confusion in the same way that early heads-up windshield displays used by jet fighter pilots did. Engineers have known about these drawbacks for decades. No company courts negative publicity, but Google didn't inform early adopters about these consequences despite its vaunted motto, "Don't be evil."[10] The brain rebels at seeing one image in front of one eye and an unrelated image in front of the other, which is exactly what Google Glass presented.

"Binocular rivalry" describes the slightly different images that each eye normally sees. Disparity between the two eyes is what gives us depth perception. The occlusion is never a problem where your nose is concerned: as with the blind spot, the brain glosses over the fact that the nose partially obstructs each eye's middle field of vision. In the natural world, biology always finds a solution. But heads-up displays such as Glass and those embedded in windshields that some automakers currently offer pose a different problem. In addition to headaches, eyestrain, and visual discomfort, the prolonged use of any gear that forces each eye to look at different images can induce an ocular misalignment called a phoria, a permanent form of crossed eyes. (Children born cross-eyed wear a temporary patch to block out the conflicting image so that the eyes can correctly align by themselves.)

* * *

By ramping up your level of monitoring and alertness, FOMO makes it more likely you will notice anything that encroaches on your peripheral vision. Fundamentally, FOMO is nothing more than an on-guard state of mind, which means you can do yourself a favor and desensitize it to be less vigilant and draining. Marcus Aurelius had pithy advice about this: "Cast away opinion and you are saved. Who then hinders you from casting it away?"[11] Fourteen hundred years after General Aurelius, the French philosopher Michael Montaigne said, "My life has been filled with terrible misfortune, most of which never happened." By this he meant that imagination alone can work us into a lather of worry. FOMO arises in the mind as a thought and in the body as a feeling, which means it is within your power to dispel the fear and replace it with more realistic, resilient, and congenial thoughts. Doing so is a basic technique of cognitive therapy, which has been around for decades.

Some people can easily change their perspective. Others find it difficult. Remember how anxious smartphone users became after only a ten-minute separation from their devices? We should not dismiss their distress out of hand; it is both substantial and meaningful, as Russell Clayton at the Cognition and Emotion Lab at Florida State University showed. He did not anticipate a strong pushback when he forbade college students from answering their iPhones while performing a word-search puzzle. Yet their heart rate, blood pressure, respiration, and perspiration all shot up, as did secretion of the stress hormones cortisol and adrenaline. Not surprisingly, the students' physiological reactions kept pace with their rising sense of panic. Even though their phone stopped ringing, the stress response did not abate until the phones were back in their hands. This indicates strong embodied cognition at work because users felt their phone to be an extension of themselves, what Clayton calls an "extended iSelf."[12]

Anthropomorphizing is nothing new, but as screen media increasingly saturate daily life, people's personal investment in and engagement with digital devices have grown proportionately. Historically, psychology assumed that anthropomorphism was a mistargeted emotional projection by the perceiver, rooted in ignorance or magical thinking. The heartbroken hero in the film *Her* illustrates the tendency taken to its logical extreme when he falls in love with, and is rejected by, the female persona of his new phone's operating system. Recent science has fortunately corrected this mistaken view, showing that conventional aspects of social cognition easily explain anthropomorphism. Socially anxious individuals may cling to their phones to relieve anxiety, and anxiously attached individuals are more likely to form anthropomorphic beliefs about them. They are also more likely to indulge in compulsive behavior such as answering the phone while driving or taking selfies in dangerous situations.[13]

For thirty years Larry Rosen, professor emeritus at California State University, studied the psychological effects of technology on adults, teens, and children. He found that "heavy smartphone users showed increased anxiety after only 10 minutes" of not having access to their smartphone, an excellent description of nomophobia. During a one-hour experiment, the anxiety among heavy users shot up to self-reported "unbearable" levels.[14] Moreover, out of sight is not out of mind. The "unhealthy connection to their constant use weighs on users," says Rosen, causing anxiety even when the device is out of view.[15]

We are social animals, and envious by nature. Joseph Reagle, professor of communication studies at Northwestern University, calls FOMO "envy-related anxiety about missed experiences." To protect ourselves we engage in "conspicuous sociality," a deliberate attempt "to convey a wealth of fun and friends" on social media. Reagle compares today's efforts at appearing continually connected to century-old concerns about "keeping up with the Joneses," a phrase that possibly originated with the 1913 comic strip by that name.[16]

Once you have fallen into this relentless (and unwinnable) striving frame of mind, it is not hard to convince yourself that you have missed something unrepeatable. But as Montaigne warned, imagination exaggerates possible consequences to unrealistic degrees. FOMO nudges you to ingratiate yourself with high-status individuals on social media. *Who is going to friend me, follow me, retweet me, or like my posts? Do I matter? Am I having an impact?* The neurotic mindset that questions whether you are posting enough can turn you into a validation junkie, constantly afraid that others are doing something you aren't and spending time with friends you wish were your friends, too.[17] In other words, FOMO has you firmly in its grip. The various hacks that exploit our biological weaknesses all seem to touch on an existential anxiety about survival: FOMO, the orienting reflex, social connections, self-worth, followers, influence, and competitive advantage on social media. Throw in television, the rest of social media, and the outdoor "street furniture" I discuss later, and it is easy to see how the screen-age deluge can catch your eye for good and leave you drained. The irony of the isolation feedback loop is that plunging into social media as a means to cope with the isolation only leads to more of it, sustaining a vicious cycle.

9

Missing Critical Time Windows Degrades Empathy

Before you can shift your attention from one salient item to another, your frontal lobes need to scan the environment. Frontal scanning is a kind of self-monitoring. As an infant develops the motor skills for language it begins to utter words and expose those utterances to frontal scanning ("What did I just try to say?"). In time the child becomes able to think, express those thoughts, and willfully shape its inner world as well as the outer one of which it is a part. The gift of evolution, particularly neoteny (knee–OT–an–ee), is the capacity to focus, reflect, hold attention, and override the Stone Age brain's ever-present susceptibility to distraction.

Early in hominoid history, homeostasis—the most fundamental drive in all organisms to maintain a stable internal milieu allowed only unthinking, unlearned behavior. Instinct ruled the patterns of daily life.[1] Mental life 7.5 million years ago was all "online," meaning that our distant ancestors could deal only with what was immediately in front of them. Anticipation or planning based on what one remembered from past experiences was pretty much nil. Only millions of years later did an "offline" capacity evolve, an ability to step back and weigh current and historical situations. The ability for self-reflection and mental time travel, projecting back into the past and forward to imagine the future, gradually made humans the shrewdest and most adaptable of the great ape species.

These developments happened thanks to neoteny, which is the retention of juvenile traits through subsequent generations. The philosopher Zoltan Torey has called the "neotenous regression to plasticity" the evolutionary breakthrough that led from hominids to *Homo sapiens*. Early plasticity linked primitive vocalizations with motor actions to beget this new, offline life of the mind. The newfound ability to imagine what-if possibilities and not respond immediately gave rise to rudimentary language.[2] This protolanguage then evolved into a sense of self, a sense that one could cause events as well as reflect on them. In time, metacognition—the ability to think about one's thinking—rendered human perception double: knowing the content of an experience and the sensation that *I* as a singular individual am experiencing it. Ever since this breakthrough, human attention has oscillated between the two strands of focus and reflection in what cognitive science calls a "global workspace."

These basic principles illuminate why digital distractions tax adults and are pernicious to growing minds. A steady dose of distractions that emphasize sensation over thought can induce the virtual autism described in chapter 1, particularly among heavy screen users. One of the downsides of plasticity is that preoccupation with mediated images competes with the laying down of circuits otherwise destined for social engagement.[3] The overall arc of primate advancements illustrates the way this interference contributes causally to developmental delays.

Neoteny stretched out the maturational timeline of both body and brain. NASA scientist David Brin calls us "the neotenous clan of apes" because we mature later than our primate relatives, more slowly, and somewhat incompletely. Thanks to the accretion principle, we retain a number of juvenile traits, such as staying relatively hairless and having larger heads, flatter faces, and bigger eyes.[4] It is telling, too, that fictional depictions of species more evolved than humans—namely, aliens—are more neotenous in having massively round heads, tiny noses, huge eyes, and no hair at all. Figure 9.1 illustrates how human infants more closely resemble chimp or gorilla infants than their respective adult forms, whose looks diverge steadily over time. Very young apes easily walk on two legs, whereas it takes about two years for human offspring to master the skill.[5] Upright walking frees the hands to manipulate objects, something else it takes longer for human juveniles to accomplish with any degree of dexterity.

The human capacity to learn throughout the life span is a neotenous trait. Even centenarians can remain flexible and open to learning, whereas in other primates only the young can adapt to the new. For them, the most consequential force is neoteny during the first years of life. For humans, argues psychology professor Bruce Charlton at Newcastle University, "the retention of youthful tactics and behaviors into later adulthood" doesn't signal arrested development but is instead a valuable trait he calls

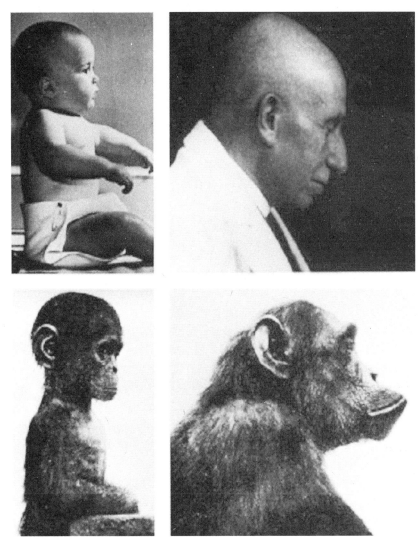

FIGURE 9.1
Chimp and human infants compared to their adult forms. Note the rounded head in both juveniles. When comparing relationships in a family tree are compared, similarities are more apparent in members close to the trunk than in members farther out on distal branches. *Source:* National Library of Medicine.

"psychological neoteny." It lets us adapt rapidly to whatever new conditions we must confront. The price for this resilience, however, is the need for ever-longer childhoods, a modern example being that formal education now extends well into the twenties in countries that prize college education.

* * *

The adult human brain contains 86 billion densely packed neurons, compared to 53 billion in a fifty-pound ape. A dense brain takes up a lot of room, at least 1,200 cubic centimeters (cc), which is too big to fit through a human mother's birth canal. Nature's solution is for babies to be born with 400 cc brains and undergo an unusually long childhood. Having a less mature and longer postnatal life than our ape ancestors shifted survival and coping from instinct to learning. Neotenous features such as the reduction in facial hair and elevation of the brow allowed more rapid and efficient signaling of social messages based on facial communication.

Human faces broadcast whatever their owners feel, think, and sometimes even intend. Many animals likewise broadcast intentional states, as Charles Darwin illustrated in his remarkable 1872 book, *The Expression of the Emotions in Man and Animals*. Our wealth of facial muscles conveys just how fundamental and evolutionarily useful the ability to read faces is. Others read us like a book just as we in turn read them. Yet only a few of our forty-five facial muscles are under voluntary control, which is why it is nearly impossible to counterfeit a smile or maintain a poker face, and why many professional poker players wear sunglasses. The upper face conveys sadness and fear, whereas the mouth and lower facial muscles convey happiness, disgust, and anger. Grandma may have told you to put on a happy face, but you can't if it isn't heartfelt. A natural smile automatically makes the corners of the mouth turn upward. We may try to imitate the movement, but the muscles encircling the eyes are strictly under involuntary control: they narrow to crinkle the outside corners. Even the Mona Lisa's smile shows this feature. The eyes give away one's game and help us tell a feigned smile from a genuine one. In technical terms the direct voluntary circuit goes via pyramidal pathways while the involuntary-emotional pathway is called extrapyramidal (figures 9.2 and 9.3).

Why do we have so many facial muscles when, in terms of utility, we need only two to open and close the eyes, and two more to do the same with the mouth? Thought of this way, the remaining forty-one would seem

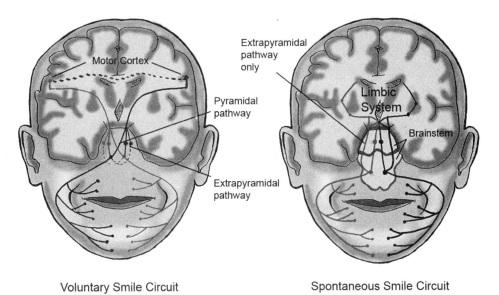

FIGURE 9.2
The authentic Duchenne smile. We have separate motor pathways that control voluntary movements and involuntary, emotionally driven ones. Drawing by Gaby Cardoso. [See also color plate 5.]

superfluous, and evolution should have jettisoned them, given the energy cost of keeping them around.[6] But evolution didn't do away with them, indicating that they likely serve an important biological purpose. Old World apes and New World moneys can make facial expressions, but our ancestral muscles differentiated away to enable greater nuance, and with it, enhanced communication. As Orson Welles says in *Citizen Kane*, "Nothing is ever better than finding out what makes people tick."

Social intelligence depends on the ability to read others. That is, we surmise the intent that lies behind an expression, decipher body language, and interpret tone separately from the words we hear. We read between the lines and notice what isn't said as much as what is. Seen from its component parts, the task seems daunting, yet we do it effortlessly. When in the film *Avatar*, the planet's inhabitants say "I *see* you," they refer to this deep perspicacity for reading others and grasping their intent. For us Earth-bound tellurians the skill was outsourced eons ago to the unconscious, where it became automatic. It isn't that modern people don't realize that others

Figure 9.3
French neurologist Guillaume Duchenne, known to his peers as Duchenne de Boulogne, electrically evoking a normal, emotionally based smile in 1852. Duchenne noted that while we voluntarily contract the zygomatic muscles that raise the mouth to say "cheese," for example, we can't willfully contract the orbicularis oculi that encircle the eye sockets. The eyes give away one's game and help us tell a forced smile from a genuine one. *Source:* Wikimedia Commons.

constantly broadcast their attitudes and feelings but that we give so little thought to this remarkable skill.

Faces figure so strongly in human perception that the brain allots more space to analyzing them than any other visual object. Six visual regions project directly to the amygdala for emotional judging, while yet others feed into prefrontal areas concerned with weighing decisions ("Do I like this person or not?"), making moral judgments ("Friend or foe?"), or resolving questions of beauty ("Attractive or yucky?"). Having multiple regions analyze a face makes us savvy to subtle distinctions in eye contact, direction of gaze, overall demeanor, even skin coloration—all while being insensitive to equally subtle variations in, say, landscapes or cloud formations. Infrared eye tracking reveals that viewer attention overwhelmingly homes in on a face when one appears in the visual field. The signals that a face broadcasts allow us to infer attitudes and states of mind, and deducing them is pleasurable, akin to getting a reward.

After practicing for hundreds of thousands of years, the brain has become good at reading faces. Defined biologically, emotions are unlearned routines for solving problems or seizing opportunities. As Harvard psychologist Steven Pinker told me, "Intentions come from emotions, and emotions have evolved displays on the face and body. Unless you are a master of the Stanislavsky [acting] method, you will have trouble faking them; in fact, they probably evolved because they were hard to fake."[7] Even one- to three-year-olds readily read faces in simple scenarios and understand what they mean. So do many animals, as any dog owner will confirm.

Growth curves comparing the brains of the juvenile *Homo erectus* ("upright man") and the young *Homo sapiens* ("wise man") are relevant to digital distractions. In each there is—or was—a time by which the brain has (had) grown enough neurons to make language possible. For the young *H. erectus* the critical age was six years. For *H. sapiens* it is only one. By the time *H. erectus* had enough neurons to engage in rudimentary language such as naming and pointing, its nonlinguistic skills were already well established. By then there was neither the need nor the plasticity to switch to vocal speech. And so *H. erectus* remained stuck at this level of development.

By contrast, the robustly plastic brain of an *H. sapiens* infant favored its budding ability to vocalize, by which it could manipulate its environment. Even simple crying summoned caretakers. Acquiring speech at around age one was the key that led to *H. sapiens* today. Slowing down maturation, neoteny repurposed what would otherwise have been strictly motor cortex destined to manipulate objects. The repurposed cortex could now manipulate words—verbal rather than physical objects—in the interior space of the mind. Even today the same modification of linking motor ability to speech must still take place in every child. Further, it must happen within a limited window of opportunity. If during this critical period a developing brain does not get the necessary stimulation needed to wire itself for verbal control of its environment, the window then closes and internal thought fails to appear, as happened in the case of Genie, who was kept locked away and isolated from most human contact until age thirteen.

As to what constitutes a growing brain's necessary stimulation, the short answer is socializing and back-and-forth banter with caretakers and the world at hand. A child needs to be spoken to and read to in complete sentences by linguistically competent adults. It needs to have new objects pointed out and named. What a child needs most is emotional engagement with an interested adult. My heart sinks at the sight of iPads hanging over bassinets, bouncy seats, and strollers or plastered on car seatbacks where they supplant a child's normally interactive view of the real world with passive, mediated images.

Screen images neither make eye contact nor provide emotional cues by way of body language that an infant can read. Despite interest groups petitioning Fisher-Price to withdraw its iPad Newborn-to-Toddler Apptivity Seat, major retailers still sell it. Its passive screen viewing interferes with socialization in all the above settings, and we know that lack of socialization blunts the development of empathy. Worst among these devices may be the iPotty, whose slogan is "Potty training meets technology." Toilet training once meant that a child learned to understand internal bodily sensations of when urination or a bowel movement needed to occur. Now it reinforces the idea that screens must not ever be relinquished, even to poop (figure 9.4). The Campaign for a Commercial-Free Childhood named the iPotty worst toy of the year. Group member Alex Reynard said, "It not only reinforces unhealthy overuse of digital media, it's aimed at toddlers. We should NOT be giving them the message that you shouldn't even take your eyes off a screen long enough to pee."[8]

Plasticity determines the brain's wiring and directs its capacity to respond to new experience. The downside of experience-driven plasticity is that any experience, not just positive ones, affects brain development.

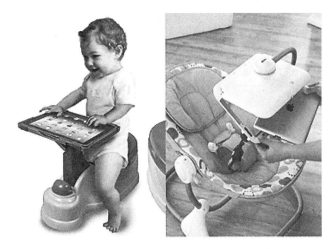

FIGURE 9.4
Screen holders on iPotties and Apptivity seats block a child's developing central vision and draw their attention away from learning how to read real human faces.

Failing to read aloud to a child is an example of not providing a necessary kind of stimulation. Preschool children who are not read to grasp language more slowly than those whose parents read to them regularly. The former have "poor lifelong literacy," according to the University of California.[9]

Aside from learning to speak, other skills take root during this early critical period: embodied cognition, imitation learning via mirror neurons, reading the emotions and intentions of others, developing empathy, and social collaboration. "Embodied cognition" teaches that there is no such thing as a disembodied mind: action influences thinking just as thinking influences action.[10] Physically based metaphors are common in everyday language; for example, affection is warmth, which language then elaborates into the concept of "warming up" to people or ideas. For these concepts to take hold requires mental flexibility and neuronal plasticity, a sound argument for putting youngsters in front of interested, responsive people rather than parking them in front of screens. Decades ago, psychologist Harry Harlow at the University of Wisconsin demonstrated the importance of emotional engagement and physical touch early in life. His series of experiments on deprivation became landmarks in the science of human attachment (formally called attachment theory), which studies how people think and behave in and around close relationships.[11]

Harlow showed that the ability to form emotional attachments could occur only during critical time windows in early life. Once the window closed it was difficult if not impossible to compensate for an earlier absence of emotional security. In one classic experiment Harlow arranged for two kinds of surrogate "mothers" to raise young chimpanzees. Both surrogates dispensed milk, but one mother was shaped from bare wire mesh whereas the other was the identical mesh form covered with soft terry cloth. Harlow observed that chimps given a choice preferred to cling to the terry cloth surrogate (figure 9.5). These chimps responded to stress quite differently from peers raised by the cold, wire mothers. When frightened by loud noises or strange objects, these chimps ran to their cloth surrogate mothers, clung for a while, eventually calmed down, and then returned to being as curious and playful as before. In contrast, frightened monkeys raised by wire surrogates never retreated to their mothers for comfort. They threw themselves to the floor, rocked back and forth while clutching themselves, and shrieked—behavior reminiscent of autistic children whose sacrosanct routines are disturbed. Harlow surmised that the first group benefited from a resilient psychological resource—emotional attachment—that was unavailable to the second group raised by wire mothers.

Harlow isolated monkeys for up to a year without social contact in what he called "the pit of despair." They were profoundly disturbed when reintroduced into society and unable to function within the group. Their lack of empathy showed itself in violent, even vicious behavior toward peers. *Screen Schooled* authors Matt Miles and Joe Clement cite Harlow's work when they reflect on the lack of empathy they witness in their own students. "Many of our children and adolescents are wasting away in 'digital pits of despair' in which they have traded human contact for a digital existence. They are isolated from peers, teachers, and parents," they say. "It is absurd that schools devote most of the day to screen viewing and then force students to spend their out-of-school homework time on screens as well. It is absurd because we know doing so reduces a person's empathy and social connectedness, and that such people are more likely to act out violently as school shooters, for example."[12]

Miles and Clement tell me that students can go through an entire school day without uttering a single word to another living being. Google Docs software allows students to "collaborate, but without ever talking to anyone." Everything mandated as "digital learning" conspires to isolate

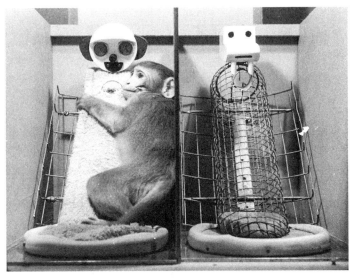

FIGURE 9.5
Top, Harry Harlow's two types of mother surrogates, with the infant always preferring to cling to the cloth-covered one. *Bottom*, Even in close proximity to a human, the infant feels secure. University of Wisconsin. Reproduced with permission.

students: Google classroom; having to watch additional videos for homework; obligatory "discussions" on the digital Blackboard platform rather than conducting them face-to-face with classmates; and so-called "flipped" videos that reverse traditional learning and replace homework by delivering instruction outside the classroom. Troubled teenagers can spend the bulk of their adolescence in online worlds, devoid of social norms, that let them sink deeper and deeper into isolation. The veteran teachers pointed me to evidence that heavy gamers are likely to identify with their fictional avatar—an idealized virtual self—rather than make the healthier choice of emulating living role models. For troubled, nascent psychopaths, this form of escapism reinforces antisocial tendencies by avoiding interpersonal encounters and anything resembling empathy.[13]

Harlow's work spoke to different types of child-parent separation and led to worldwide policy changes in orphanages and adoption agencies. Agencies began placing infants with adoptive parents as soon as possible rather than letting them languish unclaimed. Orphanages directed staff to spend more time with babies and hold them more frequently. Classic work like Harlow's informs my own wariness about exposing children to screen media during early life. He was the first to demonstrate that once a critical time window shuts, no amount of social exposure to mothers, siblings, or other caring individuals can fully undo the autism-like behavior or the emotional damage that has already occurred. When emotional bonds first become established is the key to whether they can be established at all. We are seriously in need of a new Harlow who can show us the cost of today's screens to human intimacy.

Contemporary research upholds the general sweep of Harlow's findings. Attachment theory, the most widely accepted elaboration of the issue, sees the fundamental role of parents as being available and responsive when wanted and ready to intervene whenever necessary to keep the child safe and out of trouble. The three most common attachment outcomes are "secure," "avoidant," and "anxious." Whether a primary caregiver responds to a child's needs sensitively, erratically, or not at all during the first two years of life determines whether that child grows up emotionally secure, anxious, or avoidant. Early childhood experience with one's first caregivers forms a template against which all subsequent relationships are shaped.[14] What do you imagine happens when that first caregiver is an iPad?

Critical time windows likewise exist during which our numerous visual pathways form. About 85 percent of the brain's input is visual, so organizing its extensive circuitry while it is still growing is a complicated undertaking. A two-year-old's brain contains twice as many synaptic connections as an adult's because the fetal brain manufactures an excess of both neurons and synaptic connections (two million synapses per second). The surfeit is subsequently pruned after birth, depending on the variety, intensity, and types of experiences a child has. Synaptic pruning is highly dynamic as connections fluidly form and dissolve. Which among the many millions become permanent and which wither away is a matter of use-dependent plasticity (the principle of use it or lose it). Synaptic pruning quiets down after age one or two but continues throughout childhood and well past puberty to gradually sculpt the adult pattern of neural circuits. Passive screen viewing skews the process because mediated screen images supplant a child's exposure to the real world and the social interactions that shape emotional intelligence and temperament.

Think about the ability to focus on an object and hold it there. The task seems simple, but isn't. The circuitry needed to pull it off spans all divisions of the brain—the retina, the thalamus, the midbrain, the cerebellum, the brainstem, and some two dozen areas of cortex. You have only to observe a baby in the first weeks of life to see this tracking system mature before your eyes: a newborn's eyes rove aimlessly, whereas a three-month-old begins to fix its gaze at will and hold it. Infants particularly prefer faces to anything else. The sheer complexity involved in laying down visual circuits means there are abundant opportunities for the process to go awry. One error affecting 2 percent of the population is prosopagnosia, an inability to distinguish one face from another. The impediment often plagues individuals with autism, making me wonder whether the incidence of prosopagnosia will rise as a new generation grows up having had their central vision trained largely by screens rather than by social engagement in the real world.

Nobel laureates David Hubel and Torstein Wiesel demonstrated that in cats, the critical window for vision opens during the first three or four months of life. In humans it opens during the first five to ten years, with the first year especially crucial.[15] Optimum windows likewise exist for the acquisition of critical thinking and basic social skills. The first sixteen months turn out to be optimal for developing language, the first twenty-four for general cognition, and the first twenty for establishing one of the three kinds of attachments.[16] Aristotle called man a "political animal," meaning not that we are obsessed with affairs of state but that we live in a polis, a city-society governed by laws and customs and centered on mutual social engagement. An infamous event in communist Romania shows what happens when circumstances interfere with a child's initiation into social norms.

In the late 1980s the Socialist Republic of Romania ran orphanages as if they were factories, warehousing up to 170,000 children in conditions that deprived them of human contact. Orphans were fed and changed but otherwise ignored and unstimulated during all their waking hours. Frequent staff rotations made it impossible for any child to develop meaningful relationships with a particular caregiver. A child could see as many as seventy different attendants in a week. They were rarely touched or offered rudimentary social interactions necessary for their emotional development. The situation turned into a large-scale study of what happens when unsocialized children are left to fend for themselves. Most failed to develop normal language and the ability to communicate socially—behavior similar to that seen in developmental autism.[17] When tested later as teenagers, these orphans had a variety of lingering behavioral and cognitive problems such as indiscriminately hugging strangers, awkwardness in initiating social interactions, or refusing invitations to play. Teachers graded them socially clumsy in trying to relate to others.

The Bucharest Early Intervention Project (BEIP), headed by neurologist Charles Nelson of Boston Children's Hospital, set out to rescue these children from what he called "a stunningly blank environment." In his book about deprivation and brain development Dr. Nelson and colleagues describe kids "rocking uncontrollably, heads in their hands, hitting themselves or hitting their heads against walls." The failure to show children how to navigate social relations, or even to allow them one-on-one encounters with an adult caregiver, had permanently stunted their mental and emotional growth.[18] Diffusion tensor imaging studies published in *JAMA Pediatrics* later showed that these children's brains had less white matter compared to control subjects, especially in pathways pertinent to attention, emotional understanding and expression, and executive control. The strength of their brain waves as measured by EEG was overall weaker than normal; a preponderance of low-frequency theta waves and attenuated alpha waves in the recording indicated a functionally immature brain. Reporting on the BEIP outcomes in the journal *Science,* Eliot Marshall wrote:

> A baby's brain "expects" regular stimulation, the kind that comes from an attentive caregiver. . . . If this doesn't occur, neurons don't grow properly. Abnormalities show up in neural activity and brain structure.[19]

In a baby not exposed to social engagement, the brain circuits behind theory of mind, emotional intelligence, and empathy are forced to wire themselves as best they can. Without social cues to guide them, the process goes awry.[20] For the Romanian orphans, social difficulties persisted into adolescence long after the children had been adopted or placed in foster care.

Can you imagine the level of distress these children endured? When the adrenal glands release stress hormones such as cortisol and adrenaline for sustained periods, it kills off dendrites, the cell branches that connect one brain cell (neuron) to another. Chronically elevated stress hormones kill off the parent neurons from which these dendrites sprout, too, delivering a long-term intellectual blow.[21] Leaving inadequately socialized children to fend for themselves produces these kinds of stark effects because it thwarts the most fundamental bond in human biology: one human to another. The more youngsters fail to bond with anyone, the greater their deficit of warmth and empathy later in life. Lack of empathy is the prime factor behind psychopathic behavior, and a strong correlation exists between heavy technology use and escalating lack of empathy in users. A common observation is that increasing dependence on technology reduces the frequency with which individuals engage with others, even with people in the same room. Dr. Clifford Nass, Stanford's expert in human-computer interactions and the author behind the persuasion algorithms now ubiquitous in nearly every screen app, was clear about their alienating effect: "The way we become more human is by paying attention to each other."[22] Yet his prescription has been routinely ignored.

Empathy and attention are related in a surprising way, as the philosopher Matt Bluemink lays out in his essay, "Socrates, Memory and the Internet."[23] The Latin word *attendere* means both "to direct one's mind to" and "to take care of." Its dual meaning is evident when we speak of a doctor attending to a patient, meaning that the doctor is taking care of a sick individual as well as the illness itself. To be attentive is to be compassionate. If you cannot be attentive to a point of view different from your own, then you lose the ability to empathize and take care of another. Novelist and writing professor David Foster Wallace said that learning how to think "involves attention, and awareness, and discipline, and effort, and being able truly to care about other people."[24] Yet increasingly, people, the young especially, cocoon themselves in online worlds. Rapt by their own curated image, they do not use social media to communicate but to project and promote an ideal self that has little grounding in reality. The average U.S. teen now spends six to nine hours a day engaged with online media. These are nine hours spent not engaged with others and understanding points of view other than their own.[25]

Professor Jennifer Aaker at Stanford's Graduate School of Business analyzed seventy-two studies conducted on 14,000 college students. The studies looked at their dependence on technology and their capacity for empathy. She found a sharp drop in empathy over the previous ten years as increasing reliance on technology whittled away at human interaction.[26] A 2017 study by the American Academy of Pediatrics, "Digital Life and Youth Well-Being, Social Connectedness, Empathy, and Narcissism," found that social connectedness and empathy plummet as screen time increases, while narcissism becomes pronounced.[27] When the brain is learning that two things tend to go together—"connecting the dots," which educators so often say is necessary for critical thinking—new frontal lobe connections are formed and reinforced.[28] Professor Linda Wilbrecht at Berkeley's Neuroscience Institute explains why active engagement is necessary for such connections to take hold. "You differentially sculpt your frontal circuits when actively doing something versus just passively observing it" (recall the gondola kitten). When a youngster stares into a phone or tablet screen most of their waking hours, this normal phase of development will not happen.[29]

An iPad thrust before babies presents an indirect, abstract, and disembodied two-dimensional construct cut off from firsthand experience of touch, time, and a child's sense of their own body. The experience is removed at a distance from what it feels like to imitate other people. It is time stolen from traditional games such as patty-cake, peekaboo, and "so-big," and from more imaginative ones such as "grocery store" or "restaurant" that inherently foster social skills in the young. It is time robbed from hearing and watching adults talk to them, read to them, and join in creative play with whatever is at hand. It is time taken from the practice they need to grow into engaged, self-aware social beings who can simultaneously be attentive to others. It is time that otherwise would be given to singing, retelling stories, and listening attentively to linguistically competent adults who engage them on their own terms—the natural ways that language, vocabulary, and literacy develop.

The brain's white matter bulks up during the teen years while gray matter naturally thins. Gray matter is made up of nerve cell bodies (neurons); white matter consists of the fibers connecting them. Aggregating friends and establishing social bonds aids this transitional crossover. The frontal circuits involved in self-control, planning, and the ability to anticipate consequences do not fully mature until the mid- to late twenties, and an immature frontal cortex typically gets blamed for all sorts of teenage problems, as if being a teenager were comparable with having had a lobotomy. But this is too simplistic. Early adolescence sees multiple changes in the tiny thorn-like dendritic spines. As puberty transforms the body, neurons may grow and lose 25 percent or more of their connections every week. It is the large-scale churning and turnover of connections rather than wholesale immaturity that explains why a teenager's frontal lobes are many times less efficient than an adult's; hence their iffy judgment and impulse control.

Two decades ago a new type of tissue imaging called laser scanning microscopy became available that could observe living neurons in a mouse brain before and after it had learned something new. Fast forward to today, when we can apply the technique to juvenile humans. Their frontal neurons are "wildly grasping for information upon which to form themselves" says Professor Wilbrecht.[30] But neurons cannot grasp or connect the dots of newly acquired information if young people are habitually head-down and passively hypnotized by their screens.

Neoteny dictates that intellectual skills develop only during critical time windows. Engaging with others is part of that development, and social engagement is what our Stone Age brain instinctively wants, a drive that hasn't changed in eons. What has changed is the recent invention of mercilessly powerful distractors that eat up our attention and cognitive bandwidth and outweigh the salience of everything else that may be present. Miss the developmental window, and the capacities to read others, infer intent, and empathize are permanently degraded, while the problems that result from having too little empathy, or none at all, inexorably mount.

10

HOW BLUE SCREEN LIGHT WRECKS NORMAL SLEEP

Thomas Edison napped in a chair that had large armrests. To one side he kept a pad and pen. On the other his arm rested while clutching a handful of steel ball bearings. A metal saucepan sat on the floor below. As soon as Edison began to dream his muscle tone relaxed and he dropped the ball bearings. Their clatter in the metal pan woke him up, whereupon he swiftly jotted down whatever creative ideas had been circulating in his subconscious mind.

Unlike Edison, we are part of a wired and tired nation, chronically sleep-deprived. Fortune 500 companies and law enforcement agencies now hire "fatigue management consultants" and install "fatigue management software" for their workforces.[1] But many workers still assume they can push through with an iron will and a pot of coffee, or else catch up on lost sleep during the weekend.

Both assumptions are unfortunately wrong. From an energy perspective sleep is neither inert nor passive. It is rather a metabolically active state that consumes a great deal of fuel. When habitually underrested, the brain's glucose reserves become taxed, and no evidence supports the idea that you can borrow time by sleeping an hour less tonight and an hour more tomorrow. On the contrary, departing just one or two hours from the ideal allotment of six to nine hours demonstrably degrades mental performance for several days, which makes ignoring a set bedtime ill-advised.

During sleep the unconscious mind makes connections that would not be obvious in the light of day. For instance, a good night's sleep more than doubles the probability that the following day you'll solve a problem requiring insight.[2] When actively dreaming, the brain takes what we have learned in one domain and applies it to knowledge already stored in memory.

Consider the solving of anagrams: when awakened from non-rapid-eye-movement (NREM) sleep subjects are foggy and don't do this task well. But woken from dreaming sleep, subjects solve 15 to 35 percent more anagrams compared to their daytime performance. The two conditions even feel different, they say. In the daytime or after NREM awakenings, subjects must deliberately focus on the problem; when awakened from dreaming sleep, the solution seems to pop into their head—an example of implicit knowledge coming to the fore.

The dreaming brain isn't interested in logical, step-by-step associations but instead makes connective leaps. Its advantage in solving anagrams shows how different the operating principles of the dreaming brain are relative to dreamless sleep or being awake. While awake we perceive a narrow set of possibilities. But only during dream-rich REM sleep do we exhibit an enhanced ability to link existing memories to new knowledge. Enhanced performance also happens after sixty- to ninety-minute naps that include dreaming.[3]

Digital distractions and screen viewing entwine intimately with the issue of insufficient sleep. And so, let's explore the architecture of normal sleep.

What Constitutes Normal Sleep?

Our modern notion of a normal sleep pattern differs from that occurring in most of human history, when daily life took place out-of-doors. Living in caves, castles, condos, suburban split-levels, and skyscrapers is a very recent development. For most of human existence we lived in harmony with the natural rhythms of light and dark. Gas and electric lighting became popular only 120 years ago, while the brain and body still take their cues from the Sun. We can't do otherwise because intrinsic biological clocks regulate alertness, mood, behavior, and even patterns of thought. Most of these clocks reside in the brain, while the heart and digestive system have robust timekeepers of their own.

Modern people can scarcely appreciate the force that biological clocks exert, or the cultural memory of what daily and seasonal rhythms once felt like when the mechanics of a rotating planet determined when it was night and when it was day. Schedules demanded by a newly industrialized and illuminated society forced us into an artificial eight-hour rest pattern that became standard for the purpose of maximizing worker productivity.

In the early 1990s Virginia Tech history professor Roger Ekirch unearthed historical sleep patterns that had been long forgotten.[4] By culling property records, church registries, and period literature, he discovered that earlier generations normally slept in two distinct phases. "First sleep" began shortly after sunset and lasted until about midnight. A person would then awaken and engage in activities such as prayer, sewing, diary writing, or sex before returning to bed for a "second sleep" until dawn. In Chaucer's *Canterbury Tales* a character in "The Squire's Tale" awakens in the early hours after her "first sleep," then later returns to bed. Charles Dickens referenced first sleep in his 1840 novel *Barnaby Rudge*. One can imagine how the great literature and philosophical tracts of the eighteenth and nineteenth centuries could have been written during these quiet hours, the undisturbed interlude between first and second sleep. Today, many highly productive people who don't sleep in two phases frequently claim to get their best work done before dawn. I am one of them.

When morning breaks, a discharge of hormones rouses us; when darkness falls, the clockwork secretion of a different set of hormones lulls us to sleep. Dim light of only 8 to 10 lux degrades their release; the average illumination of U.S. bedrooms measures about 200 lux. Dr. Thomas Wehr at the National Institutes of Mental Health, an expert on chronobiology, corroborates our natural inclination to sleep in two distinct blocks.[5] Chronobiology (from the Greek *chronos,* meaning "time") refers to natural bodily rhythms and the way light–dark cycles obligated by the Earth's rotation synchronize all aspects of human physiology.[6] Fall out of step, and the inevitable fatigue leads to late-hour mistakes, such as the Chernobyl nuclear disaster or the *Exxon Valdez* oil spill. Dr. Wehr persuaded volunteers to live for a month inside a windowless bunker while he controlled the periods of light and dark. Cut off from external light cues, subjects reverted to the two-phase sleep pattern described by early authors such as Chaucer and Dickens. Primitive cultures unaccustomed to artificial lighting still sleep this way, illustrating how biologically ingrained the split pattern is.[7] Once electric lighting allowed factories to run nonstop starting in the 1920s, the age-old pattern of first sleep followed by second sleep disappeared. Artificial illumination has now migrated from overhead bulbs to the palm of our hand, and this has introduced its own set of prickly problems.

There are as many patterns of sleep as there are sleepers. Any sound sleep, whether attained during an eight-hour block or a twenty-four-minute nap, physically starts clearing the brain of cellular detritus so it can function more efficiently.[8] Optimally, we should sleep no less than five combined hours every twenty-four hours and no more than nine. The benefits of napping are so well established that many businesses set aside dedicated nap rooms knowing that naps boost productivity and employee morale. People who adopt a split sleep schedule enjoy two daytime intervals of energy, creativity, and mental alertness. Longer naps are common in tropical climates such as India or Spain, where the culture favors siestas to promote efficiency. Siestas happen to coincide with a normal circadian reduction of afternoon alertness, again confirming the biological basis for two-phase sleep.

The word "circadian" comes from the Latin *circa dies*, meaning "approximately one day." Sunlight exposure resets the brain's central clock every day to the familiar twenty-four-hour rhythm, whereas volunteers kept under constant illumination fall into an otherwise natural rest-wake period of twenty-five and one-half hours. Jet lag happens because circadian rhythms adapt sluggishly to time zone changes. You may wonder how people living north of the Arctic Circle reset their circadian clocks when seasonal cycles of light and dark are so at odds with natural human rhythms. Contrary to popular belief, the Arctic doesn't have constant darkness for half the year and sunlight during the other half. The Inuit settlement of Mittimatalik (73° N) is about as far north as all but a few people live. During seventy days of winter the Sun stands persistently below the horizon, yet even a few hours of twilight are sufficient to break the monotony of darkness and thus sustain healthy sleep-wake cycles. Murmansk and Tromsø have even less time without some daily sun, whereas Shetland, situated 6° below the Arctic Circle, is never without a bit of noontime sun even in midwinter. Inhabitants south of 73° N accordingly have some sunlight or bright twilight each day in the winter.[9]

The question really, then, is how much daily light it takes to maintain circadian rhythms. Not much, it turns out. A few minutes of moonlight will do, and people adapt surprisingly well to living at these latitudes.[10] A few minutes of bright twilight without an external timekeeper (the customary term is the German *Zeitgeber*) suffice to maintain rhythms during the darkest weeks of midwinter. The Inuit people maintain their wake-sleep cycle without having any light-dark cycle to guide them, while Shetlanders strongly believe that moonlight should never fall on a sleeper's face. Unused to unexpected periods of light during the long winter nights, they may have come to notice the unsettling effect that any light had when shone on sleepers at sensitive times.[11]

Foremost among the benefits of sleep is tidying up the brain. When an office building shuts down for the night, in comes the cleaning crew. Sleep similarly ushers in a cleaning crew that mops up accumulated toxins, cellular junk, and the by-products of metabolism. Cleanup time is also the occasion for long-term memory consolidation, otherwise known as learning. While memories are created in the hippocampus, it has limited storage capacity, and so plasticity erases them during the deepest stages of sleep and transfers them to the neocortex for long-term storage.

"Sleep pressure" is a homeostatic force that builds up an irresistible urge to nod off. As it rises, so do concentrations of beta-amyloid and tau, two waste proteins associated with Alzheimer's disease (AD).[12] An association between poor sleep and AD has been appreciated for a long time, and sleep pressure is directly proportional to the length of time spent awake. At day's end we feel the urge to sleep; a good night's rest then resets the pressure level for the next day. The more sleep-deprived you are, the stronger your EEG brain waves are during slow-wave sleep, as if extra restorative forces were at work. Plaques of beta-amyloid and tau are microscopic hallmarks of AD, and plaque deposition begins destroying neurons decades before the onset of clinical symptoms.[13]

This should give every screen binger and night owl pause.

Beta-amyloid is a small piece of a larger molecule called amyloid precursor protein that normally aids neural growth and repair; tau protein maintains the cellular skeleton and regulates transportation of materials inside the axon. The first diffusion tensor imaging (DTI) study of sleep deprivation in healthy volunteers showed widespread structural degradation in brain white matter only fourteen hours after a night of sleep deprivation. DTI is a type of MRI imaging that looks at the anatomy of white matter tracts.[14] Subsequent studies confirmed the restorative power of sleep, that sleep deprivation degrades the microstructure of brain white matter, and that these changes can persist for years.[15] These findings are disturbing, and follow-up studies will tell us more. For now, the prudent course would be to limit your screen exposure, particularly in the evening.

Extracellular wastes such as amyloid and tau accumulate while we are awake; during sleep they are absorbed into the cerebrospinal fluid (CSF), from which they are then excreted. When the sleep-wake cycle is shortened or disrupted, the clearance of both is diminished and the progression to symptomatic dementia accelerates.[16] Until recently, neurologists attributed the fitful sleep of demented individuals to brain degeneration. But accumulated evidence now points to disrupted sleep as a possible contributing cause of Alzheimer's dementia rather than its end result. Two cohorts of older individuals showed that sleep fragmentation is accompanied by accelerated ageing of microglia, which may partially underlie its known association with cognitive decline; a 2021 study of eight thousand individuals followed for twenty-five years found an increase in dementia in those who slept six hours or less at age fifty and sixty.[17] Positron scanning also shows that a single night of sleep deprivation raises amyloid levels in brain regions routinely involved in AD.[18] The pattern of rising amyloid and tau levels during wakefulness and falling levels as they are cleared during sleep is especially robust in teenagers and young adults. Mothers, as usual, are right: young people need their rest, for good reason. It is disconcerting enough that poor sleep is associated with CSF biomarkers of dementia in cognitively normal adults; it is even more disturbing that apoptosis—the programmed cell death that occurs as a normal and necessary part of the brain's self-maintenance—goes haywire. An insufficiently rested brain will actually start to cannibalize itself by killing off neurons and synapses, a possible explanation for why it is difficult to think clearly after late-night binging on screen media.[19]

The brain routinely encrypts a staggering amount of information every day, yet sleep commits only a mere fraction of it to memory. Normally we experience four to five ninety-minute sleep-dream cycles each night. Most dreaming takes place during REM sleep, while sequential involvement of the four stages of NREM sleep consolidates memory. Shortchanging shut-eye by staying up late to binge on Netflix dulls thinking the next day because you never experience the last one or two REM sessions. Let's say you've had two nights of good sleep and then stay up late the third night, thereby depriving your brain of expected REM sessions in the latter half of the night. What happens? Your ability to learn new material the next day suffers. In fact, staying up late cancels out subsequent memory consolidation for up to three days. An incorrect mixture of brain waves and chemical transmitters overwhelms any attempts at catchup sleep as the shortchanged dream cycle tries, and fails, to resynchronize itself with the hard-to-budge circadian clock.[20]

A more effective strategy when you must study or rehearse a presentation is to front-load your sleep. Instead of staying up late to cram, turn in early and set an alarm to wake up seven hours later. That way you will achieve the deepest, most beneficial sleep during the early part of the night. The deep slow-wave stages III and IV normally drop away two or three hours after you hit the pillow, while REM episodes progressively lengthen as the night wears on. The golden minutes of slow-wave sleep are the most restorative and yield superior memory consolidation compared to the lighter stage I and II sleep. It might seem that staying up late would merely time-shift your standard sleep cycles. But not sticking to a regular bedtime also prolongs your exposure to artificial light, which further inhibits optimal rest and learning. There is no cheating your circadian clock.

* * *

Infants and children are a special case because brain learning algorithms have their heaviest workload during early life, as two observations should make clear: infants spend most of their time asleep, and sleep duration shortens with age. With respect to memory, childhood sleep is three times more effective than that of an adult who sleeps the same amount of time, and memory consolidation is in full swing even during the earliest months of life. Following a ninety-minute nap, babies will retain words learned during the hours prior to falling asleep better than unrested controls. What is more, they awaken knowing how to generalize. On learning the word "doggie," for instance, they first associate it with one or two specific examples. But after having slept they can suddenly associate it with canine varieties they have never seen before. Studies in preschool and kindergarten children likewise show that an afternoon nap strengthens the memory of material learned in the morning. The benefit occurs only in regular nappers and when sleep takes place within a few hours of new learning. Adolescents show similar gains in learning and the capacity to generalize. Regular sleep boosts good grades in school as well as self-esteem at having mastered new lessons.

In all cultures and geographic locales, the hormonal changes of puberty shift the timing of the suprachiasmatic nucleus progressively forward in ways every parent recognizes. Whereas younger siblings may be sound asleep at 9 p.m., teenagers don't feel tired. They stay up late and then have a hard time arising for morning classes. Even at 11 p.m. it may be hours before their brain begins to wind down.[21] Caught between a biological drive for later bedtimes and societal pressure to rise early, teens and young adults may curtail their nightly sleep to unhealthy levels (the Guinness World Records did away with who could stay awake the longest after realizing the damaging effects of sleep deprivation). The American Academy of Pediatrics advises that delaying school start times improves students' attendance and classroom attention and bolsters their grades.[22]

How Blue Light Ruins Sleep

The bluish light emitted by LED screens—found on TVs, smartphones, tablets, laptops, desktops, wearables, and seatback infotainment—powerfully disrupts the pattern of normal circadian sleep (tip: don't turn on the seatback screen during long-distance flights if you want to rest, and use an eyeshade). Recall that all life forms from bacteria to humans beat in time to circadian clocks that impose a twenty-four-hour rhythm that is in step with geophysical time. That is, all internal clocks are connected to the Sun.[23] All human physiology is rhythmic because internal clocks are present in nearly every cell in the body, influencing everything from hormone levels to blood pressure to appetite. A central timekeeper in the suprachiasmatic nucleus (SCN) of the hypothalamus acts like an orchestra conductor who keeps all the players in sync. The SCN and the tract from retina to hypothalamus are particularly sensitive to short-wave light, which outwardly looks like a bluish green called cyan.

Because of the way sunlight scatters on entering Earth's dense atmosphere, blue light predominates just before sunrise and again around sunset. The SCN registers the time of these peaks, while the primary force that resets the circadian clock every day is overall light exposure. No drug can shift the circadian phase to a new time zone. Melatonin-based medications can speed up adaptation a bit, but larger trials are needed to see how effective they really are. Meanwhile, new research reveals that being out of sync from jet lag, Daylight Saving Time, or other reasons impedes the formation of new neurons in the hippocampus (neurogenesis). It also interferes with spatial cognition, explaining why we can easily get mixed up when inadequately rested.[24]

The tech titans have reluctantly admitted, but only in the glare of scrutiny, that years of overexposing us to excess nighttime illumination in general and short-wave light in particular has had consequences, particularly for brains that are still maturing. Teenagers have a naturally delayed circadian clock; theirs is also roughly twice as sensitive to the wavelength mixture of natural light that predominates at the end of the day. Spending a single hour after dusk in front of a laptop or other display suppresses melatonin production in fifteen- to seventeen-year-olds by 23 percent, much more than it does in adults.[25] When a teenager stays up late staring into a phone or gaming console, their natural hypersensitivity to light pushes sleep onset into the early morning hours and leaves them shortchanged just when they need to get up for work or school.[26] Many U.S. teens wake in the middle of the night to check their phones because they either have received a text alert or want to check social media.[27] But even a few seconds of screen light during these hours undermines the benefits of switching off and turning in.

Before the advent of iPhones and iPads held close to the face, the largest source of short-wave light was the Sun. Photopic vision is what we experience outdoors during daylight and indoors with good lighting. Our eyes then are most sensitive to the green part of the spectrum, around 550 nanometers (nm) (refer back to figure 7.1). The retina changes from photopic to scotopic vision as darkness falls, and we lose the ability to discern colors. At night the eye's peak sensitivity shifts to 498 nm, toward the blue end. But within the range of blue only a small slice of spectrum between 460 and 490 nm can affect the retina's melanopsin receptors. Melanopic vision acknowledges that more occurs in vision than merely seeing, namely, that light has a broad biological impact on us. The hormone melatonin (hence "melanopic" vision) released from the pineal gland regulates the circadian pattern of wakefulness and sleep. During the day, naturally blue-rich sunlight peaking at 464 nm suppresses it. Light energy from LED screens is considerably greater than that contained in natural sunlight, which is why all nighttime screen displays adversely affect sleep (figure 10.1).

The only way to meaningfully reduce the light energy hitting us from this spectral range, aside from shutting devices off after sunset, is to turn down the brightness of the device's blue channel (the tall peak on the left of the figure). A number of apps reduce blue brightness, but only slightly. To actually matter the screen would need to look uncomfortably yellow. Inexpensive blue-blocking glasses claim to filter out the "bad" light rays, but to be effective such glasses would need to have a high optical density

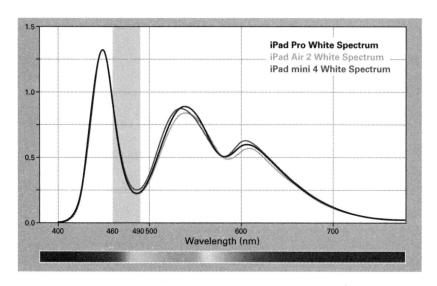

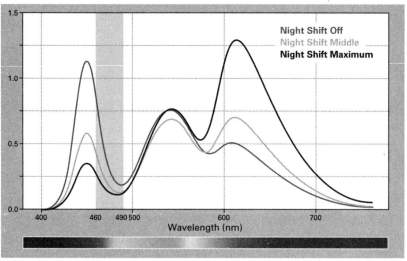

Figure 10.1

Top, Spectra from typical handheld devices. The proportion of light energy in the 460-490 nm range (vertical gray bar) is much greater than that in the solar spectrum, which is why exposing yourself to nighttime screens can degrade your sleep. *Bottom*, Typical screen spectra using Night Shift, with the melanopic 460–490 nm range shown by the gray bar. *Source:* DisplayMate Technologies. Reproduced with permission. [See also color plate 6.]

that rendered them dark orange or red and would result in very low visibility. One study from the University of Houston noted an upsurge in melatonin levels of 58 percent in participants who wore dense glasses that successfully filtered out blue light—many times what supplement pills can achieve. Study participants fell asleep faster, slept more soundly, and slept twenty-four minutes longer compared to controls.[28] Dr. Lisa Ostrin, the lead researcher at the university's College of Optometry, says that "by using blue-blocking glasses we are decreasing input to the photoreceptors, so we can improve sleep and still continue to use our devices. That's nice because we can still be productive at night."[29] While true, it downplays the fact that subjects wearing glasses that are effective enough can barely see what they are doing. It also dances around the root problem of screen devices being so addictive, distracting, and wasteful of mental energy, something no spectacles can rectify. The free software f.lux turns a user's screen color warm at night and cool during the day.[30] Apple doesn't permit f.lux in its App Store, and while Apple's Night Shift setting does decrease the blue channel, it superimposes an orange cast beyond yellow by increasing the red output to garish levels. There is no known biological benefit to doing this other than perhaps the placebo effect of making users feel they have done something good for themselves.

* * *

Context dictates whether light that is predominantly blue or red would be best for a given circumstance. For decades, doctors have used light of specific wavelengths (phototherapy) to treat conditions ranging from psoriasis and neonatal jaundice to macular degeneration and seasonal depression. Building on this, Flinders University in Adelaide, Australia, invented the Re-Timer, an LED device worn like ski goggles that beams cyan-colored light into the wearer's eyes. Donned either in the morning or evening, depending on whether the user has insomnia or habitually wakes up too early, it effectively resets the circadian clock.[31] The intensity and duration (dose) of artificial light exposure has society-wide implications for school starting times, elder care, and anyone who habitually uses a screen device after sundown. A good night's sleep depends on getting the right amount of the right kind of light at the right time of day and none when the body expects it to be dark. Manufacturers such as Phillips and Lighting Science have responded by introducing "smart bulbs" that adjust their wavelength output throughout the day. Their Sleepy Baby and Good Night light bulbs claim to not interfere with melatonin secretion, whereas their Awake and Alert bulbs, intended for morning use, output a greater proportion of short-wave light.[32]

Short-wave light emitted from digital screens matters because it often blasts the retinas with the wrong kind of light at the wrong time of day. Interestingly, the ability of short-wave light to reset circadian rhythms is not a function of its color: any photoreceptor could have evolved to register oncoming daylight and approaching darkness in the brain's suprachiasmatic nucleus. But more than other bandwidths humans can perceive, short wavelengths better penetrate the oceans where life and photoreceptors first evolved. Every animal has melanopsin photoreceptors, the blue-sensitive retinal ganglion cells that reset the daily clock. Even blind catfish and similarly sightless creatures have them, testament to how fundamental sunlight exposure is to sustaining life and well-being. Circadian timing makes skin wounds sustained during the day heal faster than those incurred at night. Because all cells answer to the conductor of our central *Zeitgeber,* skin fibroblasts involved in healing are metabolically more active during daylight hours. "We consistently see about a two-fold difference in wound healing speed between the body clock's day and night," says molecular biologist John O'Neill, who led a study at Cambridge University showing that daytime healing of burns is 60 percent faster than healing measured at night.[33]

Light is vital enough to human survival that it fits the definition of a nutrient: any substance an organism must obtain from its surroundings that provides nourishment essential for growth and maintenance of life. Yet most people don't spend much time bathing themselves in the bright morning light that the body evolved to respond to. In addition to perturbing the circadian clock when exposed at the wrong time of day, high-energy visible blue light (HEV) packs more energy than longer wavelengths do. It penetrates the cornea all the way to the back of the retina (figure 10.2). Some research suggests that HEV blue light, especially from LEDs, because they are so bright, leads to vision loss similar to that of macular degeneration. But even ordinary screen light affects young and old eyes in other ways, too. Age differences are especially meaningful in preschoolers.[34] Just a short exposure to the typically bright light from a tablet suppresses melatonin in youngsters by 90 percent. An hour afterward, melatonin remains

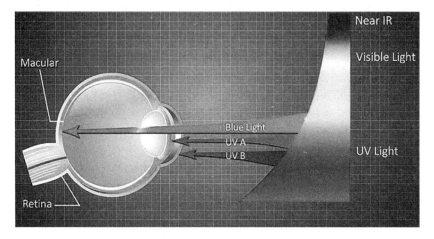

FIGURE 10.2
Short-wavelength blue light packs a higher level of energy than light coming from other parts of the spectrum. It can reach all the way back to the retina, where it may cause damage. [See also color plate 7.]

suppressed and not back to even 50 percent of the previous day's level. The chemical upset wreaks havoc with youngsters' bedtimes.

Aside from having more transparent lenses, preschoolers are more sensitive to ambient light levels because they have larger pupils that let more light fall on the retina and boost the signal to their internal clock (pupil size decreases with age, especially in those with blue or light-colored eyes). "To help children sleep, go dark" advises pediatrician-author Perri Klass.[35] Dr. Judith Owens, director of sleep medicine at Children's Hospital in Boston, recommends that given a preschooler's "circadian profile and melatonin release is relatively earlier compared to adults . . . parents should avoid having children exposed to very bright light before bedtime." Screen light, including television, is typically a household's brightest and bluest source. As for nightlights, says Dr. Owens, "have them low near the floor. You don't want anything shining directly in the eyes."[36] Her advice is remarkably like the Shetlanders' caution of not letting bright moonlight fall on a sleeper's face.

Exposure to strong light after sundown encourages late-night habits that perturb circadian rhythms and leave you chronically tired. The circadian clock will synchronize to any natural cycle if given the opportunity, as a clever experiment in Colorado demonstrated. Participants camped outdoors during a midsummer week that measured 14 hours 40 minutes of daylight and 9 hours 20 minutes of darkness. The study banned flashlights and electronic devices, making campfire the sole evening illumination. Fire radiates only long wavelength light. After one week, participants' circadian clocks had aligned themselves with solar time and the campers' biological night began at sunset and ended at sunrise. All were astonished that their weekly light exposure had increased more than four times compared to what it had been in the city, an illustration of how little we appreciate the chasm between light levels in home and office compared to those occurring naturally outdoors.[37]

Just three generations ago people got an average of ten hours of daily daylight exposure. Interior office light today measures about 300 lux and the average home even less, whereas outdoor light intensity can measure 50,000 lux. If we don't allow ourselves vigorous daytime exposure to sunlight, the SCN becomes overly sensitive to the short wavelengths that predominate at twilight, mistakenly telling us to "wake up" when exposed to the bright, short-wave light thrown off by the devices we use in the evening and the bright LED advertisements and storefronts that we encounter on the street. Dr. Mariana Figueiro at Rensselaer Polytechnic Institute, who studies the health effects of light exposure, says that while digital screens flood users with melatonin-suppressing short-wave light at night, for those who need it, screens "could act as invigorating light therapy during the day."

At an Albany nursing home Dr. Figueiro repurposed a flat-screen TV into what the staff dubbed the "miracle table." It transmitted only light, no picture. Elderly residents congregated around it every morning for two hours, and soon it began to alleviate the fragmented sleep so common in older individuals, particularly those with dementia.[38] The device worked as well as two hours' exposure to outdoor sunlight, making it useful for those prone to wandering or too frail to go outside. Dr. Figueiro likewise introduced luminous ceilings in intensive care nurseries, transforming hospital units from a twenty-four-hour environment uniformly lit by fluorescent tubes to one that delivered an artificial sunrise and sunset matched to the circadian rhythm of incubator babies.[39] A handful of international hotels currently offer smart lighting to help guests arriving from far-flung time zones reset their circadian clocks and alleviate jet lag. In 2019, Crowne Plaza introduced circadian lighting at its Atlanta Airport hotel. It slated the

Healthe bedside lamps for deployment across numerous properties of the InterContinental Hotels Group.[40]

Smart-lighting techniques are also effective in dimly lit industrial settings where workers sometimes spend hours with little light exposure. For example, Dr. Figueiro explored how much light submariners need to stay sharp and healthy. These sailors typically spend three to six months in dim fluorescent lighting, often following an eighteen-hour watch schedule to which the human circadian clock will never adapt. Studies by the U.S. Navy and the National Science Foundation in Antarctica are currently exploring how well blue light entrains individuals in addition to keeping them more alert (entrainment is the process that synchronizes one's internal clock with external cues). Because blue light suppresses melatonin, the issue is also relevant to factory shift workers. The National Toxicology Program, an interagency program within the U.S. Department of Health and Human Services, thinks that a relative lack of blue light exposure may explain the increased risk of cancer, cardiovascular disease, and depression that has been documented in this population.[41] We already know that long wavelengths boost alertness without suppressing nocturnal melatonin, and indeed, compared to current industrial lighting in a closed environment, exposing shift workers to more red light at night promotes alertness and improves performance.[42]

Having to choose among blue light, red light, colored filters, or special goggles seems like a lot of effort. Daily life would be easier too if the growing panoply of backlit gadgets were more circadian friendly. Instead of one more technological solution for a problem caused by screen technology in the first place, what if we just took a brief walk outside? Our internal clock is most sensitive to the benefits of light during the first two hours after waking up. But going outside at any time of the day is beneficial. Leave the sunglasses and ordinary glasses behind. Look up at the sky. Full-spectrum sunlight will brighten your mood and make you more alert. Even on cloudy days, outdoor light exposure is much greater than what you would get in an office building, even sitting by a window. Dr. Figueiro's prescription is: "Wake up and go for a walk every morning." Without your phone in hand, it is a bright idea to get your lux in for the day.

11

HOOKED IN THE PURSUIT OF HAPPINESS

Like the Sirens of Greek legend whose song lured Odysseus, modern technology bewitches us. It's fun. It's cool. It's amazing. We can do previously unimaginable things. But its "infinite scroll" can make us waste an hour swiping up or sideways. No need to say "just one more" because there is always one more, just as there are always eager opponents in multiplayer video games that seize participants so firmly that some wear diapers to avoid having to leave the action.[1] Obsessed gamers have been known to forgo food and sleep. Two adolescents in Chongquinq, China, exhausted after gaming nonstop for two days, passed out on the railroad tracks and were killed by an oncoming train.[2] Equally obsessed players have died while sitting in front of their screens, brain and bloodstream roiling with stress hormones.[3]

These examples are admittedly extreme yet they are indicative, reason enough to take them seriously. Like Odysseus, we too are easily lured by siren song. When a 2019 Stanford University study reconstructed adult screen experiences from recordings, it found that digital lives weave in and out among multiple applications and information sites in less than twenty seconds. Many subjects couldn't resist switching unless they metaphorically lashed themselves to a mast as Odysseus did—hence the numerous paid apps that promise to limit time online but that users typically circumvent.[4] We circumvent them willfully because our wanting system does what evolution designed it to do. Chasing novel stimuli may make us feel more productive, but we don't accomplish more merely because we are busier. The bias for seeking novelty and a high sensitivity to change that helped our Stone Age ancestors secure adequate food, shelter, and safety work to the detriment of modern people. Putting more on our plate makes it less

TABLE 11.1

Wanting and Reward	Liking and Pleasure
Dopamine	Opioid endorphins
More extensive	Smaller in scope
Easy to engage	Hard to activate
Nearly impossible to satiate	Satisfaction is short-lived
Impulse largely unconscious	Consciously felt

likely that we will stay on task. Pull your phone out to check something, and you will likely find yourself distracted by matters extraneous to what you initially intended to do.

Within the brain's neural networks are two feel-good circuits: one concerned with pleasure and liking, the other with wanting and reward.[5] The pleasure circuit mainly uses opioid neuropeptides, the brain's natural narcotics, as its transmitter. Of the two, the pleasure network is smaller in scope but harder to activate.[6] The resulting satisfaction is short-lived, whether from a good meal, good sex, or a good night's sleep. The glow of applause or social admiration, for example, quickly fades, as does the satisfaction of getting likes, pins, and retweets.

The more extensive wanting and reward network, however, relies on dopamine, the neurotransmitter with a ticket to every part of the brain, and is the fulcrum of addiction whether physical (drugs and alcohol) or behavioral (gambling, compulsive sex, gaming, shopping overeating, anorexia, and constant online checking).[7] Reward deficiency syndrome, now an official diagnosis in the DSM-5, defines physical and behavioral addictions in terms of biology. Their physiological and chemical features are similar, as shown by scan images that map their sites of action. As a person's wanting network ramps up, it becomes increasingly harder to satiate. Because the bulk of wanting operates unconsciously, it is useless to address addiction with pleading, punishments, or lectures about willpower. Addiction specialists now seek to identify genetic, anatomical, neurochemical, and external influences that predispose some to addiction, including the rising incidence of screen addiction.[8]

Compared to pleasure, wanting is easier to trigger and sustain but nearly impossible to satiate. One can never satisfy desire because as soon as we get what we want, we want something else. This psychological feature is the basis of advertising. Commercials promise greater satisfaction—cleaner floors, sexier relationships, better-smelling laundry—if only we use their new and improved products. But advertisers know perfectly well that the satisfaction of desire is ephemeral and that we will go shopping again.[9] Perhaps that is why the only thing new and improved about most products is the packaging and the ads.

In the Declaration of Independence, Thomas Jefferson called "the pursuit of happiness" one of our "inalienable rights." Stirring words, but as a nation, how well has that pursuit paid off? In *The Geography of Bliss*, NPR correspondent Eric Weiner writes, "Americans are three times wealthier than we were half a century ago, yet we are no happier. We are the richest country on earth yet rank merely twenty-third on the list of happy nations," despite an explicit mandate to seek happiness in the nation's founding documents.[10] Since 2012 the United States has steadily dropped in rank, according to the UN World Happiness Report, despite our ever-rising standards of living.[11] Some of the blame can be laid to our ancient feel-good circuits. They goad us into wanting big homes, flashy cars, bragworthy salaries, and other status symbols because our cave-dwelling ancestors sought the Stone Age equivalents of these things. The Greek legend of Sisyphus is a metaphor for this endless struggle, while in modern times the French speak of *la chasse au bonheur*—"the hunt for happiness"—while social psychologists Philip Brickman and Donald T. Campbell, mindful perhaps of America's own cultural hedonism, have coined the term "hedonic treadmill." They compare the failure to satisfy desire by acquisition and status with running on a treadmill from which we can never escape.[12] Climbing aboard this treadmill is unwise, like playing a rigged game we can never win, because it entices us to pursue happiness in the wrong places. Connect the dots if you will: if leading search engines and social media giants rely on targeted ads for their revenue, and if people increasingly rely on screen-mediated technology to govern their social interactions, and if technology giants spend millions of dollars lobbying Congress for a favorable regulatory environment, and if politicians rely on political donations to win elections, then doesn't it follow that these moneyed interests stand to benefit by promoting a culture of addiction among people who turn to their screens for happiness?

Humans did not evolve over millions of years to pursue happiness. Our Stone Age incentive system couldn't care less whether we find happiness or not. It cares only that we survive long enough to procreate and pass on our genes. What did evolve was a mindset now transmitted through cultural values and priorities that anticipates the kind of circumstances that will make us happy. How well it has achieved that goal is debatable.

Ambitious people by definition are never satisfied. Despite always striving and racking up tangible rewards, they have measurably lower levels of happiness compared to their less-driven counterparts. Materialistic individuals consistently rate themselves as less satisfied than those happy to have fewer possessions because external factors cannot alter their basic disposition or bestow long-term satisfaction. Some rich people are joyless thanks to their own personalities and inner thoughts, not because money makes them glum.[13] Physically attractive or wealthy people are no happier than their less fortunate counterparts even when their looks or money inspire envy in others because emotion operates by comparison, making them compare themselves upward to even better-looking or wealthier individuals. Anthropologists say that status seeking likely evolved ages ago among tribes whose size numbered a few dozen at most. Today we are exposed to the most gorgeous, talented, and successful people on a planet of seven billion. No matter what we do, someone will always surpass us in every measure.[14]

You can never win if you habitually compare yourself upward to those better off. And here is where Facebook is especially insidious in wreaking havoc on self-esteem.[15] For peace of mind, try a time-tested strategy and instead compare yourself downward. A case in point was my ninety-two-year-old mother and her unsightly varicose veins. "At least I have legs," she said. "They don't hurt." Her downward comparisons stopped her from wasting energy on something she couldn't do anything about. It freed her to focus on more upbeat things, and she remained upbeat and optimistic until the end. In the same way that singing aloud when you feel blue improves your mood (try it and see), the simple expedient of downward comparison can profoundly change your perspective and help alleviate the FOMO that you aren't living as fabulous a life as the people you obsess about on social media.

Granted, attempting to short-circuit your hedonic desires may be hard because the wanting system never pauses. Aristotle touched on this when observing, "It is the nature of desire not to be satisfied, and most men live only for the gratification of it." People around us strongly influence what we want (keeping up with the Joneses, again), and most people want the same things despite insisting that their individual taste is unique. The delusion makes modern advertising possible.

If in doubt, ask yourself: If you were the only person left on Earth, would your wants be different from what they are now? If yes, then your challenge is to understand how you have been conditioned from birth to think the way you do. The term "conditioned" comes from the Indian philosopher Jiddu Krishnamurti, who said, "We become what is done to us, apprehending everything through the screen of conditioned thinking instead of with a fresh pair of eyes." He meant that parents, teachers, society, religion, relationships, ambition, and one's own life experience unavoidably shape us.[16] For those who fancy themselves rational, experiments conducted in both rats and humans show how easy it is to make us want rewards that we don't actually like, such as an addict scoring a hit that nonetheless renders him sweaty, nauseated, and unable to sleep; or a gobbled-down meal that leaves you bloated and with diarrhea. Unconscious parts of the brain motivate us to pursue things that aren't pleasurable or don't pay off yet make us race after them like, well, rats.

Think about a midlife crisis. Whatever its outward manifestations—getting a facelift, buying a snazzy red sports car, embarking on an affair—the moment inwardly reflects dissatisfaction at how poorly the years spent rewarding your desires have paid off. Such a poor showing after decades of striving only verifies your enslavement to the brain's reward network. That such a network exists and is not a social construct or the result of conditioned thinking is vividly demonstrated by a Parkinson's patient who underwent a brain implant intended to stop her tremors. When the current to her implant switched on, this otherwise happy woman flew into a crisis of despair. "I don't want to live anymore," she wailed. "I've had enough. . . . I'm disgusted with life!"

To her doctors' dismay, reasoning was futile. Morbidly depressed, she felt empty and worthless, her incentive to live crumpled along with the pleasures she had previously enjoyed. But once the electrical current was cut off, the woman regained her cheery disposition as quickly as she had lost it. The cycle time from her usual frame of mind to despondent and back again took all of eight minutes. She joked about what had just happened, and gamely volunteered to repeat the experiment enough times under varying conditions that no other explanation was possible but that her doctors had serendipitously stumbled on a neurological circuit involved with desire.[17]

Arko Ghosh, who revealed that screen swipes change the brain's representation of the hand in sensory cortex, has continued his research into exactly how smartphone behavior in naturalistic settings (as opposed to the laboratory) alters specific brain structures.[18] His downloadable "tap counter" app measures how many times a user touches, or "taps," their screen. It also tracks larger arm and hand movements via the phone's built-in accelerometer. Measurements show that people can generate up to 40,000 taps a day. "Surprisingly," Dr. Ghosh tells me, "people still play with their screen while they are supposedly asleep." The astonishing number of times that people touch their screen, consciously or not, seems eerily like the behavior of frenzied rats that furiously press a lever to self-stimulate the pleasure center in their brain.[19]

In 1954 James Olds at McGill University first demonstrated the existence of a pleasure center in the brain, a serendipitous discovery that secured his place in neurological history. Rats that have electrodes implanted in a specific deep nucleus will repeatedly press a lever to give themselves a sensation that must feel good because they forgo food, sleep, and sex for the opportunity to self-stimulate five thousand times an hour until they keel over and die (there are only 3,600 seconds in an hour).[20] Olds and colleagues showed that deep-brain stimulation turns on neural networks ordinarily activated by everyday rewards such as food and drink. Today's digital distractions show how easily artificial stimulation can elicit complex behavior like that observed in obsessive smartphone users. Pavlov would marvel at how modern people have willingly conditioned themselves.

Humans who have had analogous implants to treat Parkinson's disease, epilepsy, or other neurological conditions confirm that the sensation produced is indeed pleasurable. Early recipients often experienced orgasm during stimulation. Robert Heath, who succeeded James Olds and founded the Neurology Department at Tulane, described patient B-19, who stimulated himself 1,500 times over a three-hour period. When a session was over he pleaded with the doctors to let him stimulate himself a few times more.[21] Others more recently have reconfirmed Olds's and Heath's conclusions by showing that infusing the same region with dopamine evokes the same kind of pleasurable self-stimulation as that triggered by electrical stimulation alone.

These driven actions seem suspiciously like those of the screen dependency that plagues today's users. Screen engagement is engineered to make users feel good and its use is designed to be a reward unto itself. Unfortunately, reinforcement renders it nearly impossible to convince users that attention is an easily squandered and limited resource. Reinforcement encourages and strengthens a pattern of behavior via encouragement or reward; it shapes habits and influences decisions ranging from food, finance, and clothing to religion, relationships, and career. As a rule:

> The more time you spend in front of a screen
> the more reinforced the behavior becomes.

The correlation might sound obvious, but few people heed it, mainly because we are oblivious that our behaviors are being reinforced. By obsessively checking or exchanging texts and photos you condition yourself to keep doing these things merely to feel normal. The term "reinforcement" refers to any stimulus that sustains or increases a certain behavior. For example, approval (the reinforcer) given immediately after a child puts away their toys (the response) makes them more likely to put away their toys in the future. Reinforcement does not require an individual to consciously grasp the effect a particular stimulus elicits. Having set yourself up by repeatedly checking your phone makes trying not to look at it uncomfortable (aversive conditioning) in much the same way that withdrawal from physically addictive substances causes intense discomfort—recall Professor Larry Rosen's experiments with nomophobia, in which subjects couldn't bear to be separated from their phones for even a few minutes.

Reinforcement is an evolutionary tool to modify behavior and maximize a desired outcome. Food, water, and sex are all primary reinforcers because they satisfy strong biological desires. As we evolved, reinforcers became more numerous and sophisticated. A parent today has experience that a child lacks and cannot yet intellectually comprehend. But children can learn from reinforcement. Sensitivity to reinforcement is a neotenous trait that existed long before psychology gave it a name. We used to call it animal training or child rearing because both trainers and parents instinctively reinforce good behavior and dissuade the bad whenever they mete out praise or disapproval. Positive reinforcements shape behavior more effectively than negative ones, although people typically use a combination of the two. Pavlov's dogs are the classic example of positive reinforcement: the pairing of a stimulus (a ringing bell) to a reward (food) that makes the dog salivate in anticipation. Once Pavlov conditioned the dog, it salivated to the sound of the bell alone. The fact that most animals are susceptible to reinforcement testifies to its biological power. It is not necessary to consciously grasp the link between a stimulus and effect for reinforcement to work.

Reinforcement is not a tangible thing but rather a relationship between a behavior and whatever reward propagates it. A dog looks up at the dinner table, so you give it a morsel of food. Instead of retreating the dog looks up again, this time with an endearing expression because dogs have evolved to excel at inducing positive emotion in their human companions.[22] Pushover that you are, you give in and feed the dog another scrap, thereby unwittingly reinforcing its begging behavior. Relevant to compulsive screen checking: any reinforced behavior can sustain itself for a long time without needing to have the reinforcement repeated. This means trouble because breaking the cycle will require sustained effort. Most people conditioned to screen distractions are either unwilling to commit or else don't understand how to break their conditioned behavior.

One doesn't need brain implants to manipulate group behavior: positive intermittent reinforcement does the job nicely. Consider "love bombing," showering a group of recipients with flattery and affirmation to make them feel welcome and safe. Doing so invokes the hormone oxytocin, normally released during childbirth, breast-feeding, sex, and even simple hugging. It is profoundly pro-social and associated with empathy, trust, and relationship building. It instills a feeling of closeness when we are with the people we care about, including online groups we belong to. The flip side is aversive conditioning, sometimes called brainwashing because it stokes trust and group cooperation at the expense of distrusting those outside the group. One example is today's extreme political polarization. The film *A Clockwork Orange* (1971) shocked audiences with its violent depiction of aversive conditioning, images not so different from contemporary political altercations.[23]

* * *

But what exactly are rewards? They aren't just dopamine squirts but objects, activities, physical states, or frames of mind that have value. They create feelings of pleasure and well-being and reinforce the frequency, vigorousness, or shortened latency of a behavior that achieves its objective. The interaction between person and environment is obviously complex, so neural networks must not only sense the presence of rewarding or repellant stimuli but also predict whether they will occur based on past experience. Pavlov's classic conditioning study (associative learning) revealed a great deal about rewards and why animals and people acquire classic conditioning so easily: the dopaminergic network responds more strongly to the prediction of a reward than to the rewarding stimulus itself.

Decades after Pavlov demonstrated conditioned reinforcement, scientists stumbled on intermittent reinforcement, a far more powerful force in shaping behavior than straightforward rewards or punishments. Both people and animals engage in a reinforced behavior more frequently when they are rewarded only some of the time. The influence becomes maximal as the uncertainty of getting a reward or not approaches fifty-fifty. Intermittent reinforcement relieved our distant ancestors from having to make the kinds of agonizing decisions we face today when surrounded by uncertainty. Marketers know this rule and use it to make decisions for us. Game and app developers similarly use it against us. To see how only-some-of-the-time reinforcement works, let's say that the next time your dog begs at the table, you decide not to give in. You shoo it away. Yet the dog sits there and waits, making entreating expressions because you have previously reinforced its begging behavior. If you refuse to give in, then continued refusal can, in time, extinguish the reinforcement, and the dog will stop begging.

But it will take weeks to snuff out your mistake if you give in only once again. If you remain resolute most of the time but reward the dog now and then, you will have made the situation infinitely worse through positive intermittent reinforcement. Children, like dogs, are clever animals. They quickly learn that sometimes no means maybe, or at least not yet. A frustrated child keeps pestering to make sure that no really means no. They may escalate by throwing a tantrum, but woe to the parent who doesn't stand firm because once they cave in, the child will have learned that nagging and tantrums work.

The principle of intermittent rewards makes slot machines addictive and profitable for casinos. Although the probability of hitting the jackpot is the same on every spin, the number of spins needed to hit it is variable. This induces gamblers to keep playing despite knowing in advance that casino games are unpredictable by design. All machines keep a small amount of the money fed in and pay out the remainder in irregularly spaced winnings. If, for example, the payout odds for a one-dollar bet were predictably eighty cents, then players would lose interest. What hooks them are the slot's small but relatively frequent payouts, followed by sporadic, medium-size ones that yield a proportionately larger jolt, then the unpredictable and thus even more thrilling win of bonus features, and finally the joy of hitting the jackpot itself. Intermittent reinforcement induces players to plow modest winnings back into the machine or else move on to a different one they hope will be more genial. Either strategy is irrational, but players don't care, just as those who complain about being addicted to their phones fail to turn them off or put them face down. Gamblers may keep at it until they go broke, which is why casinos make between 65 to 80 percent of their revenue from slots.[24] Online multiplayer games entice participants to purchase "loot boxes" that grant the player a chance to win paltry game accessories or nothing of value. Loot boxes are another kind of screen distraction. Unsavvy players who buy them can wind up maxing out their parents' credit cards.

Repeatedly checking your phone strengthens the same reinforcement tactic as playing the slots: you checked it once and got something useful, funny, important, or worth reposting to garner social affirmation—so you check it again. And again. And yet again. An unpredictable schedule of payoffs keeps you rooted to your screen while the emotion of anticipation is busy releasing dopamine, as is hope that the next screen will deliver something rewarding. The downside of staying vigilant for another hit while wading through a mountain of drivel is that you easily deplete your mental energy and attention capacity. An analog of screen obsession might be fishing. Perhaps you catch a fish in the first ten minutes on the water. But then you wait an hour or even all day before another nibble. The unpredictable reward schedule faced by naughty children, hungry dogs, eager gamblers, hopeful anglers, and compulsive screen checkers will keep them engaged in tantalizing, energy-draining anticipation of a reward that may or may not come.

Screen devices are addictive precisely because their unpredictable rewards are more exciting than fixed ones we can foresee. By putting animals and people in brain scanners we know that dopamine release is proportionately bigger for intermittent rewards than for predictable ones. Each time our smart device disgorges a reward like a text message or a social alert, it excites us. We want it to happen again, even though most of the digital flux that washes over us costs time and energy to register, categorize, and decide whether it is worth the trouble or not. Every single screen notification forces us to expend resources to evaluate its relevance, salience, and ultimate value. Maybe we get a relevant hit once every thirty messages, but that is enough to keep us hooked. Social anxiety also drives us to check our devices hundreds of times a day even though studies show that frequent checking reduces feelings of well-being and makes the user feel less connected.[25]

When I spoke with NPR correspondent Eric Weiner about the hedonic treadmill, he said no one had ever asked him why it exists in the first place. From an evolutionary perspective, he thought it made sense. "If we were satisfied with X then we would never strive for Y." He also saw self-preservation at work since the treadmill "works to blunt suffering as well as happiness. Perhaps we are not very good at predicting what will make us happy and what will not. We eat that third hamburger thinking it will taste as good as the second or the first because we are fixated on the short-term pleasure of the recent past, and not the wisdom of hamburgers past."[26]

I was surprised to read the *Oxford English Dictionary*'s definition of happiness:

> HAPPINESS. Good fortune or luck in life or in a particular affair; success; prosperity. The state of pleasurable content of mind, which results from success or the attainment of what is considered good.

The definitive English-language dictionary paints happiness as a passive result of attainment. If you have clothing, food, and a roof over your head, then the dictionary defines you as happy. Contentment, by contrast, has to do with not letting desire disturb you:

> CONTENTMENT. Having one's desire bound by what one has (though that may be less than one could have wished); not disturbed by the desire for anything more, or of anything different; satisfied so as not to repine.

The difference between these two states suggests that we would feel less stressed if we didn't chase desires so vigorously. But can we ever hop off the hedonic treadmill? Possibly, if we appreciate the three reasons why we mount it in the first place:

1. We typically miswant things that, once attained, fail to satisfy the craving that led us to seek them in the first place.
2. We overestimate how long a desire, once indulged, will gratify us.
3. We adapt and take what we get for granted, quickly becoming dissatisfied yet again.

It follows that the worst way to overcome dissatisfaction is by trying to gratify the never-ending stream of desires that well up within you. The whole purpose of Zen meditation is to address the existential dissatisfaction inherent in being alive. Because desire is insatiable, we cycle endlessly from craving to action to discontent and back again, never satisfied for long with what we already have. This insight is the basis of Zen's prescription for detachment. Some individuals wake up to realize the futility of chasing after their wants. Prince Siddhārtha in India relinquished his palace twenty-five hundred years ago to become the Buddha. Francis of Assisi in the twelfth century and Thomas Merton in our own lifetime likewise renounced their personal possessions to become legendary examples whose simplified lives held deep meaning and were examples to all. But most of us are not saints. We must settle for what enlightenment we can get while living in the bustling modern world, a smartphone glued to one ear, an earbud stuffed in the other.

"Miswanting" is the culprit behind impulse purchases and buyer's remorse because we adapt to new circumstances and return quickly to our temperamental set point.[27] The euphoria of lottery winners wears off after a month or two as recipients return to their prior levels of satisfaction. Those who were previously happy resume their sunny outlook while curmudgeons go back to being sour. Interestingly, the return to one's idiosyncratic set point applies from both directions. In comparing lottery winners with those left paralyzed by an accident, psychological studies find that lottery winners are initially euphoric and paraplegics despondent, as expected. But within a few months the newly rich revert to their typical mood, while paralyzed individuals "rebounded to happiness levels only slightly lower than before their accident."[28] In other words, resilience lets us adapt to what life throws at us. A quotidian example comes from a reader of *Psychology Today* who shared his surprise at not liking something he thought he wanted:

> I recently had an increase in salary, which I thought would make me happy. What I found was quite the opposite. I became quite depressed. Reading your column made me aware why. I have for a substantial period of time been very content with life in general. Receiving this increase in salary brought to the forefront of my mind all the "wants" in my life. My desires once again were disturbing my contentment.

Quick fixes often involve what I call "someday thinking." We say: I'll be better off when I'm ten pounds lighter . . . older . . . richer . . . married . . .

have my degree . . . get that promotion . . . whatever. The list is as infinite as the hedonic treadmill's belt that cycles incessantly under our feet. By focusing on a future that may never arrive, someday thinking pushes away the chance to be content right now.

All emotions are based on comparisons, as I've said, which is why "keeping up with the Joneses" is a familiar motivation, and miswanting a common trap. The urge for social approval only confirms that other people exert more influence over us than we wish to admit. If we did not value social status, or if neighbors laughed at rather than complimented our expensive cars, latest tech, and upscale homes, we would not strive so hard to acquire them. If we didn't value social approval, we wouldn't knock ourselves out to cultivate a public face and spend inordinate time, energy, and money maintaining it. Our Stone Age ancestors placed a premium on social approval because inclusion in a group conferred huge benefits compared to being ostracized and left alone in the wild.

Among the eight billion people in the world, some are spectacularly accomplished. Trying to compete against top winners is guaranteed to sap your energy, so no wonder the term "social media fatigue" has entered our lexicon or that the pressure to conform feeds one's fear of being judged. Gregory Berns, professor of neuroeconomics at Emory University, has shown that conformists have less brain activity in their frontal executive areas and more in areas associated with perception, indicating that peer pressure literally changes the way we weigh conditions and makes the opinions of others matter more than our own internal compass.[29]

Susan Cain concurred as much in *Quiet*, her in-depth study of introverts. "Groups are like mind-altering substances," she says. She cites Warren Buffett as someone who "divides the world into people who focus on their own instincts and those who follow the herd." The pressure to conform, even when slight, impedes creative thinking. Ms. Cain finds introverts to be better at making a plan, staying on task, ignoring distractions, and not giving in to emotion, whereas extroverts frequently seek emotional approval—and constantly check their phones.[30] Extroverts are suckers for positive intermittent reinforcement, and thus easily addict themselves to digital distractions.

12

Pandora's Box: How Ambivalence Keeps Us Hooked

Ambivalence means being of two minds (Latin *ambo,* meaning "in two ways," + *valentia,* meaning "strength"). A cardinal feature of physical and behavioral addictions, ambivalence describes the state of mind of those who want to stop but discover that they cannot.

Smokers in the 1940s had no reason to feel ambivalent about nicotine addiction because smoking was socially acceptable, even glamorous at the time. Films such as *Now, Voyager* (1942) illustrated romance by having Paul Henreid light two cigarettes in his mouth, one for Bette Davis and one for himself. Not until 1964 did the U.S. Surgeon General implicate cigarettes as the principal cause of lung cancer.[1] Until then, movie stars smoked on screen, in fan magazines, and later in television appearances. Growing up in the '50s and '60s I remember the *Journal of the American Medical Association* arriving every week on my father's doorstep, the back cover portraying a doctor, white-smocked and typically feet up, relaxing with a Camel cigarette. What could be healthier the ad implied? When smartphones first came out, no one imagined they might be addictive. We greeted them as miraculous devices sure to usher in a beneficent era of untold wonder.

At some point during their dependency an addict wants to stop but discovers they cannot, despite such consequences as getting arrested or losing a job, a family, a home, their health, or even their life. Remember my acquaintance who jabbed a needle into his thigh while calling out for help to stop? He was a highly educated, successful physician, but neither success nor his education shielded him from addiction to narcotics. An alcoholic judge told me how, before he became sober, he'd go to the liquor store determined to have "only two drinks" that evening, and then, to his dismay, would empty the bottle in a matter of hours. More familiar are the quotidian struggles of smokers and dieters. Yet no matter the addiction, brain plasticity changed by exposure to an addicting agent alters the wanting network so that short-term cravings win out over considerations of negative consequences in the future. David Eagleman at Stanford University calls this relentless force "the power of now." Rats work hard to continually "self-administer drugs at the expense of food and drink," he says, two pleasures essential to their survival. They make this senseless choice for the same reason humans do, "because the drugs tap into fundamental reward circuitry in their brains. The drugs effectively tell the brain that this decision is better than all the other things it could be doing."[2]

Regaining sobriety entails support from fellow addicts and talk therapy that, over time, aims to reverse whatever unconscious learning established the addiction in the first place. Addictive drugs and behaviors, including screen addictions, rewire the brain in the same way that the plasticity behind learning lays down new memories. While common medications for seizures and depression also target the brain, they are not addictive because they don't release dopamine or alter the motivational circuitry even after years of continued use (suddenly stopping some SSRIs after prolonged use can cause withdrawal symptoms, but they are not addictive in the traditional sense the way alcohol and recreational drugs are). Illicit drugs do alter the circuitry within a few hours of ingestion and can keep it altered for weeks.[3] Such rapid rewiring is a form of learning via the formation of new synapses, but that kind of learning is unconscious and leaves users ignorant that their brain has been altered. One's first exposure always marks the high experience as deeply salient, and the feeling can push the vulnerable to use again and again.

When someone tries to quit after repeatedly indulging in a drug or reinforced behavior, an insidious form of plasticity called "incubation of craving" occurs. Cravings typically develop during early abstinence. As the weeks pass they intensify until they stabilize and become something like a long-term memory.[4] Anything once associated with the addiction—people, places, objects, settings—can instantly trigger a craving. Enduring changes wrought in the pathways of motivational circuits are just one reason why quitting and staying sober takes continual effort. It is worth repeating that the Latin word *addictum* described the length of time an indentured slave, or addict, had to serve their master. The word's etymology means "bound to." And experience shows we can be easily bound to all sorts of substances and behaviors, including possibly the screen in our hands.

* * *

The pertinent question is whether incubation of craving applies to screen dependency. It isn't possible to say definitively until we conduct more research in naturalistic settings like the kind Dr. Arko Ghosh has done with smartphone swipes. But the parallel is clear enough to raise concerns.[5] The logic goes like this: Addictive drugs and addictive behavior both commandeer ordinary motivational circuits by strengthening the synapses that drive dopamine cells and their downstream targets. Instead of encoding new memories, these synaptic changes impel an individual to repeat the behavior or take the substance again. Evolution has optimized motivational circuits over eons so that attaining the materials for survival feels good (recall how food, water, and sex are all primary reinforcers because they satisfy fundamental biological desires). Rewiring these basic circuits has ruinous consequences because an addicted brain will subordinate health, employment, relationships, money, and life itself as if scoring a hit were the most crucial thing for its survival.[6]

When it comes to excessive screen checking, ambivalence is self-evident in laments of how much time you are wasting scrolling through your usual feeds. Perhaps you feel the internal tug between having promised yourself to cut down and then, without a second thought, violating that promise. This is the real-life paradox of wanting-but-not-enjoying: we fail to tear ourselves away despite repeated evidence of negative consequences.

In *Ten Arguments for Deleting Your Social Media Accounts Right Now*, Jaron Lanier recounts his journey from pioneering computer scientist who helped shape digital culture to an internet skeptic dismayed at how toxic and isolating social media have become. He writes, "If triggering emotions is the highest prize, and negative emotions are easier to trigger, how could social media not make you sad?" Positive intermittent reinforcement feeds addiction not through rewards but by never letting you know if or when a reward will come. Lanier frames it this way: "The algorithm is trying to capture the perfect parameters for manipulating a brain . . . but the brain isn't responding to anything real, but to a fiction. That process—of becoming hooked on an elusive mirage—*is* addiction."[7]

It is easy to yank the brain's wanting and liking systems in opposite directions. An experiment from Stanford University called "Lusting While Loathing" surreptitiously prevented volunteers from winning a prize they had previously coveted. Rating the prize afterward revealed the contradiction the researchers had set up: losers desired it more even as they simultaneously judged it less desirable compared to their initial assessment. The ability to disdain something yet still want it illustrates the magnificent human capacity to simultaneously hold contradictory thoughts and feelings. In the experiment the greatest emotional disparity between wanting and liking occurred in participants who weren't strongly self-aware of their feelings as a rule. The experimenters emphasized that emotion plays a big role in the "relative harmonization of wanting and liking."[8] Individuals will be happiest if they can strike a balance, but a good number of people cannot, thanks to ambivalence. Lusting while loathing is an example of the larger category of choice blindness that shows just how malleable decisions can be. Once subjects are shown two pictures of unknown individuals and asked to indicate which of the pair they find more attractive, the researcher then asks why the subject picked that individual. When the researcher surreptitiously swaps out the chosen picture, participants then end up explaining why they like a person they actually never choose.[9]

What may superficially appear irrational or contradictory preferences are in fact valid reflections of assessments that we weigh automatically. As circumstances change, we implicitly assign shifting valences to people, places, and things. Because wanting and liking are weighed unconsciously, we end up doing things we rationally know won't be good for us, whether skipping the gym because we don't feel like it, eating an entire bag of cookies, opening a fresh pack of cigarettes, shopping online for things we don't need, or snorting cocaine in lieu of eating a decent breakfast.[10]

A few years ago Dr. Kent Berridge at the University of Michigan challenged prevailing wisdom when he insisted that dopamine was not the reward neurotransmitter but the one related to desire (think of the automatic urge to pick up and check your phone). Time has proven him correct: dopamine is responsible for the positive reinforcement that makes us crave another donut even when, rationally, we feel sated or wish to lose weight. Initially it was hard to explain why, after years of abstinence, drug addicts could still crave cocaine, alcoholics liquor, smokers nicotine, dieters carbohydrates, and gamblers a chance to take a bet. Dr. Berridge found that addictions that successfully commandeer the dopamine network, as screen addiction does, change it permanently. "The enduring change is the basis for ongoing addiction," he explained to me.[11] This is unhappy news because it applies to screen and other behavioral addictions as much as it does to substance addictions, and so seriously indicts our uncritical embrace of technology. For example, by the time the average American child is eight years old, says pediatric advocate Dr. Aric Sigman, "they will have spent more than a full year of 24-hour days on recreational screen time."[12]

It gets worse: If addiction can permanently change the wiring for desire, then it exposes the limits of plasticity. On one hand, neuroscience has revealed the brain as marvelously plastic; on the other it has not adequately discussed limits on it. Consequently, people hold on to the notion of free will without realizing that the majority of their brain's operations must take place outside conscious awareness or willful control.

All addictions sensitize the wanting network throughout its extensive course in the brainstem, limbic system, and cortex, causing it to release more dopamine than it otherwise would. Receptors sprout for the glutamate transmitter, which in turn results in even more dopamine being released.[13] Cues associated with getting a reward consequently become more salient and leave individuals hypersensitive to whatever triggers a particular craving—drinking scenes in films, the sight of someone lighting up, the street corner where they used to score, or the Krispy Kreme sign flashing "Hot Now." To Dr. Berridge, this trigger of intense wanting constitutes heightened "incentive motivation." Yet here is the odd part: a conditioned Pavlovian response makes individuals (and animals) come to prefer the cue over the actual substance. Weirdly, anticipation supersedes attainment, becoming many times more rewarding than securing the actual object of one's addiction.

Following a three-month rehab, a different medical acquaintance of mine continued to inject himself—but with sterile water. He had become a so-called needle junkie, attracted to the ritual and sensation of injecting his veins. "I have no idea why I still do it, but I get antsy if I don't," he told me, illustrating how permanently embedded the drives behind wanting can be even as they remain involuntary and unconscious. In alcoholics, for example, electrode recordings from deep limbic nuclei reveal the involuntary nature of the wanting that drives them: subliminal drinking cues flashed on screen too briefly to consciously register still trigger cravings that the addict consciously feels.[14] A brief image of a film or TV character biting into a slice of pizza can similarly set off a food addict because incentive sensitization underlies eating disorders, compulsive gambling, pornography obsession, and the like.[15] It is reasonable to question whether the same might apply to screen addiction. Recall that it takes a long time to extinguish intermittent reinforcement and that a single instance of giving in makes a conditioned behavior worse. Given the permanent brain changes wrought by addiction and new cravings that are easily reignited after years of abstinence, you can still crave your smartphone after being away from it in much the same way a former cigarette or pot smoker suddenly resumes the habit after being exposed to a seemingly innocuous cue.

A recovering addict can successfully resist cues to indulge again for a long time only to relapse when stressed, emotionally riled, or exposed to the formerly alluring substance (this speaks to the hard question of how to balance the need to interact with screens and avoid becoming addicted to them). One does not need to consciously feel the force of emotional coloring for it to impel such behavior. By comparison, we can prime a rat's wanting system with dopamine and not see any difference in its behavior so long as the reward cue remains out of sight.[16] As soon as the cue becomes apparent, however, the primed rat frantically tries to obtain the associated reward. As Dr. Berridge puts it, "It is the cue multiplied by the level of dopamine interaction that determines the degree of wanting." And, I would add, habitual conditioned behavior.[17]

We know that both novelty and anticipation activate dopamine release. Drop your keys once—a change in prevailing conditions—and dopamine neurons fire to alert you of the change. But drop them a time or two again and your neurons habituate: they don't provide anywhere near the first-time jolt because the cue has become less salient with repetition and the consequence increasingly predictable. But the thing about digital distractions is that they fail to habituate. They remain endlessly novel and attractive because we have made screen flashes and the whoosh of notifications highly salient. Through repeated use, reinforcement conditioning has made them anticipatory cues for imminent emails, texts, videos, and potential social connections, the latter being perhaps the most potent and meaningful influences there are.

* * *

A large divide exists between those who question the unintended effects of ever-present screen exposure and those who root for more technology starting in infancy. In the latter group are people who plop infants in front of iPads either because it is the path of least resistance or because parents

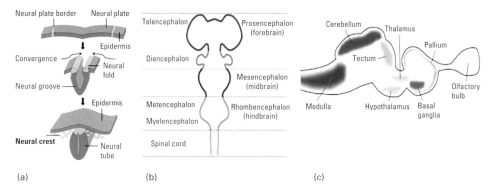

PLATE 1
The nervous system of every vertebrate embryo starts out as a line of cells on its surface called the neural streak. [See figure 4.1.]

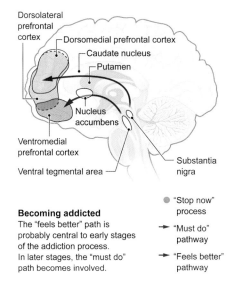

Becoming addicted
The "feels better" path is probably central to early stages of the addiction process. In later stages, the "must do" path becomes involved.

● "Stop now" process
→ "Must do" pathway
→ "Feels better" pathway

PLATE 2
The hypothesized main neural circuits in addictive behaviors. [See figure 5.1.]

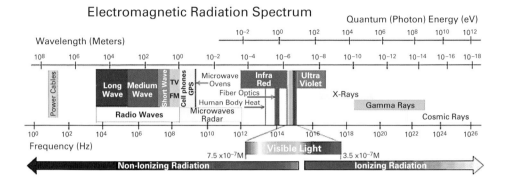

PLATE 3

The visible light to which humans are sensitive is less than a ten-trillionth slice of the universe's objective energy spectrum, which covers a billionfold span. [See figure 7.1.]

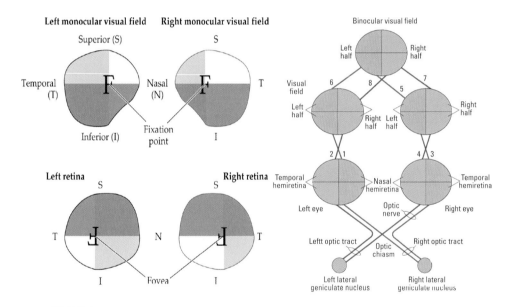

PLATE 4

Left, The four quadrants of the visual fields (top) and their representation on the retina (bottom). *Right*, The second row illustrates the outer quadrants where most distractions catch our eye and seize our attention. The upper quadrants connect most strongly with the emotional brain. [See figure 8.1.]

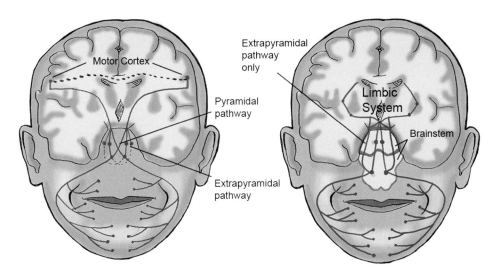

PLATE 5
An authentic Duchenne smile. [See figure 9.2.]

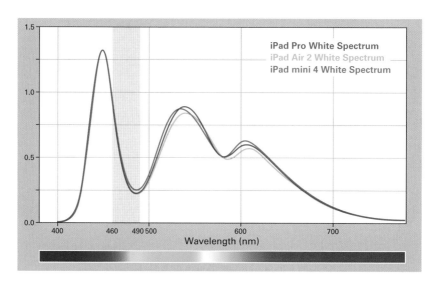

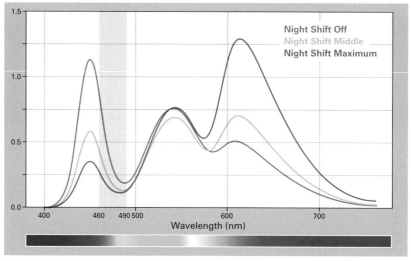

PLATE 6

Top, Spectra from typical handheld devices. *Bottom*, Typical screen spectra using Night Shift, with the melanopic 460–490nm range shown by the gray bar. [See figure 10.1.]

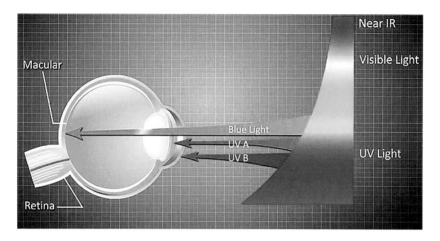

PLATE 7
Short-wavelength blue light packs a higher level of energy than light coming from other parts of the spectrum. [See figure 10.2.]

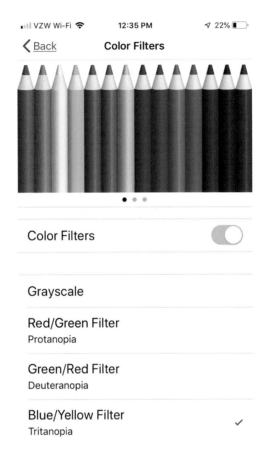

PLATE 8
Recommended adjustments of the iPhone color filters. [See figure 15.2.]

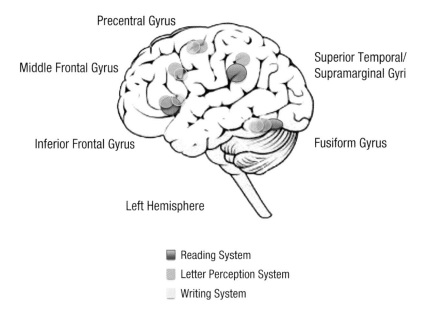

PLATE 9
There is considerable overlap of visual-motor letter perception and letter production. [See figure 17.1.]

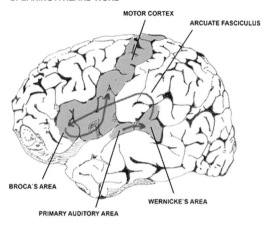

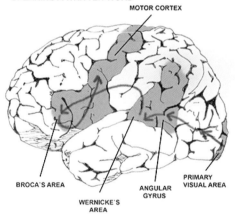

PLATE 10

Top: The sensation from a heard word reaches the primary auditory cortex but is not understood until the input is transformed by Wernicke's area. *Bottom*: A word that is read first registers in primary visual cortex (V1) and then transfers via the angular gyrus, which associates the visual form of the word to its corresponding sound pattern (auditory lexicon) in Wernicke's area. [See figure 17.2.]

believe that the experience, often labeled "educational," is enriching even when purveyors offer no evidence that their peddled wares are.

Arguing for the view of no restraint is Professor Annette Karmiloff-Smith, a British authority on early child development (about whom I'll say more in the next chapter). "Everything we know about child development," she argues, "tells us that tablet computers should not be banned for babies and toddlers."[18] Voicing the counterargument are such authorities as the American Academy of Pediatrics, which for more than twenty years has advised against exposure to screens of any kind before the age of two. But even the AAP has begun to crater in response to popular pressure from schools, sidestepping questions of harm by calling for "more research." But this copout fails to acknowledge a common conflict of interest in having tech purveyors participate in and fund such reports (it is exceedingly rare for any journalist to report that funding for a study about discretionary screen time originates from a digital media corporation, including Meta and Google).[19] If we wait for more research to yield an unequivocal answer, the present level of screen saturation may already have permanently altered millions of developing brains. This is one reason I call the uncritical adoption of screen technology the largest social experiment ever conducted and the largest undertaken without its participants' knowledge or informed consent.

A call for "more research" sounds well-intentioned, but looked at critically is one of a number of calculated obstructions that tech proponents and their media allies level at anyone who dares urge precaution in questioning the negative effects of increasing hours of screen exposure on children. Vested interests dismiss the judgement of highly trained clinicians who actually work with children as scaremongers while simultaneously abusing the concept of "evidence-based medicine." In so doing, says Dr. Aric Sigman, a member of Britain's All-Party Parliamentary Group on a Fit and Healthy Childhood, they urge us to err on "the wrong side of caution." No one has countered the establishment's viewpoint more clearly than this credentialed child advocate:

> We would all like the luxury of formulating public health guidance on the basis of comprehensive neatly quantified data from prospective randomised controlled trials. . . . [Discretionary screen time] is obviously not a pharmaceutical substance but a complex, multi-factorial lifestyle behaviour. Therefore, producing definitive proof of causation in the many domains of study, from neurobiology to psycho-social, will be a long time coming.[20]

In his defense, critics of the precautionary principle often misunderstand science and the precautionary principle itself. "People must humbly acknowledge that science has limitations in dealing with the messy complexity of the real world . . . thus, there is no contradiction between pursuing science and taking precautionary action," says the World Health Organization (WHO).[21] That is prudent advice, given findings from the Department of Pediatrics at the University of Michigan Medical School that "manipulative and disruptive" advertising was the norm in popular apps aimed at preschoolers aged five and under. This vulnerable group has become conditioned to peddling and sales pitches. Of the 135 apps from the Google Play and Amazon stores examined by the Michigan pediatricians, 95 percent contained at least one advertisement. Even apps categorized as "educational" by their vendors or available only by purchase featured a great deal of colorful, deceptive advertising.[22] For decades, the federal government has regulated TV advertising to young children, yet similar guidelines for digital apps do not yet exist. Online pop-up videos typically interrupt a child's play with nudges to buy the "full version" app to avoid ads, enable higher levels of play, or open to more characters (table 12.1). By design, children cannot succeed in these play apps unless they make a purchase.

Over three thousand marketing messages confront the average American each day. While ads may merely annoy adults, children cannot distinguish an ad from content. They cannot grasp the motives behind entreaties that screen characters make. Their ability to focus attention and their impulse control are far from mature, and the enrichment that apps claim to deliver has proven questionable. In "Doctor Kids," an app marketed as "educational" and intended for six-year-olds, the user becomes engrossed in playing doctor in a children's hospital until a pop-up ad invites them to purchase a mini-game for $1.99 or the full version for $3.99. If a child declines, a screen character shakes its head, looks disappointed, and cries. The game is structured so that the decision to not buy anything shames the child into thinking they have done something wrong.

"Given how few adults are able to resist the attention-sucking powers of app designers and advertisers, how can kids possibly stand a chance?" asks *Atlantic* columnist Joe Pinsker.[23] Using cartoon characters for commercial purposes extends to reputable companies such as Walt Disney. The Dole foods website invites users to "Celebrate 90 Years of Mickey Mouse with

TABLE 12.1

Advertising Code Category	Frequency, N (%)
Any advertising	129 (95.0)
Use of commercial characters	65 (48.2)
Full-app teasers: Offers/reminders to buy "full" version of app to avoid ads, have access to more characters or levels, and make gameplay easier, or a design that will not let a child succeed unless they buy the full version	62 (45.9)
Ad video interrupting play	
• Pop-up ad automatically appears without clicking anywhere, when idle, or when one level ends and before the next level begins	47 (34.8)
• Prompts to watch ad videos or to try out other apps to unlock play items or levels	21 (15.6)
In-app purchases to buy tokens, lives, or items to level up, make play easier, or have access to more characters or levels	40 (29.6)
Prompts to share	
• Rating on app store	38 (28.2)
• On social media	19 (14.1)
Distracting and deceptive ads	
• Banner across top, bottom, or sides of screen	23 (17.0)
• Camouflaged ads (e.g., a bouncing present that when clicked takes child to ad video)	9 (6.7)

Note: Advertising approaches documented in 135 apps marketed to or played by children aged five years and younger. From M. Meyer et al., "Advertising in Young Children's Apps: A Content Analysis," *Journal of Developmental & Behavioral Pediatrics* 40, no. 1 (2019): 32-39, table 1.

Dole" by directing them to recipes on Pinterest, where yet more ads aim to capture their attention as they scroll through screen after screen. A recent Disney-Dole initiative, "There's Beauty in Healthy Living," features licensed Disney characters who ostensibly market produce to children. But Dr. Charlene Elliott, Canada research chair for Food Marketing, Policy, and Children's Health at the University of Calgary, examined the ethics behind this and concluded that "Beauty and the Banana" is a blatant sales commercial, not a public health campaign.[24]

Starting in late 2018, the biggest names in tech began trying to atone for the damage they wrought by releasing "digital wellness" tools such as

Apple's Screen Time to show users how much tech consumes their daily life. These powerful corporations tried to come across as genuinely concerned, but people questioned whether their so-called wellness tools were in fact a cynical ploy. "Popular apps for preschoolers are rife with marketing that takes unfair advantage of children's developmental vulnerabilities," says Josh Golin, executive director of the Campaign for a Commercial Free Childhood. "Disguising ads as part of gameplay and using cartoon characters to manipulate children into making in-app purchases is not only unethical, but illegal."[25]

That same year Tristan Harris and Jim Steyer from the Center for Humane Technology in San Francisco began a yearlong tour across the country to explain to parents how social media keep kids tethered to their smartphones. At the same time, they spent hours covertly meeting with like-minded tech designers who were beginning to worry about the ethics of their work. Harris has characterized big tech's business model as "a race to the bottom of the brainstem," while Steyer says, "We will hold the industry accountable." Apps can either promote positive behaviors such as exercise or can make kids waste their time, the way Snapchat does. Its "snapstreak" feature assigns a number and a fireball symbol alongside every name in a user's contact list. It then counts consecutive days that the child has sent a "snap" to that friend. If they skip a day, the streak disappears, making the young user feel anxious and bad. In response to criticism, Snap redesigned its app, although the feature still remains and the company's sparse home page now says, lamely, "We contribute to human progress by empowering people to express themselves."[26]

Former *New York Times* technology columnist Farhad Manjoo points out the absurd ways in which tech giants contradict themselves. Google and Apple market their latest phones as irresistible while simultaneously wanting to appear virtuous by promoting software that lets you use it less frequently.[27] They do this, says Manjoo, because "we have reached peak screen" saturation. "Tech has now captured pretty much all visual capacity." When NYU professor Clay Shirkey observed that screens in the classroom acted like secondhand smoke, he banned them. Adrian Ward, marketing professor at the University of Texas at Austin, similarly found that a smartphone visible within "glancing distance" measurably reduced cognitive capacity. Phones are so irresistible that our conditioned Stone Age brain has to exert energy resisting the temptation to look whenever they catch the corner of our eye.[28] Realizing this, tech giants had to stop fighting over merely winning eyeballs to find other ways of holding users' attention hostage. Their solution was to weave a nonvisual digital net around us made of wearables such as watches, fitness trackers, sleep monitors, headphones, eyeglasses, head-mounted displays, GPS navigation, geofencing, and smart speakers—the Internet of Everything.

Voice applications turn out to be surprisingly good at manipulating us because they can read emotions carried in a user's voice. Facial recognition software combines psychology with big data to discern shifting emotions as people shop, watch ads, and move through public spaces. Their accuracy is extraordinarily good. Affectiva, an emotion-detection software maker in Waltham, Massachusetts, has amassed an enormous database by measuring seven billion emotional reactions in 2.4 million face videos taken in eighty countries. That is nothing, however, compared to what tech companies glean from your voice alone. Algorithms can deduce your state of mind, so be mindful of your tone the next time you are on the phone. Your health insurer can't see your face when you call, but they do have your voiceprint and don't hide the fact that they are recording you, purportedly "for training purposes." Call centers all over the world do this. What they never tell you is that they are creating millions of customer voice files that they then sell to behavioral analytics companies. Your voice is scored, combined with your demographic information, and then further sold to marketers who can target you with alarming specificity.[29]

*   *   *

Why do millions find it nearly impossible to momentarily put their screens aside (preferably face-down or out of sight)? Doing so is hard because we have come to associate screens with pleasure, and pleasure is a fundamental drive for survival and well-being. Pleasure is integral to homeostasis, which functions like a thermostat to stabilize our internal physiology—not just the chemical soup inside us but also how we think, feel, and act. Without having to think about homeostatic balance we act out pain and pleasure routines written into the fabric of our DNA and backed by powerful reinforcers.[30] Dr. Berridge says, "Human brains notice, remember, think about, anticipate, and plan for pleasure. It kept us alive for eons."[31] Our appetite for pleasure is why we always have room for dessert when we say we are stuffed, and why "one more for the road" sounds like a good idea.

Within the liking and wanting networks Dr. Berridge discovered "hedonic and motivational hotspots" and elucidated how natural opioid-like transmitters such as endorphins and enkephalins (the molecules released in the runner's high) amplify pleasant sensations to make them even "more likable." Doing so primes us to repeat actions or opt for choices that produce more of these pleasure-inducing neurotransmitters. When we are hungry the brain releases the neuropeptide orexin into a hedonic hotspot and makes food taste better (after a few weeks in rehab, alcoholics and addicts are apt to notice this). Similarly, flooding another hotspot with anandamide, a natural brain-based version of the intoxicant found in marijuana, amplifies any pleasurable experience.[32] Because the psychology and physiology of pleasure and wanting are separate and distinct, screens can tantalize us without providing much pleasure at all. No matter how many hours we spend swiping up, right, left, and down we are unlikely to score that satisfying but ephemeral euphoria we are driven to seek. Yet we continue to swipe because the possibility of that euphoria tantalizes us. The ancient Greek myth of Tantalus, from which we get the word tantalize, tells of how wanting in isolation can be agonizingly unpleasant. According to the legend, Tantalus had stolen ambrosia from the gods and revealed to human mortals secrets he had learned on Mount Olympus. Banished to the underworld, he was forced to stand in water that receded every time he bent down to take a drink, and beneath overhanging fruit trees whose branches lifted beyond his grasp whenever he tried to reach them.

Ubiquitous screens offer tantalizing tidbits that prod us to pick up our phone while standing in line, waiting at a stoplight, or even using the toilet—as if a moment of silence or a pause in activity has become intolerable. Perhaps it has, thanks to ever-present digital media that have reshaped and reconditioned our brains as if we were Pavlov's dogs. In 1654 the philosopher Blaise Pascal said, "All of humanity's problems stem from man's inability to sit quietly in a room alone." Yet does the anxious checking of your screen today deliver satisfaction, make you happy, or give life meaning? More likely it feels as if you were nervously waiting in front of a slot machine hoping for some kind of payoff. Relief from boredom is nothing like pleasure, even though we have conditioned our brains to think that it is.

Physicist and essayist Alan Lightman rarely ponders the material world without weighing its human significance. In both *Searching for Stars on an Island in Maine* and *In Praise of Wasting Time*, he shares his thoughts on imagination, creativity, and the value of solitude that I discussed in describing silence as an essential nutrient. Experience quiet, he says, and you will find something larger than yourself. The difficulty of slowing down in today's accelerated world is the very reason to purposely waste time. Doing so is good. Essential, in fact. "That's when the mind has a chance to think about what it wants to think about" without bombardment by outside opinions and demands.

After twenty-five summers spent on Maine's Pole Island in Casco Bay, Lightman discovered that silence "perpetuates itself" and reveals new ideas and feelings. It still boggles his mind that nearly all the volume of an atom is empty space. Yet what do we make of this extraordinary emptiness, this void in the reality that encircles us? Mulling it over, the scientist and humanist recollects moments of change on the island when suddenly he saw things, and himself, differently. "We must honor our inner lives. Otherwise, we are prisoners in the modern world we've created."[33]

Amen to that.

13

IPADS IN THE NURSERY OR NOT?

This is the question that two well-known child advocates battled out in the pages of *Nursery World*, a British publication devoted to "best practice for everyone working in childcare and early years education."

Favoring a full immersion of infants in screen-based technology was Dr. Annette Karmiloff-Smith, an award-winning cognitive scientist with expertise in developmental disorders such as autism, Down syndrome, and Williams syndrome—conditions that result from errors caused by both inborn (genetic) and external environmental factors. One of Dr. Karmiloff-Smith's studies demonstrated that a "tiny difference in the style" of how a mother interacts with her child (either "a little bit more directive or a little bit more sensitive") affects when that infant achieves specific cognitive milestones. Yet she insists that "everything we know about child development tells us that tablet computers should not be banned for babies and toddlers."[1]

Opposing early exposure to screen technology was Dr. Richard House, fellow of The Critical Institute and editor of *Too Much, Too Soon? Early Learning and the Erosion of Childhood*.[2] Dr. House argued that proponents have not thought through "the consequences of their enthusiasm. . . . We have no idea what these devices are doing to the younger brain, so perhaps we should exercise a degree of caution when we sit the younger generation in front of iPhones and iPads."

In the end, the two experts remained at odds, so how is a parent supposed to choose which route to follow when the experts can't agree?[3] Perhaps getting past the binary logic of whether screens are either inherently good or evil is a helpful place to start. A better-framed question asks whether phone and tablet use disrupts the social and emotional relationships between adult and child. Young children are exquisitely sensitive to the emotional tone and actions of their parents (the basis of such sayings as "That kid doesn't miss a trick" or "They soak up everything like a sponge," because, well, they do). We know that simpatico behavior in human pairings creates a social connection between the two, and that mutual gaze aligns brainwave patterns in parent and child as well as activations that are measurable on functional magnetic resonance imaging (fMRI) scans. These alignments create a transient but powerful network that facilitates communication and helps a child understand, via implicit learning, how to talk to and socially interact with other individuals.[4] Emotional attachment has its roots in the central nervous system, and the mother-child bond is the closest form of human attachment there is. Biology wires the pair of them to connect and hormonally bond via the attachment molecule oxytocin, which also doubles as a neurotransmitter. Coupled in this way, a child can embody the emotional experiences of familiar adults, forming a template on which to build intimate relationships throughout their life span. Psychology calls this the Russian doll model of empathy, after the wooden matryoshka in which dolls of decreasing size are nested one inside another, each relating to the outermost one just as the first connection between mother and child colors all subsequent connections in that child's life.[5]

Because parents and children naturally attune to one another, behavioral psychologist Sarah Myruski and colleagues at Hunter College asked what happens from the child's point of view when the parent's use of a smartphone interrupts this expected attunement. Myruski compared parental preoccupation to a classic psychological paradigm called "still face" because using smart devices in front of their children renders a parent temporarily unavailable. Psychologically speaking, parents who engage with their phones "disappear and become invisible," and this virtual disappearance affects the child's eventual emotional attachment. The amount of brain space on the fusiform gyrus and elsewhere devoted to recognizing and analyzing faces is considerable, and a child's ability to read faces (which they can do long before they can speak) is crucial to their forming a theory of mind: the ability to think about one's own mental state and infer the state of others.

In children aged seven to twenty-four months, the still face paradigm consists of three phases: mutual free play between mother and child; the still face phase, during which the mother is physically present but stares down at her phone and neither responds to nor initiates any gambits for attention; and a reunion phase that restores the parent as fully engaged and emotionally available. When gestural or vocal attempts to gain a parent's attention go unanswered, children become distressed. In the lab, the mother is instructed to scroll, type, and focus on her phone for just two minutes—the still face phase. On average we pick up our phones about ninety times a day, so Mommy may think that checking takes only a second. But kids see it differently, as shown by a video of the experiment in which one boy tries to get his mother's attention:

Come on, mommy. [no response]

We have another thing to do, mommy. [no response]

Mommy. [no response]

You are not listening to me, mommy. [no response]

Listen to me, mommy. [no response]

The toddler repeats his plea seven times during the short two minutes of the still face phase. In another video a different toddler looks to have already learned that when forced to compete with her mother's phone, she might as well give up. Resigned, the two-year-old sits silent and motionless, waiting while Mommy ignores her during the still face interval. The study's supervisor, Dr. Tracy Dennis-Tiwari, urges parents to understand that face-to-face engagement is the most important time during which to give children emotional feedback and tune in to their needs. "These quality interactions aren't just icing on the cake in terms of how young children learn about themselves and the world. It is the cake."[6]

In previous generations telephones were either bolted to the wall or tethered to the kitchen counter (and they didn't have an attention-grabbing screen). Mommy could signal with a look or a gesture that she'd be with you in a minute. She made eye contact and tried to reassure you while she dealt with whomever was on the line. But today when a child sees Mommy's head looking down, there is no telling how long it will be before she looks up again. The final reunion phase of the still face experiment provides an opportunity for mother and child to reconnect emotionally. Unfortunately, research shows that the more habitually preoccupied a parent is with their screen, the less successful reunion is at repairing the emotional breech created by the device.[7]

Despite efforts to shift away from either-or approaches, binary thinking about screen media still predominates. Forces pro and con ask whether iPads really do accelerate learning, as advocates claim. Do they bestow cognitive benefits on infants who haven't yet learned to speak and somehow give them a leg up in life? Does using them free up analytical space by offloading facts that previously had to be committed to memory, and what is the evidence that youngsters are now more "analytical" as a result? Or do digital devices overload and confuse developing sensory, motor, emotional, and social networks that are forming just as children struggle to make sense of the actual world of human interactions into which they have been born? For me, the most interesting question is:

> To what extent do electronic screen media divert the development of the Stone Age brain from the natural course it has followed for 200,000 years?

An iPad cannot react to a child's smiles, gestures, or babblings; nor can it sense a child's sudden shifts in attention and mood the way a human caregiver can. What it can do is compete with long-established biological processes for laying down key neural networks that serve social interaction. Multiple studies confirm that children younger than thirty months do not learn from television and videos as well as they do from hands-on instruction and face-to-face engagement.[8] Even knowing this, some might still believe that viewing iPads is no different than watching *Sesame Street* or *Mister Rogers' Neighborhood* was in years past. But it is.

From their inception, televisions have been fixed pieces of furniture. Viewers gathered around, making TV sets a social hub. They chose from a limited menu of programs available only at specific times. A digital device, on the other hand, is an extension of oneself—in your hand, going wherever you go, filled with personal data. Televisions were turned off for the night, whereas digital screens disgorge inexhaustible viewing material and exist in multiple iterations as videos, games, social media postings, podcasts, and interactive apps. Cumulatively, Apple has sold more than 200 billion apps, millions of which are targeted to children. As of 2017 well over half of all infants in affluent countries had played with an iPad before they learned to speak. Mobile devices now number in the billions, and for much of the world they are the main portal to the internet, perhaps boosted by the spread of Elon Musk's Starlink satellites that encircle the Earth. The global audience for YouTube alone is over two billion, or more than 26 percent of a world population of 7.9 billion.[9] The negative effects of screen exposure, if any, should therefore be more evident in the youngest children whose brains are developing robustly. Organizations such as the AAP and Common Sense Media have issued guidelines, but no authority has yet systematically looked into the consequences, pro or con, of heavy screen exposure in early life.

We could ask if influences are cumulative and proportional to the amount of time a child spends on screen. Common sense says they should be, but definitive proof of causation will be a long time coming, owing to the multiple factors involved in measuring discretionary screen time. Yet why aren't parents, teachers, pediatricians, and policymakers more concerned about how seldom anyone asks the question? If lead or similar toxins were at issue, people would be outraged. Until we can say whether screens are toxic or not, shouldn't the precautionary principle prevail? How do we know that any damage wrought at the start of a child's life won't persist for years, or that ill effects from early screen exposure might not appear until puberty, when the flood of hormones causes massive brain reorganization for the second time in that individual's life? The first reorganization happens after birth and dwindles until about age three, during which time synaptic pruning eliminates the excess neurons and connections formed during gestation (the fetal brain makes two million synapses per second). Peak pruning then dies down until the hormonal storm of puberty initiates a second round of synaptic reorganization that every parent knows as typical teenage behavior. The most pertinent confounding variable affecting the consequences of digital screen media, however, is that the brain is not fully developed until age twenty-five. To be fair, one must admit other confounding variables, especially the disparate methodological ways of defining "screen time," let alone measuring it. One scholarly group calls this problem "the conceptual and methodological mayhem of 'screen time.'"[10] Analysis of thousands of studies over the past decade reveals major weaknesses common to them. Aside from poor conceptualization of the issue, studies use nonstandardized measurements that are almost always self-reported. Context and content examined over time—what people actually see and how they interact with their screens—are not adequately controlled for, "especially in relation to the problems that doctors, legislators and parents worry most about."[11] Whatever chosen measurements of screen time are used, tracking how these variables change over time can at least discriminate between objective quantities of screen exposure and subjective self-reported estimates. And while I have focused mainly on the untoward aspects of screen time and social media, existing studies point to the beneficial effects of digital technologies on well-being; some address the conceptual and methodological distinctions mentioned above.[12] Still, I am sticking to the precautionary principle of less is better for all involved.

* * *

Dr. Richard House called early exposure to screen technology "unnecessary, inappropriate, and harmful" because in his view, children will learn whatever skills they need to operate digital devices "much more rapidly when their motor coordination is already well developed." Besides, any software they become adept at using will be outdated by the time they are older. He also regarded early engagement with technology at the expense of unstructured free play as "inappropriate" because it rests on the false assumption that "learning can be 'accelerated.'" Thinking is not an activity you can speed up: living cells chemically fatigue when pushed past their inherent Stone Age limits. Trying to do so increases the secretion of the stress hormones cortisol and adrenaline, which overstimulate the brain and lead to impulsive errors and clouded thinking. Attempts at multitasking likewise burn through oxygenated glucose, the fuel that helps us concentrate and stay on task. We become frazzled if we force ourselves to push ahead because multitasking demands that we continually make multiple unrelated decisions. Little ones consume as much energy as the big ones, and the first thing to go is impulse control: after making numerous decisions that don't matter much, we are primed to make really bad ones about something that does.

It has always taken time to acquire new knowledge and skills. From the brain's perspective, learning equals the physical formation of new synapses and the laying down of pathways that will become strengthened through repetition. Repetition works the same way that a footpath becomes worn and easier to navigate the more times people walk it. You cannot rush learning any more than you can accelerate thinking. (Actually, ingesting enough amphetamines does induce so-called "racing thoughts," but such individuals neither think rationally nor function very well.)

Three studies undertaken by the Alliance for Childhood found that typical kindergartners had less than thirty minutes a day set aside for play. "Most of the activities available during choice time," meaning time officially reserved for free, unstructured play, were in fact chosen by teachers and involved "little or no free play, imagination, or creativity." Long-term studies confirm than any gains attributable to early learning acceleration vanish by the fourth grade. By contrast, children placed in more play-based early education surpassed their peers at age ten "in reading, math, social and emotional learning, creativity, oral expression, industriousness, and imagination."[13] In other words, self-directed play bolsters a child's capacity to self-regulate and focus on sequential tasks. These are important for establishing emotional intelligence and fostering creativity, imagination, resilience, self-competence, and most of all social skills (figure 13.1). Earlier I cited the benefits of letting children roam as opposed to smothering them in helicopter parenting.[14]

FIGURE 13.1
Listening to a story or reading a book engages the imagination, while screen-based pictures preselected by third parties kill it. In earlier generations children saw what they heard in their mind's eye, and with repetition this developed a rich imagination. Do we want kids to think for themselves today or to have others do it for them? *Source:* Vanessa Vietes, Shannon M. Pruden, and Bethany C. Reeb-Sutherland, "Childhood Wayfinding Experience Explains Sex and Individual Differences in Adult Wayfinding Strategy and Anxiety," *Cognitive Research: Principles and Implications* 5, no. 12 (2020), https://doi.org/10.1186/s41235-020-00220-x.

Screen media supplant face-to-face communication. The definition of "medium" is any mass communication device interposed between an individual and that person's direct experience of reality. Early screen exposure is tantamount to "playing Russian roulette" with the natural unfolding of a child's development, says Dr. House. When remote third parties choose, edit, and package information on our behalf, "then imagination withers." Heavy screen use unbalances early development compared to everyday human rhythms and natural experiences, which engage all of cognition, emotion, physicality, and temperament.[15]

I have spoken with teachers at big schools and small schools, religious and secular schools, private schools and inner-city public ones, with young teachers fresh on the job and with seasoned instructors who have years of experience. From all of them I heard educators disgruntled at being micromanaged by higher-ups and expected to hew to rigid lesson plans preloaded on classroom tablets. Many wondered if the trend of offloading teaching from in-person instruction to tablets—far cheaper than hiring qualified teachers and retaining seasoned ones—wasn't an administrative solution to overcrowded, understaffed public schools. Some teachers welcome screen-based technology in the classroom while others see it as a flashy distraction. I found no obvious fault line between pro and con. I did find that the number of U.S. teachers, principals, and education policymakers who champion more technology use at ever-earlier ages outnumber those who caution prudence and waiting until students are older before they engage with screen media. Their enthusiasm persists despite being aware of the hypocrisy of Silicon Valley parents who forbid their own children to use smartphones and tablets.[16]

Theirs is a curious disconnect because the mission statements on the education pages of websites for Apple, Google, Microsoft, Dell, and Blackboard all say "children learn differently" and imply that students should use their company products to learn "at their own speed." Or at least they did. Corporate websites change constantly, and one can find the exact quotes only by searching using the internet's Wayback Machine, which archives older versions of web pages.[17] Prior to the pandemic, the contention that kids learn differently appeared in the root URL of Educational Technologists Limited (etlearning.com/resources/todays-kids-learn-differently). Claims like this sound good and appeal to common sense. But no evidence has ever supported the premise that allowing students to work at their own pace while using screen-based software improves educational outcomes. And yet Apple's "iPad in Education Results" boasts rosy outcomes, which the company attributes to using its products.[18]

The company claims that "Apple was not involved in the gathering or analysis of the data reported, nor has any knowledge of the methodology used." Yet how can it be unbiased when Apple's logo appears throughout the publication? Patently unbelievable assertions include the following:

> Academic achievements of students at Archbishop Edward A. McCarthy, as evidenced by the number of National Merit acknowledgments, continued to rise. Commendations rose from three to eight and finalists rose from one to eight—an increase of 200 percent since the introduction of iPad in the classroom.

This is an egregious misrepresentation of data based on a minuscule sample size: claiming that a 200 percent improvement—printed in an enormous font—was due to the introduction of iPads. One cannot assert correlation here, let alone causation, insofar as other variables can easily account for a one-year increase in National Merit finalists among such a small number of students at a single U.S. school. An analogously meaningless claim would be that giving $5 to a homeless person who had only $1 put an end to poverty because his wealth had increased by 500 percent. Many parents and policymakers have trouble judging these kind-of-statistical-sounding but specious arguments. Cathy O'Neill dissects this numerical flummery in her book, *Weapons of Math Destruction*.[19]

Despite lack of evidence that iPads and kindred devices deliver anywhere near the kinds of benefits that their manufacturers and advocates claim, education policymakers appear to be in lockstep with the tech giants, and this coziness isn't a recent development. During his tenure as education secretary, Arne Duncan laid out a National Education Technology Plan for Congress "for applying the advanced technologies used in our daily personal and professional lives to our entire education system to improve student learning, accelerate and scale up the adoption of effective practices, and use data and information for continuous improvement."[20] He cited no rationale for adopting such a policy or any evidence that it had worked in small trial programs. But he did say that the Education Department had "incorporated input received from hundreds of industry experts."[21]

Back in his 2016 State of the Union address, President Obama announced a $4 billion proposal to teach computer science in schools, especially coding. It "isn't an optional skill," he asserted, "it's a basic skill, right along with the three R's." Aside from the president being unqualified to render such an opinion, Obama assumed that knowing how to code would "make sure all our young people can compete in a high-tech, global economy" without first asking what young people should know to compete in the wider world before they know how to code or not. Putting the how before the why confuses ends with means, but that always happens with technology. People rarely first ask, should we do it? Instead, we spend a fortune on equipment and then make teachers figure out what to do with it.[22]

Based on forty years of combined classroom experience, *Screen Schooled* authors Joe Clement and Matt Miles tell me that education policymakers "are most concerned with looking good and pleasing the public." Bureaucrats want to "fix education quickly," so they buy "tons of these gadgets and force teachers to use them. If teachers don't use enough gadgetry," their performance evaluations are downgraded, even though their lessons may be excellent and their students raptly engaged. Large public school systems like those that employ Miles and Clement in Fairfax, Virginia, dread losing federal funding if they don't have enough technology on board. Because of this, "Every school in our massive school district has a technology specialist (who earns more than a teacher, by the way) who has only one function: get teachers to incorporate more technology into their lessons."[23] When I asked Clement about the reception of their book, he said, "Fairfax County has tried to ignore us at least and silence us at worst. They have no interest in hearing the truth" or in the experience of seasoned educators.[24]

Writing in *The Atlantic* in "How the New Preschool Is Crushing Kids," and amplifying her observations at book length in *The Importance of Being Little,* early childhood educator Erika Christakis echoes the observations made in *Screen Schooled.* She laments the overtaking of early education by bureaucrats and eager-to-please politicians who don't know the first thing about how children actually learn and frankly don't appear to care. Christakis calls one-on-one conversation "the most efficient early-learning system we have." The beginning of literacy, she says, "is that children have heard and listened. They have spoken and been spoken to. They have asked questions and received answers." No matter how feature-rich screen media become, what they offer is nowhere close to achieving the same kind of intimate, mutual exchange. Nonetheless, reading, writing, and cognitive tasks that have traditionally been introduced in the second, third, or fourth grades are now being shoved forward to preschoolers, who, according to Christakis, "lack the motor skills and attention span to be successful."[25]

Dr. Karmiloff-Smith's Birkbeck Centre for Brain and Cognitive Development in London, published a survey one year after its endorsement of early iPad use, looking for any association (not causation) with developmental milestones in 366 toddlers aged nineteen to thirty-six months. It found that "earlier touchscreen use, specifically scrolling the screen, was associated with earlier fine motor achievement (e.g., stacking blocks)." But "no significant relationships were found between touchscreen use and either gross motor (walking) or language milestones (producing two-word utterances)."[26] The conclusions were largely circular: scrolling a screen early on enables a child to do similar actions later. A subsequent study from South Korea confirmed that there is no evidence that tablets facilitate earlier language acquisition.[27]

After reading my essay, "Your Brain on Screens," in *The American Interest*, a concerned father wrote to say he was about to move to Europe to work for "an educational company" that specialized in preschool apps.[28] During the months before his scheduled move, he had let his young daughter "play these 'educational' games from the company."

> Well, let me just say that the only things these apps did for her was to get her better at playing the apps. Not only was there no "carry over" of skills to other parts of her life, as they said there would be, but we observed over weeks a visible degradation in her attention span, as well as a slowing down of her fine and gross motor development.
>
> I tried at first to chalk this up to something else. As time went on, this became harder to do. These things are worse than entertainment, much worse. I cannot in good conscience go to work for them now. In fact, I have been spreading the news about your findings to others I know, in the hope that they will make the right choices for their children and their students.[29]

What babies learn to recognize best during their early visual apprenticeship with the world are facial expressions. Without effort they distinguish one look from another, understand what different looks mean, and judge smiles of different intensities. They can differentiate surprise, anger, happiness, sadness, disapproval, and fear—an aptitude that watching screens can never instill.[30] Babies learn to read others long before they can speak, making the capacity a key part of emotional intelligence. The ability to read faces is necessary for human empathy and shared subjectivity. Up until now we have been overly optimistic about how transformative iPads would be and have not given enough thought to the price such technology would exact. Newborns effortlessly adapt to all the unfamiliar newness around them. They have never seen, heard, tasted, or felt anything like their new multisensory world—so different from the oceanic oneness felt inside their mother's womb. Rather than empty vessels waiting passively to be filled, babies are active explorers of their surroundings, other people, and what their own bodies are capable of.

Babies take in, organize, and interpret the energy flux coming from multiple sources, including proprioception of their own bodies (the awareness of its limb and core position, movements, and equilibrium) as they learn to distinguish inner sensations from outer ones. Through repetition they learn the links between perception and action, which thoughts produce the movement they intend, and which looks, expression, and vocalizations bring about a desired result. Infants are superbly motivated to explore, soak up the environment, attend to this and that, and, most of all, engage socially.

Recall our discussion of the retina's fovea, the matchhead-size spot responsible for sharp 20/20 vision. The fovea does not fully mature until age four, which is reason enough not to occlude the visual field with an iPad during this critical window of growth. A newborn's visual acuity typically measures 20/400. Acuity sharpens in the first months after birth. Color vision becomes functional around age four to six months, as do other networks for decoding the perceptual complexity of different kinds of movement. Despite their underdeveloped vision, newborns are astute social creatures: locking onto an adult's dilated pupils is a common sign of interest and pleasure that makes an infant smile.[31]

Visual attention similarly takes time to develop. The requisite eye muscles and brainstem circuits are functional at birth, whereas learning how to attend to and track an object takes practice. Smooth tracking steadies itself around two months. Command of the quick, jerky saccades that shift the visual frame from one point to another, as they do in reading or looking, is a different skill that takes longer to mature. Object recognition requires visual attention—learning to understand edges, boundaries, textures, and partly occluded objects; learning to calculate shape, size, relative movement, velocity, and distance from the way edges intersect; and learning that objects maintain a constant shape and size despite changes in distance and the viewing angle. It is an enormous undertaking; no wonder babies sleep so much. While they sleep, pertinent synapses and network circuits are being laid down and consolidated so that these skills gradually become automatic. A curious child learns these things naturally. Sticking an iPad in the way impedes the process as well as the evolutionarily ancient neuronal development that leads to their attachment to fellow humans.

With respect to hearing, infants are partial to high-frequency sounds. The ability to discriminate low-frequency sounds does not reach adult levels until the age of ten. Infants are more sensitive to frequency differences (changes in pitch) than intensity (loudness), meaning that locating the source of a sound in three-dimensional space is difficult for them. Infants also have a hard time separating speech from other sounds even though they are universally sensitive to the individual sound units (phonemes) of all languages. They must surmount the difficulty of categorizing sounds into meaningful groups, a tough cognitive task that is yet another reason to spare them the interference from screen media.

A synesthetic cross-sensory connection occurs naturally during the first year of life, especially close in time to the incipient emergence of speech. That is, all infants see speech as well as hear it. Normally, sight, sound, and movement are tightly coupled, which is why even bad ventriloquists can convince us that the dummy is talking. Cinema likewise persuades us that dialogue comes from the mouths on screen rather than surrounding speakers. Without realizing it we all lip-read, too, and the noisier it is the more we are forced to look at the speaker's face to see what they are saying. Cross-sensory couplings are so automatic and compelling that even two-year-olds succumb to their effects. They fall for the McGurk illusion in which a listener hears a different sound from the one the speaker actually makes. For example, listeners hear the sound of "ba" as "da" when looking at the lip movements associated with saying "ga." (You can find illustrations of the McGurk effect on YouTube.) Visual lip movements and vocal sounds influence one another even before sound and visual perceptions become assigned to a specific phoneme or word category.[32]

While we cannot get inside an infant's mind, we can draw on more than a century of research about how perception and thought develop in order to prudently weigh the eager but evidence-free claims of overenthusiastic tech advocates.

14

Human Contact Traded for a Googlized Mind

How do Google and the internet influence the malleable mind of the Stone Age brain? Nicholas Carr first addressed the question in his essay, "Is Google Making Us Stupid?," and later at book length in *The Shallows: What the Internet Is Doing to Our Brains*. Carr wrote:

> Over the past few years I've had an uncomfortable impression that someone, or something, has been tinkering with my brain, remapping the neural circuitry. . . . I'm not thinking the way I used to think. I can feel it most strongly when I'm reading. . . . Now my concentration often starts to drift after two or three pages. I get fidgety, lose the thread. . . . The deep reading that used to come naturally has become a struggle.[1]

This is the challenge for the Googlized mind. In the wake of *Fifty Shades of Grey*, a fashion columnist for *ELLE* magazine described how she commandeers her son's attention so she can protect her personal fantasy time while driving.

> In the car, I begin by neutralizing the attention of our son, age eleven, by selecting the app Angry Birds on his iPad. . . . There is nothing like it to make a young person blind, mute and above all deaf.[2]

This supposedly educated editor can't see that her son's already Googlized mind has gravitated like a magnet toward games and screen media to assuage the loneliness she inflicts on him by failing to pay him attention. Indeed, she celebrates the way her son can't resist the screen's pull, the way it stupefies him to the point that he cannot pierce her selfish, private bubble. Unwittingly she is contributing to her son's desocialization, unaware that the earlier compulsive screen consumption begins, the harder it will be for her son to overcome it. Games like Angry Birds and Pokémon

GO are clearly addictive to adults, teens, and preadolescents. Yet without considering the consequences, parents give fun "toys" like these to toddlers whose brains should otherwise be undergoing sensorimotor development. It is precisely the first years of life that establish the shape of one's personality. And it is this that screen distractions interfere with the most.[3]

* * *

An outsized example of Googlized minds comes from China, the first country to officially declare internet addiction (*wangyin*) a clinical disorder and the most recent, as of 2021, to enact weekly limits on gaming for citizens under the age of eighteen. In 2015 the Chinese government called wangyin "the number one public health threat" to its teenage population after realizing that boys were evading home and school to spend all day playing video games in internet cafés. They hammered away at keyboards, playing unseen far-away opponents as they tried to forge the social connections that eluded them in real life. As Baroness Susan Greenfield framed these sterile encounters, "No one looks them in the eye and they look no one back."[4]

The Chinese Communist Party typically views the internet as politically destabilizing. To neutralize the threat it established military-style boot camps to "rehabilitate" wayward teenagers who prefer the anarchy of a freewheeling virtual world to the constraints imposed by Communist life. Stern treatment regimens combine punitive measures such as solitary confinement with coercive practices typically found in the mental institutions of repressive regimes. For the documentary *Web Junkie,* two Israeli filmmakers were given access to the Daxing boot camp program in Beijing, which lasts three to four months and is designed to rehabilitate thirteen- to eighteen-year-olds.[5] Battles among parents, patients, and staff can be epic. Children complain that parents "are brainwashed by the psychiatrists" even as they deny they have a problem. Parents, for their part, are at their wits' end. "I'd unplug the computer and hide the cables," says one mother. "Nothing we did mattered. The games affected his mind. He wouldn't wash. He changed into a different person."

Colonel Tao Ran, Daxing camp director and resident psychiatrist, says, "All of them feel zero degree of emotion. They know the internet inside out, but nothing about human feelings toward another person." The absence of empathy for want of social connections eerily echoes Harry Harlow's "pit of despair" and the observation of teacher Matt Miles, who lamented that some of his adolescent students in Fairfax, Virginia, were wasting away "in digital pits of despair in which they have traded human contact for a digital existence." Miles watched as compulsive gaming usurped socialization and the formation of emotional attachments in some students during these critical years of adolescent development.

Late one night in Daxing, staff members discover some of the boys missing. Out goes the alarm. A count is made. Seven have escaped over the walls and hailed a cab to an internet café. Like salmon leaping over waterfalls to reach ancestral grounds, addicted gamers return to their screens for relief from psychological pain, a reprieve similar to what junkies in withdrawal feel when they finally score a hit. Guards round the boys up. The ringleader is sentenced to ten days in solitary confinement. After his release no one speaks to the offender. When Dr. Tao asks what makes him determined to go online despite the consequences, the boy says, "Another lonely person who sits on the other side of the computer."

* * *

After blaming gaming for poor academic performance and increasing nearsightedness across a broad swath of it 190 million youth, to say nothing of their poor social skills and alarming levels of teenage depression, the Chinese government issued new rules in late 2019: there would be no playing of video games after 10 p.m. and not more than ninety minutes of gaming on weekdays. In 2021 the edict expanded to no online video games during the school week, and one hour a day on Fridays, weekends, and public holidays.[6] In 2022 it banned minors from tipping influencers on livestreamed platforms such as TikTok. In 2023 the Cyberspace Administration of China moved even further ahead of other countries in regulating how much time youngsters spend online. Speaking of her thirteen-year-old son, one mother said, "It would be great if there was a way to force him not to spend so much time . . . scrolling and playing video games. He's been nearsighted since he was very young."[7]

In the West, experience has taught that coercion is counterproductive and only leads to anger and more antisocial behavior. We may question the cultural wisdom of China's approach even as we applaud its recognition of gaming addiction as a bona fide mental health issue. South Korea takes the precaution of banning young players from online gaming portals between the hours of midnight and 6 a.m. Its government also subsidizes gaming addiction treatment clinics. The United States is slowly catching up: in 2019 Purdue University blocked Netflix, HBO, Pandora, and other streaming services from 7 a.m. to 9 p.m. in academic spaces across campus. According to IT Communications Manager Greg Kline, its aim was to prevent students from hogging wireless bandwidth during class hours. At any given time, 55,000 devices were drawing on the university's wireless network.[8]

The American Psychiatric Association currently labels gaming disorder a "condition for further study," whereas the World Health Organization (WHO), which first recognized compulsive gambling three decades ago, acknowledged "gaming disorder" as an illness only in June 2018. The following 2022 edition of their *International Classification of Diseases* (ICD-11) includes it in the same category as drug abuse. Yet its inclusion begs the question as to why there is no "smartphone disorder," which surely is a bigger problem. Perhaps the concept is age-related, or perhaps adults who easily find fault with gaming among the young are blind to their own phone obsession.[9]

American treatment centers are few and gentler in approach than their Asian counterparts. One of the earliest, run by Dr. Hilarie Cash, is a digital detoxification clinic in Washington State called ReSTART. A place to "unplug and find yourself," it restores the idea of a Sabbath, a day of rest to reconnect with what gives meaning, feeds your soul, and rebuilds your capacity for attention.[10] The Dutch have a cultural practice called *niksen,* deliberately doing nothing that might entail staring out a window for a few minutes or momentarily putting life on pause.[11] At ReSTART, residents engage with nature and one another to break the habit of losing themselves in screens and game consoles. While screen-mediated engagement may give the appearance of a connection, it is superficial and counterintuitively pushes users into competitive, pathological isolation. "Launch your life, not your device," Dr. Cash tells patients, who "don't really know how to build and maintain intimate relationships." Typical symptoms that lead to her door are academic decline, termination of employment, ultimatums in relationships, and failure of young individuals to launch. Heavy media users lie and feel guilty about the amount of time they spend online. One boy received a $26,000 text messaging bill he had to justify to his parents. It did not go well.

Even after unplugging for only two weeks, young adults show salutary improvements. Preteens improve even more rapidly, which the UCLA

Children's Digital Media Center discovered when comparing two groups of sixth graders over five days. The study group had no access to screens of any kind, whereas the control group was allowed its usual amount. It is difficult to find kids willing to cut themselves off for five hours, let alone five days, but the Pali Institute, an outdoor educational camp for public school students located seventy miles outside Los Angeles, persuasively arranged for an absence of phones, TVs, and computers. Both control and study groups came from the same public school and shared similar demographic backgrounds.[12]

After only five days face-to-face with peers and camp staff, children in the preteen study group improved their ability to read nonverbal cues (tested by having them infer emotional states expressed in photographs of various facial expressions and videos with the sound turned off). The result is encouraging because social aptitude heavily depends on an ability to read facial expressions, tone of voice, gesture, body language, and gaze. Emotional intelligence is well known to correlate with an individual's degree of personal satisfaction in academics, work, athletics, and relationships. It is worrisome, then, that teens resort to text messaging more than any other form of communication. They prefer texting and social media to reading books, which at least require readers to speculate about motives and intentions, and that in turn fosters empathy.[13] Every text sent and received is time not spent socializing in person or developing empathy by reading linear narratives.[14]

It took just five days in the UCLA study to measure statistically significant improvements in preteens' emotional aptitude and social engagement. Does the rapid reversal imply that there are no long-term consequences to developing minds? We don't know conclusively, but that it no reason to abandon the precautionary principle.[15] The film *Wall-E* portrays a human population comically blind to the tangible world outside the mediated images with which it has surrounded itself. Residents are obese, passive, focused on their screens, and entirely dependent on mediating technology. Referring to the film, the UCLA study cautions, "Before screens become the only thing we look at . . . let's devote some resources to study the costs and benefits. The stakes are high, and our children are worth it."

The film *Celling Your Soul* and the companion book of the same title follow high school and college students as they embark on a weeklong "digital fast," only to confront the stark realization that their generation, the most technologically advanced in history, has also become the most socially awkward as it has steadily replaced interpersonal engagement with digital simulations of it. The French government forced a partial digital detox on students fifteen and under by banning cell phones during school hours and encouraging them to read a book or play outside. "Children don't have the maturity" to use smartphones, one French mother says. "Some adults don't either."[16]

The consequences of developing a shallow, internet-saturated brain (decreased attention span, depression, stress, irritability, addiction) will not dissipate until general users unplug often enough and long enough to effect change. The five-day UCLA experiment suggests that the washout period need not be long. On the other hand, a single reexposure to unwanted behavior makes reinforcing it infinitely worse (think back to Pavlovian conditioning and giving in once to the begging dog). Opposing arguments make it nearly impossible for readers to know what to do, and so the precautionary principle might remain the best practical guide, as is remembering that ambivalence, positive intermittent reinforcement, and sensitivity to reward cues are powerful determinants that make screen dependency challenging to shake.

Two decades of cumulative research have demonstrated the many ways in which the internet affects behavior, thinking, and feeling to say nothing of the neural substrates behind them. The digital world offers remarkable benefits, but problems as well that we can't ignore. The challenge is not to let drawbacks overwhelm the positive, which seems currently the case. The internet's precursors, Darpanet and Arpanet, date from *Sputnik* in 1957 and from 1969, respectively. They came online as research tools limited to a small number of universities and government institutions. Not until April 1993 did the public gain access to it. In historical terms the internet as we know it is relatively young, and we still have much to learn about its enduring effects on our psyche.

Recently, Dr. Kep Kee Loh at the Stem Cell and Brain Research Institute in France and Ryota Kanai at the Sackler Centre for Consciousness Science in Britain reviewed the large body of research on internet access to assess its current influence.[17] They began by acknowledging that cognitive networks in the brain have always changed in response to whatever tools we invent, whether stone flints and shovels or language and high-level mathematics. Through plasticity, the acquisition of every new skill encroaches on and gradually reshapes existing circuits. The acquisition of music, literacy, and arithmetic each led to repurposing brain areas originally devoted to visual recognition, sound analysis, or spatial perception. Brains are infinitely adaptable. They are adapting even now to the steady bombardment of screen exposure, but are they doing it in a good way or a bad way?

A change in response to experience is the definition of learning; plasticity achieves change either quickly or, through practice, over time. In string musicians, the sensory cortex devoted to the nonbowing hand typically encroaches on and expands into neighboring areas. Likewise, in newly blind individuals learning Braille, the brain map devoted to the reading finger invades newly unused visual cortex, changing its functional assignment from deciphering retinal signals to comprehending the tactile meaning of Braille. At first the newly blind just feel embossed bumps, but soon meaning emerges. Their finger begins to read words instead of feeling raised bumps in the same way that sighted individuals read meaning on a page rather than needing to decipher text letter by letter.

In a similar vein, sighted volunteers who are blindfolded for only two days remap their primary visual cortex, V1, so that it responds to touch, tones, and spoken words—an acquired type of synesthesia or coupling of the senses.[18] The demonstration of plasticity in blindfolded individuals is important because two days is far too short a time for axons from elsewhere to grow into new visual targets and establish fresh synapses.[19] The volunteers' newfound skill must therefore depend on existing connections among the senses that were hitherto dormant. To confirm this, removing the blindfold for only two hours reverts V1 so that it again responds solely to visual input. The brain's sudden and reversible ability to see with the fingers and ears depends on connections from other senses being already there but suppressed so long as the eyes input a signal.[20]

I have laid out evidence that, through plasticity, swiping smartphones hundreds of times a day induces a remapping of the brain's sensory cortex. Teachers around the country tell me that smartphone saturation has penetrated even the youngest grades. More than 60 percent of first graders have phones, and ownership skyrockets by the second grade. Because digital technology not only consumes attention but also plastically reshapes a growing brain, I wonder whether parents would be as permissive if screen technology came with an obligatory manufacturer's release of liability affidavit that they had to sign in case a piece of gear or app irreversibly affected their child's intelligence, social skills, and capacity for empathy. The software user agreements we are typically forced to accept can be longer than Shakespeare's *Macbeth*. Would tech firms peddle their wares as vital for success without evidence aside from self-funded studies, or would corporate lawyers panic at the possibility of catastrophic liability exposure?[21] Why do we not discuss these scenarios when an innocuous-seeming but ubiquitous action that physically reshapes the brain, a screen swipe, is already proven to change an individual's psychology? We already know that repeated screen swiping physically changes the cortical representation of the moving hand in direct proportion to the amount of time users spend swiping.[22] The research specifically asks which aspects of smartphone use change the brain and which are negligible. Some argue that does not prove causation; then again, as Edward Tufte, emeritus professor of statistics at Yale, says, "Correlation is not causality, but it sure is a hint."[23]

Back in 2015 the American Academy of Pediatrics noted that there were 80,000 apps in the Apple store labeled "educational." Today that figure is well over two-and-a-half million. Yet it is nearly impossible to tell which might be useful and which are a colossal waste of time or even harmful. The majority of so-called educational apps are untested and unregulated, and no established rating criteria exist. Most reveal nothing about their developer's credentials or whether third parties have conducted any peer-reviewed research as to quality or whether an app accomplishes what it purports to achieve. Today the AAP lamely cautions that "interactive media requires more than 'pushing and swiping' to be educational."[24]

It is tricky to compare what we already know about brain remapping in musicians and blindfolded volunteers to the way swiping a screen hundreds of times a day might alter a user's brain. Musicians hone their skills by practicing scales and études for hours a day, whereas people use their smartphones all day long in a variety of ways: checking the time, the weather, sending and receiving texts and emails, setting and responding to timers and alarms, looking up facts and definitions, reading the news, and engaging with various apps. To an outside observer, different swipes may look alike, but brain wave recordings show that some yield meaningful rewards while similar-looking gestures prove meaningless. More research like Dr.

Ghosh's is lacking because not until late 2018 did science and society even consider the topic of digital overuse worthy of study. The gap is reminiscent of the years it took to convince people that watching television changed the physiology of viewers' perceptions.[25] Just as learning to play the piano or speak a foreign language alters thinking and behavior through plasticity, Silicon Valley giants only recently admitted that prolonged, repetitive exposure to digital screen media may alter the structure of brains via plasticity, too.[26]

As to the Googlized mind, let me share a writer's observation about attention in relation to critical analysis. There is something oddly different about reading printed text compared to reading the same words on a screen. Ostensibly they are identical. Yet it is easier to catch mistakes and see glaring errors in logic and narrative flow when I work from the printed page. It is less exhausting to do so, too, perhaps because there are no hyperlinks. On a screen, hyperlinks, underlined and colored blue, stand out from normal text and take an extra byte of available attention. They exact a metabolic cost within the two dozen areas of the visual brain, forcing readers to decide at every break whether to follow a link or not. These small decisions cumulatively consume more and more of our limited bandwidth. Following an underlined blue link forces you to expend additional chemical energy and mental effort to integrate what you freshly encounter in the hyperlink jump with the details of what you were reading in the first place, assuming you can hold it in your mind (recall the limits of working memory). You then must navigate back to where you were amid multiple open windows, yet another energy drain. The number of items anyone can hold in working memory, such as phone numbers or a grocery list (the equivalent of a sticky note in the brain), is seven ± two.[27] This limit explains why it is taxing to juggle loose ends from the links you encounter while deciding which ones, if any, are worth following. As the Oscar fiasco illustrated, we are not capable of handling two tasks at the same time when both require the use of working memory. Switching attention from one task to the other always costs us in terms of slowed reaction time and mistakes made.

Almost every writer colleague I have asked likewise prefers revising from a printout to editing on a screen. The difference in perception is difficult to explain, but the difference is there. As to how the internet affects cognition, Drs. Loh and Kanai conclude that we have collectively moved to "a mode of shallow learning," echoing Nicholas Carr's point of view of "The Shallows" as typified by:

- Skimming, nonlinear reading, and a reduced ability to sustain attention;
- Rapid shifts in attention that lead to distractibility and multitasking that impedes rather than sharpens mental focus;
- A decline in metacognition, or the ability to analyze one's own thinking;
- Decreased memory retention, especially when caught up in hypertext and multimedia that steal bandwidth from the resources needed for deep learning; and
- Depleted ability to think critically and solve problems by connecting the dots because shallow learners have fewer dots—culturally shared facts and experiences—that they can connect.

The high prevalence of screen dependency among digital natives compared to older cohorts reflects changes that digital media have induced in reward networks. A shallow mind results when you have effortless access to online information and little incentive to register it in temporary working memory. If it never makes it to working memory, it can never be stored in long-term memory as a dot you might later connect with some other meaningful dot. When we offload facts to the cloud or always rely on Google, we have nothing in the back of our mind ready to fall into place and suddenly make sense (we call this insight) when a seemingly stray datum enters our consciousness. Two examples of a mind primed to connect what look like disparate dots are Kekule's dream of a snake eating its tail that led him to visualize the unique ringed structure of the benzene molecule (a problem he had fruitlessly pondered for years), and Poincaré stepping into a bus and suddenly seeing the solution to a mathematical theorem that had long eluded him. In both cases a prepared mind grasped the relevance of a seemingly stray fact and implicitly made a meaningful connection to related details acquired long ago and waiting, metaphorically, for the other shoe to drop.

Googlization of the mind begs the question: If future generations don't have a grasp of common facts, then how can they know what facts to look up? The question is not theoretical. Pat Sajak, the *Wheel of Fortune* host, says that as shared culture becomes less common in America, inventing puzzles for the show has gotten harder. "We rarely do books anymore, because fewer and fewer people read them." The irony of the iPhone, he says, is that "there is nothing in the history of mankind that you can't find out in about 15 seconds." Yet people have less and less factual information at their command with which to inform their thoughts. "It's all in the devices," Sajak laments.[28]

Transactional memory refers to the transfer of knowledge from your head to an external device such as a calculator, an appointment calendar, or a notebook. In theory, say advocates of offloading, not cluttering your mind with unimportant details frees up mental space and bandwidth, which should make you sharper. Depending on the context this might be true, but over time, habitually offloading what could be meaningful content comes at the expense of the deep reading and retention of knowledge that normally undergird a lifetime of learning. The skill to see connections, an aspect of wisdom, simply atrophies. Another drawback is that transactional memory works in only one direction, turning us into passive receptacles and relieving us of any initiative to retain relevant information. To expand on the drawbacks of hyperlinks, third parties presume to tell you what is important and what isn't. But what interests one person won't pique the curiosity of another. Compare its one-size-fits-all perspective to underlining or writing your own marginal notes in a book. Something catches your eye. It triggers a train of thought. The thought train connects new material with something you already know and are interested in. Your imagination wanders, and suddenly you are deeply engaged with the material, perhaps in a surprisingly satisfying way.

Whatever you jot down will be personally salient in a way that hypertext links can never be. With the latter you are nowhere in the loop. But when engaged by your own curiosity you will retain details far better than if the same information had poured in passively through a screen. Writing in the margin with a fountain pen engages me with my own thoughts, my own interpretations, and keeps me focused on them instead of being pulled away by someone else's perspective.

15

THE CONSEQUENCES OF FORCED VIEWING

In Moscow I'd been invited to give a keynote speech. Although it was October, I didn't need any of the winter clothing I had packed. People were walking around in shirtsleeves. But my first shock on arrival was the traffic: worse than London, New York, and Washington, D.C., combined. What an *omáčka*, as my grandmother would say. A gravy. A lumpy gravy, *hrudkovitá omáčka*.

I decided to walk to that evening's opening since it was unexpectedly pleasant outside. It wasn't far from where the hosts housed conference participants at the Marriott Royal Aurora, near the Bolshoi Theatre, to the St. Regis Nikolskaya, where the opening dinner was being held. My stroll down Petrovka Ulitsa would take me past the Kremlin in the ancient part of the city, a fortunate up-close look. I had heard that Moscow is the second brightest city in the world even though it gets only a daily average of five hours of daylight. #MoscowLights on Instagram shows why the city is brilliant, even from space.

Lights, lights, everywhere lights—overhead, underfoot, outlining buildings, pushing up through the sidewalk around kiosks. Wires overhead ran along the street dangling multiple ribbons of sparkling orbs, icicles, and cascades that rippled like waterfalls and swayed in the mild evening breeze. Sidewalk kiosks beckoned with ads, some for luxe baubles from familiar names like Prada, Cartier, and Swarovski, others from houses I could only guess at since I couldn't read the Cyrillic alphabet. It was painful to glance at them for more than a few seconds, and after passing twenty of them I lost count. I hadn't even made it to the TSUM shopping mall, dazzling and recognizable by its Gothic façade and pointed gables.

Across the way the Bolshoi Theatre stood out as a lavish neoclassical affair, its frontage a bank of uplit Roman columns bracing a pediment topped with five rearing stallions. The ground-level portico shone with a dazzling white. A short distance opposite stood a long, covered display of six larger-than-life lumières depicting scenes from Bolshoi opera and ballet productions. Lights, lights, everywhere—eye-popping lights. The foreground plaza with its twin fountains were a maze of walk-through ornaments three times human height: globes, snowflakes, archways, and arabesques made of glittering white, yellow, and blue lights. Next to the Bolshoi and in front of the butter-colored Maly Theatre, likewise awash in halogen, sat an enormous open Spanish fan sculpted in blue and yellow LEDs. People stopped and gawked.

It went on like this toward Red Square. Merchandise and info stanchions. Bus stops and the stations they serviced illuminated by three-foot discs visible a good block away. Promenades of trees wrapped in white lights, their delicate branches wafting in the night air. Store windows and arches one after another outlined in lights. Cobra-head streetlights and tent poles holding halos of a dozen blinding metal halide lamps that shone down on pedestrians.

Just before the Karl Marx Monument I turned left onto Teatralny Proyezd and continued on to Lubyanka Square and my destination, the St. Regis Nikolskaya. Upscale stores lined Nikolskaya Ulitsa, as expected. But then I saw what its attraction was as a tourist must-see: thousands upon thousands of lights. Ten-foot-tall spirals of light curtains shaped like drapery jabots wound themselves around street poles. Arched in the thoroughfare leading to the hotel entrance hung more roundels, orbs, icicles, starbursts, and clusters. I passed through this luminous artifice as if strolling through the Milky Way. Dazzling beautiful, but glaring and hard to look at directly. Like stepping into a dark movie theater, I had to wait for my eyes to adjust once I got inside the hotel lobby.

*   *   *

My "Moscow Nights" excursion was part of an emerging phenomenon I call "forced viewing," in-your-face displays and digital signage that seize your attention. Its rapidly increasing presence in cities worldwide is cause for concern. From New York, Shanghai, and Paris to Madrid, Hong Kong, and London, the world's busiest metropolises are awash in lights, even powerful lasers, and plastered with ultra-bright LED screens, or what the outdoor advertising industry calls "street furniture." Whether situated at bus stops or subway entrances, in airport and shopping concourses, or scattered along busy sidewalks, they broadcast commercial advertising twenty-four hours a day. Passersby cannot avoid looking at them.

In 1800, a mere 2 percent of the world population lived in cites. By 1900 the proportion had grown to 15 percent. Today more than 50 percent are city dwellers, even though cities cover only 2 to 3 percent of the planet's surface. For 200,000 years, humans lived on grasslands and in forests. In short order, but relatively recently, they had to contend with 90-degree corners, glass windows, ever-taller buildings, and street lighting that extinguished the natural nighttime darkness. Artificial light spreads for miles beyond its intended target, and the displacement of darkness has had biological consequences. One-third of vertebrates and two-thirds of invertebrates are nocturnal. Bats have abandoned steeples "because the churches all glow like carnivals in the night," writes Swedish bat scientist Johan Eklöf in *The Darkness Manifesto*. Hong Kong citizens live under a night sky 1,200 times brighter than an unlit one. In some large cities inhabitants' eyes have never adapted to true night vision, which takes thirty minutes of absolute darkness to make the shift.[1] Light pollution keeps more than a third of the world's population from seeing the Milky Way (figure 15.1). The National Park Service calls artificial light "an intrusion." Its night sky team tries to raise awareness about the importance of darkness. Its slogan: "Half the park happens after dark."[2]

To see one plausible outcome of today's collaboration between municipal governments and media corporations, one need only look at the video ads depicted in *Blade Runner 2049*. Aside from the relentless forced viewing, ads appear for simulacra such as Joi, the digital wife of Ryan Gosling's character. How different are these fictions from what is done by companies such as JCDecaux, which calls itself "the Number One outdoor advertising company in the world"?[3]

From orbit at night, Las Vegas is the brightest city on Earth. The Strip is awash in brilliant, flashing LEDs, as are the Fremont Street Experience and Glitter Gulch, all of them affecting pedestrians' circadian clocks. But visitors to Vegas aren't primarily interested in sleep, so a better example might be New York City on an average evening. With a population density of more than 27,000 people per square mile, 67 million annual visitors, and more millionaires per capita than any other city in the world, New York City represents an irresistible market for advertisers and anyone else clamoring

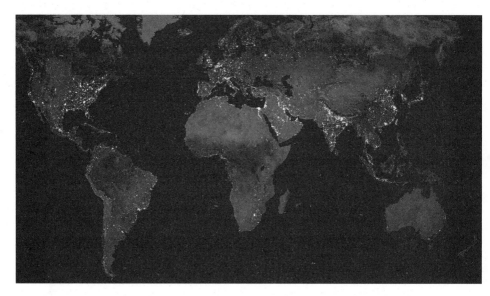

FIGURE 15.1
An image of Earth's city lights created with data from the Defense Meteorological Satellite Program. The world is now comprehensively illuminated, affecting countless species. One-third of vertebrates and two-thirds of invertebrates are nocturnal. Humans need a full thirty minutes of darkness to adapt to night vision. *Source:* NASA, *Earth at Night.*

for viewers' eyeballs. While public advertising does help maintain infrastructure by generating over $1 billion in revenue for city government,[4] it is now almost impossible to walk down a busy thoroughfare without encountering multiple high-resolution video advertisements on a timed loop.[5] Some intersections, such as Eighth Avenue and 57th Street, feature more than twenty dazzling screens that span the entire block. Pedestrians stop and stare, oblivious that they are obstructing foot traffic as they watch the ads cycle.

Advertisements in smaller spaces are now commonplace, too, such as elevators and waiting rooms, airplane seatbacks and escalator handrails, the sides of buses and buildings. Street-facing storefronts play dazzling dynamic advertisements many times larger than sidewalk street furniture. A good number are two to six stories high, and even taller around Times Square. City taxi tops, now digitalized, are brighter than the earlier signage they carried. In Miami, boats first started trawling the waterways towing huge digital screens created by Ballyhoo Media that touted its "water-based advertising." Similar displays plied New York rivers until the governor banned them in August 2019 as "a nuisance that blights our shores."[6]

Earl, a millennial New Yorker, tells me, "Daily life has become a constant interruption of commercial breaks. The proliferation of street screens has diminished the quality of pedestrian life. I can no longer take a walk, especially at night, without getting a screen headache. You can't look away either, because it risks collision with a fellow pedestrian, a car, a truck, or a bicycle."

JCDecaux sees it differently. It "wants to improve your outdoor experience and make cities more attractive, more intelligent, more responsible, more connected and more engaging." In 1964, Jean-Claude Decaux invented the advertising bus shelter in Lyon, France, and the concept of street furniture advertising caught on worldwide. He offered to provide cities bus shelters free of charge, maintained and managed by his company and paid for by advertising. Currently the global leader in outdoor video screen advertising, the company focuses on transportation advertising, billboard advertising, and street furniture. Operating in 4,033 cities with populations greater than 10,000, it boasts: "Every day, JCDecaux reaches more than 410 million people on the planet."[7] While the company's goal is to "engage consumers," pedestrians and even drivers pay for that engagement with their time and by involuntarily relinquishing part of their brain's limited allotment of attention.

The industry calls this paradigm "out-of-home" (OOH) advertising, intended to surround audiences in their day-to-day environment. Given the "growing challenge to catch and maintain the attention of consumers," says JCDecaux's marketing literature, "[it] can reach them . . . when they are less distracted by other media. We can evaluate different media using a common currency of attention." It deems an eye that alights on an ad for more than 100 milliseconds (one-tenth of a second) as having "viewed" it, while an average dwell time of 1 to 2 seconds is enough for OOH to have "impact." Data from the Outdoor Advertising Association of America show that digital adverts consistently engage more pedestrians than users of mobile or desktop devices, making it financially worthwhile to attract consumer eyeballs by this means.[8]

To understand the benefits and downsides of this practice and to get a corporate perspective, I spoke with Gabrielle Brussel, executive vice president of JCDecaux North America, and Mike Naclario, its head of digital advertising.[9] The Eighth Avenue and 57th Street location I mentioned earlier in fact features LCD displays from four competing companies. This is in addition to the Hearst building located there, "which is quite robust itself," Ms. Brussel says. "Everyone wants the best location." LinkNYC kiosks, provided by City Bridge, "are popping up all over the city," with 7,500 units contracted for placement throughout the five boroughs. LinkNYC provides maps, directions, and access to city services, along with free WiFi, phone calls, and rapid charging ports. OUTFRONT holds the Metro Transportation Authority concession, which has displays throughout a system that includes the Long Island Railroad, Metro North, and the New York City subways, with more displays slated to come. All this signage is not just a vehicle for advertising, she explained, but "part of communication systems that cities use quite effectively."

At any of its 50,000 locations, OUTFRONT'S networked signs can post real-time train and bus arrivals. "In terms of being able to deliver relevant information," says Mike Naclario, "you can send an amber warning alert to every screen and billboard in minutes. You can send weather updates, flu shot reminders, or notices of cultural events. The city of Chicago has posted more than 200 million municipal messages in the five years they have operated their digital billboards, and the city has emergency generators to keep the messaging network going in the case of a power failure."

The average person can walk three hundred feet in sixty seconds, and obviously fewer on a crowded city block. The experience of walking past LED displays, however, is different from staring into your phone, tablet, or TV screen. One study JCDecaux commissioned, titled "Using Eye Tracking to Understand the Reality of Attention to Advertising Across Media," found that "pedestrians pass OOH panels quite slowly and so usually can see most of the ads on the loop. Drivers naturally move faster and may be exposed to only one or two of the six ads on rotation."[10]

"We sincerely believe we can only capture a person's attention for a few seconds," Naclario says, "and that's how we design our maximum loop before you see a repetition." Loop duration is further bound by contract and the concession that shows it. Ms. Brussel adds, "I don't think a digital ad versus one printed on paper is any more effective in leading to a consumer purchase. It just makes it easier for advertisers to change content and for cities to communicate better."

That may be, and there is no denying the municipal benefits of advertising revenue and public alerts. But I wonder about the level of distraction that increasingly dense signage causes and whether the intensity and wavelength composition of their emitted light upsets the circadian clock or reinforces existing screen dependencies. Getting answers has been quite hard because many interests are involved, and technical measurements are not available for cross comparison.

JCDecaux highlights three groups of urban consumers it identifies as young, affluent, and informed, all of whom spend a large part of their day outside the home: shoppers ("mobile, affluent, influential and in a positive mindset receptive to visual messaging that helps them make purchasing decisions"), daily commuters ("professional, affluent, heavy tech users"), and airport passengers ("tech savvy, early adopters, less concerned about necessity or price than they are when at home").

All groups respond "well enough to dynamic OOH," the company says, to make sophisticated eye tracking at a distance worthwhile. This real-time technology has been used for years by science and industry, including travel and tourism industries.[11] Companies can tailor an ad in different languages or appeal to a particular individual's sensibilities and interests thanks to enormous databases amassed on millions of consumers. A "dynamic ad campaign" envisioned by JCDecaux resembles the in-your-face infomercials that swarm Tom Cruise in the film *Minority Report* or Ryan Gosling in *Blade Runner 2049*. They raise Orwellian privacy concerns in light of corporations already vacuuming up a wealth of information about consumers, such as airline passengers, whose movements IP sensors track as long as their phone is turned on.[12]

The burgeoning presence of screens and cameras in public spaces has elevated our general level of mental stress. Earl, the New Yorker quoted above, spoke of frequent headaches from nighttime screens that he felt were too bright. Neurologically, intense light affects mood and causes visual discomfort. Individuals with migraines, high blood pressure, glaucoma, cataracts, and nasal allergies are among those most likely to be sensitive to overly bright illumination and the inherent glare of LEDs, along with their high output of blue-rich light. The American Medical Association notes that birds, insects, mammals and even the rhythms of budding plants are affected by LED lighting, which is generally too bright and too blue at the wrong time of day. It recommends an intensity threshold that minimizes blue-rich light and encourages lamp shielding to minimize glare.[13] In September 2021, Pittsburgh became the first U.S. city to rethink urban lighting. It passed a "dark-sky ordinance" not only to improve visibility of the night sky but also to help plants, animals, and people maintain a crisper day-night cycle and reduce glare for better night vision.[14]

The physiology of visual discomfort induced by flicker, brightness, and strain is well understood in terms of brain energy consumption and its increased demand for oxygen uptake. Earlier I described visual discomfort as a protective reflex that kicks in whenever oxygen reserves in visual cortex approach their limit. I also cited striped patterns, especially lines of text, as the worst offenders. LED advertisements frequently feature moving stripes along with the stripes of text that are likewise in motion. As for the quality of the light itself, its spectrum falls predominantly within the short wavelengths known to be biologically problematic for modern city dwellers.[15] Once there was a time when we could find some green space, take a mental break, or go for a quiet stroll to clear our head. Now, mediated advertisements greet us at every turn, and animated shop windows badger us to look at them and attend.

I call this "forced viewing" because the high saturation of these digital images makes it nearly impossible not to look at them. Trying not to look forces you to expend mental energy and brain fuel. Its mere presence is like secondhand smoke in the way it exacts a hidden, serious cost. I explained the evolutionary basis for why all brains respond reflexively to novelty, especially why things caught in the corner of our eye not only snare our attention but do so with an emotional charge. Gabrielle Brussel directed me to current studies that examined billboard distractions, especially during the years printed ones were transitioning to digital. Until now, distractibility rather than luminous intensity has been the greatest safety concern, and so the question I posed must go unanswered at present. The question is important, even a moral one, because it asks whether the increasing radiance of digital screens is innocuous or not. What current research shows is that digital billboards negatively influence glance duration and driving behavior. Short display times of three seconds are the most distracting and correlate with higher vehicle speeds when pedestrians are nearby.[16] When a driver's eyes are dark adapted at night, the glare from bright signs temporarily blinds them. Researchers characterize digital advertising as "clutter that adds demands to the driving task" and impairs "expected patterns of eye movement"—otherwise known as looking where you're going.[17]

A 2018 compendium of studies about electronic billboard distractions in ten countries concluded that LED billboards were more detrimental to driving performance than their static counterparts in terms of "speed control, braking, and lane position maintenance."[18] When a pedestrian is present, braking reaction is 1.5 times slower and overall crash rates 30 percent higher than at control sites. Crashes at some study sites declined by 60 percent after the offending billboards were covered, only to return dramatically when they were once again visible. Studies conducted in other locations show that advertisements with eight or more words cause drivers to drift from their lane. Drivers also gaze at digital billboards more than twice as long as they do at nondigital ones.

Screens do not merely hook the corner of our eye. They divert wholesale attention and degrade concentration. While blocking an infant's central vision with an iPad interferes with its developing vision, seizing an adult's central vision with mediated images impedes their ability to follow a line of thought, hold a thought, or perhaps even think at all.

Nighttime TV and Reading from Tablets

Understanding the consequences of exposure to undesirable wavelengths or intensity of light after sundown requires knowing some technical measurements pertaining to brightness and luminous intensity. One is the term we use, the other the term we mean. Brightness is what the brain perceives, whereas we measure intensity with instruments: one is psychological, the other a measurement of radiant flux reflecting off a surface or emanating from a point of light. Add to that the viewing distance, viewing angle, screen size, and other aspects of the optical geometry between your eye and a screen plus the physiology and psychology of your response, and you have a complicated issue thanks to the many variables. Nonetheless, I promise to conclude with clear recommendations for nighttime viewing.

To start with, the surface brightness an object casts on the retina is independent of viewing distance. While the size of a retinal image appears smaller the farther away you move, its surface brightness stays the same. This is measured in candelas per square meter (cd/m^2), also called nits—the most relevant number that specifies the perceived brightness of televisions, handheld screens, and display signs of any size.

Imagine looking at a candle one meter (3.2 feet) away. That's a candela, a measure of luminous intensity coming from a specific direction.

The candela is not a practical unit because it applies only to a theoretically ideal point of light. Optical scientists instead use lux (short for "luminous flux") for the amount of light spread out over a surface: 1 lux = 1 lumen per square meter. Confusingly, total brightness is the surface brightness times the image area, which does decrease with the square of the distance as well as the viewing angle. Increasingly, light bulbs are no longer rated in terms of electric watts they use. To recap:

- Lumens measure how much light a point source emits.[19]
- Lux indicates how bright a work surface such as a desk will be.
- Candelas measure the visible intensity of a light source. 1 cd/m^2 = 1 nit, the unit typically used to specify the luminous intensity of a display.
- Brightness is the subjective sensation by which we distinguish differences in luminance. It describes a perception, not the physical intensity of the illumination itself.

To put these terms in context, I spoke with Dr. Raymond Soneira, head of DisplayMate Technologies, a company that specializes in optical display calibration and mathematical screen optimization.[20]

Televisions are without question the brightest source of nighttime light exposure and require different screen settings than laptops or handheld devices. Standard cathode ray TVs may output approximately 300 nits, while HDR (high dynamic range) LED TVs go up to 2,000. HDR produces dazzling images by amplifying the peak brightness of the highlights in a picture. Whereas the 4k standard yields enhanced resolution, HDR really ramps up the light output. Screen size also influences the nits-to-lumens relationship. A sixty-five-inch TV emitting 500 nits will have about four times the lumen output of a smaller, thirty-two-inch TV that emits the same 500 nits.[21]

That said, people typically have their TVs set way too bright for nighttime viewing. Turn it down—a lot. While today's TVs can crank out a lot of light, Dr. Soneira says that you only need 100 nits or less at night. You don't need a light meter; just lower the brightness as much as you are comfortable with. This will drop the total light energy emitted, including the blue spectrum, by a factor of four or more. In the picture settings menu, chose an average picture level (APL) closer to 20 percent rather than the 100 percent that produces a bright white screen. Doing so gives an automatic five-to-one reduction in intensity. By making these two adjustments, you have reduced the light output by a factor of twenty or more. Finally comes color adjustments, including the white color temperature. Most people prefer warm tones that range from greenish yellow to orange-red, which will further cut the blue light level by another factor of two. The ultimate reduction factor of forty or more from these three steps may be all you need to fall asleep naturally after watching TV at night.

Don't forget that room lighting affects circadian rhythm, too, and that fluorescent and LED lamps both emit more blue than incandescent lighting does. Ambient lighting increases screen reflections also. In ideal viewing conditions that are dark, such as cinemas and home theaters, a relatively low picture brightness is all you need.

* * *

An active mind takes a while to unwind. Or, better said, an active mind needs to unwind. Looking at smartphones, tablets, e-readers, laptops, and desktop monitors at night is much different from watching TV because texting, doing work, and social media engagement require active participation. We typically view TV from eight to ten feet away. The constellation of smaller devices is usually set to be very bright and is viewed at much closer distances: eight inches or less for smartphones, ten to twelve inches for tablets, and eighteen inches for laptops and monitors. These two factors alone are guaranteed to impair your sleep.

The initial study that correlated nighttime use of digital screens with negative effects on sleep, circadian timing, and next-morning alertness was conducted jointly by Harvard Medical School and the German Aerospace Center in 2015. Their representative survey found that "90 percent of American adults used some type of electronics . . . within 1 h before bedtime." Late-night viewing "phase-shifted the circadian clock," "delayed the onset of REM sleep, and reduced alertness the following morning." Even the use of e-readers such as the Kindle or Nook had "unintended biological consequences," they found. "Compared with reading a printed book in reflected light, reading a light-emitting e-book in the hours before bedtime decreased" the most restorative delta and theta EEG waves of sleep. Participants who used an e-reader were sleepier in the morning compared to their counterparts who read a paper book, even though the light levels were "identical for both reading conditions."[22]

I said that Apple's Night Shift reduced light coming from the 460–490 spectrum. The bottom half of figure 10.1 showed a reduction of 57 percent for the middle setting and 42 percent for the maximum, both of which produce a distinctly orange cast. Yet we still don't have crucial and convincing evidence that these manipulations can prevent circadian disruptions. On Apple devices these settings are buried within multiple-level menus. The single most effective change you can make for nighttime reading is switching from black text on a white background to white text on black. It will reduce the light hitting your face by a factor of ten or more. You can further raise that factor to forty by turning down the brightness.

Lastly, I recommend permanently engaging a color filter. On an iPhone, "Invert Colors" and "Color Filters" are found under Menu Settings\General \Accessibility\Display Accommodations (figure 15.2). Under color filters select the blue-yellow tritanopia filter to diminish melanopic emissions. All these adjustments help, especially with the newer, larger phones. The light emitted from the more recent phone's OLED screen is well into the blue-green melanopic spectrum, which tells your body, "Wake up, it's daytime." This latest phone produces a peak brightness of 8,850 lumens (705 nits) when Automatic Brightness is turned on.[23] According to Dr. Soneira, the iPhoneX and beyond is "well beyond the range of early iPhones to delay sleep. It is not a 'little' screen anymore, and it's very good at making light that your non-visual photoreceptors will see." Actually, all iPhones do this, which is reason enough to turn down everything to the edge of your comfort level.[24]

Oftentimes in science different pieces of evidence contradict one another, as illustrated in 2019 by the Lighting Research Center (LRC) at Rensselaer Polytechnic Institute. Researchers there say we should be more concerned about our amount of nighttime light exposure than about the proportion of blue light in it. Their experiments assumed the consensus view that light containing the normal amount of blue radiation would suppress melatonin more than light that had a lower energy output of between 475 and 495 nm. Technically, this waveband is known as the cyan gap. However, it turns out that blue-scant light suppresses melatonin just as much as light containing a normal amount of blue energy, even when subjects are exposed to it for less than an hour. Others will have to replicate this finding to see whether it holds up.

"Both light amount and its spectrum are important," says Mariana Figueiro, director of the LRC. "Blue light [alone] does suppress melatonin . . . but if enough energy is emitted at other parts of the spectrum, we will

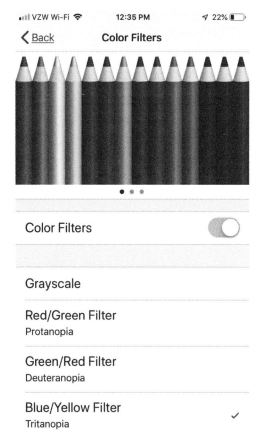

FIGURE 15.2
Recommended adjustments of the iPhone color filters. This applies to all adjustable screens. [See also color plate 8.]

see a response, because the circadian system uses all types of photoreceptors to respond to light for melatonin suppression."[25] In an earlier study of iPad users, the LRC found that brightness settings, not color settings, were the main factor that suppressed melatonin.[26] Nonetheless, lighting manufacturers continue to market bulbs for nighttime use that strip out the cyan gap, including the portable JOURNI circadian lamp, which is based on a patented spectrum technology developed "to support the circadian rhythms of astronauts on the International Space Station."[27]

One application that favors blue light is the nighttime dashboard display. Instrument panels with a strong blue component could theoretically improve automobile safety by keeping drivers alert and reducing the likelihood that they will fall asleep at the wheel.

TOO LITTLE SLEEP AND NOT ENOUGH LIGHT

People regularly commit to treating their high blood pressure yet ignore sleep pressure despite similarly fraught risks to their health. Perhaps we should make sleep pressure a new vital sign in addition to temperature, pulse, and respiration. Adults who sleep just six hours or less compared to eight or nine have a higher prevalence of diabetes, a larger waist size, a fatter body mass index (BMI), lower levels of good cholesterol (HDL), and an elevated risk for stroke and Alzheimer's dementia.[28] The National Sleep Foundation recently revised its age-specific recommended sleep durations (table 15.1).[29]

Morning light exposure influences body fat and the appetite-regulating hormones leptin and ghrelin that the stomach secretes. Forty-five minutes of morning light at an intensity of 1,300 lux between 6 and 9 a.m. for three weeks reduced body fat and appetite in a study group of obese women (0.77 pounds over three weeks). To put this in perspective, indoor daylight is typically dim, at 150–500 lux, with shorter wavelengths predominating in the morning. Even moderately intense light exposure beyond 500 lux early in

TABLE 15.1

Age	Recommended	May be appropriate min. and max.	Not recommended
Newborns 0–3 months	14–17 hours	11–13 < > 18–19 hours	< 11 > 19 hours
Infants 4–11 months	12–15 hours	10–11 < > 16–18 hours	< 10 > 18 hours
Toddlers 1–2 years	11–14 hours	9–10 < > 15–16 hours	< 9 > 16 hours
Preschoolers 3–5 years	10–13 hours	8–9 < > 14 hours	< 8 > 14
School-aged 6–13 years	9–11 hours	7–8 < > 12 hours	< 7 > 11 hours
Teenagers 14–17 years	8–10 hours	7 < > 12 hours	< 7 > 11 hours
Young adults 18–25 years	7–9 hours	6 < > 10–11 hours	< 6 > 11 hours
Adults 26–64 years	7–9 hours	6 < > 10 hours	< 6 > 10 hours
Older adults ≥ 65 years	7–8 hours	5–6 < > 9 hours	< 5 > 9 hours

the day alters one's basal metabolism. The authors conclude that "exposure to moderate levels of light at biologically appropriate times can influence weight, independent of sleep timing and duration."[30]

The reduction in daily light exposure compared to that experienced by earlier generations also coincides with the rising incidence of nearsightedness (myopia). The COVID-19 pandemic confined school-aged kids indoors and focused them on screens, which sped up the trend of worsening adolescent eyesight.[31] In 2023 the trend continues unabated. By contrast, time spent outdoors reduces the chance a child will develop myopia, which typically begins during elementary school years when the normally growing eye grows too much along its front-to-back axis. It is not possible to shrink the eye, so prescription glasses or contact lenses are necessary, although laser surgery is sometimes possible. A child with two nearsighted parents has a 60 percent chance of also being myopic, whereas spending just two hours a day outdoors effectively neutralizes that genetic risk. Bright sunlight stimulates dopamine release from a class of specialized retinal cells that are not involved with vision. These then trigger a cascade of chemical signals that retard the elongation of the eyeball.[32]

Princeton University researchers examined 1,567 nine-year-olds, whose recommended nightly sleep is nine to eleven hours. Those who slept fewer hours than half the group displayed more rapid cellular ageing in the end caps of their chromosomes. These structures, called telomeres, are yet another kind of biological clock that starts ticking down from the day we are born. They represent biological age as opposed to chronological age and prevent degradation of the base pairs in our DNA and the unwanted recombination and fusion of chromosomes. All of our telomeres shorten with time and affect the pace of ageing, age-related diseases, and our individual life span. The nine-year-olds' telomeres were 1.5 percent shorter for each hour they slept less than their comparable peers. In an understatement, the authors conclude, "Starting off life with a shorter supply [of telomeres] is unlikely to be a good thing."[33]

Many people turn to sleeping pills for fear that they will otherwise toss and turn. Alcohol, another common resort, may make you initially drowsy, but it distorts the normal progression through stages I, II, III, IV, and REM sleep, thus worsening the very insomnia you hope to cure. Your brain's sleep patterns can deteriorate after using sleeping pills for just one week because, like alcohol, they disrupt the cyclical EEG patterns during stages I, IV, and

REM sleep. Sleep lab studies of Lunesta and Ambien, for example, show that they provide volunteers all of eleven minutes of extra sleep. Because a hazy memory is normal in just about everyone on awakening, none of us ever has an accurate measure of how long we've slept unless total sleep time is objectively measured via EEG in a sleep lab or by wearables.

While my point has been the ways screen exposure affects restful sleep, it is not off topic to sum up with a prescription for assuring sound and restorative slumber.

- Rise and go to bed at roughly the same time every day, including weekends.
- Make your bedroom dark or wear an eye mask. Quiet eyes make for a quiet mind. People are surprised at what a difference this makes. An eye mask will also help you conk out quickly.
- The two most restful sleep postures are three quarters prone or on one's side with the legs straight and a pillow between the knees. Back sleeping is restful and can be good for those with back pain, but back sleepers tend to ruminate about past and future events once in bed, and so stay awake longer and make themselves anxious.
- Keep your bedroom cool—ideally 65 to 68 degrees. Body temperature normally drops 1 to 2 degrees Fahrenheit during the night, so don't bundle up. If you wake up during the night to throw off the covers, then your bedroom is starting out too hot. Make adjustments. You'll be more pleasant to be around the next day. Consider a cooling gel mattress pad or pillow, too, or the ingenious Eight Sleep cooling pod. A corollary is not to bundle up watching TV in the hours before bedtime. Allow your body to cool down instead.
- Put low-wattage LED nightlights, preferably red ones (never blue), like those made by LOHAS, in the bathroom.[34]
- Dim household lights an hour or two before bedtime. All electric lighting gives off short wavelengths within its overall mixture of illumination. Discover the sensuous pleasure of candles. Try reading by one the way Benjamin Franklin did and see how you like it. It's a new kind of cozy.
- On the flip side, why not reacquaint yourself with the dark? Our Stone Age ancestors had only the Moon and the stars. Take a walk outside before bedtime and look up at the sky, assuming you can get away from

streetlamps and the light pollution common to big cities. Entertain and educate yourself with the Pocket Universe app or existing books and pamphlets to identify stars and planets.

- If you must use a digital device after sunset, make the three adjustments I've outlined and do the same for your TV. Install the free f.lux software. Avoid mentally stimulating work such as paying bills or debating emotionally laden subjects before bedtime. Postpone arguments until the morning. Few bothersome things are worth bothering about, and everything will feel less important in the morning.

- Never look at your phone when in bed to check the time. The brief blast of brightness—a mere second—will undo all the above. Better to get a digital clock with red numerals, but even looking at that will get your mind going about how much time you have left in bed. If you use your phone as an alarm, put it on the other side of the room. Turn off the whoosh of push notifications, too, and activate the Do Not Disturb function.

- Open the curtains and blinds on awakening. Prioritize exposing your brain to natural light over privacy. Nobody is looking at you anyway.

- A brief morning walk without sunglasses or even regular glasses is an excellent way to start the day. Look up at the sun. The preponderance of blue wavelengths early in the morning will energize you and keep your circadian clock well synchronized. The Re-Timer light therapy glasses at re-timer.com can reset your circadian rhythm to your preferred wakeup times. They are also useful for those who hate the twice yearly clock changes.

- Nature is restorative, so spend time outdoors in your own ecotherapy spa. People happily discovered this during the COVID-19 lockdowns. When you take a walk, look at the trees and flowers. Look up at the sky instead of down at your phone. Stash your earbuds and listen to the birds and your surroundings. This will allow the mind to wander naturally. Wandering invites unexpected connections that are the springboard to creativity. If a terrific thought strikes you, you can record a voice memo.

- Socialize a little, too, even at six feet. Face-to-face engagement is what your Stone Age brain evolved to do and what social media compete with. Don't let them win.

16

DOES HEAVY VIEWING INDUCE AUTISM-LIKE SYMPTOMS?

The subject of virtual autism is both complicated and fraught because it is easy to conflate the behavior of autistic individuals, who have a developmental brain disorder, with similar-looking outward behaviors resulting from something else entirely. A parallel from neurology's trove of clinical vignettes is Parkinson's disease versus parkinsonism. The "shaking palsy" described by the English surgeon James Parkinson in 1817 results from the degeneration of cells in the substantia nigra (Latin for "black substance"). Situated in the brainstem, it is the sole source of brain dopamine. This tightly defined condition is Parkinson's disease. Parkinsonism, on the other hand, manifests in other neurological states as muscular rigidity, slowed movement, and tremor. The boxer Muhammad Ali did not have Parkinson's disease despite numerous popular reports to the contrary; he had parkinsonism stemming from pugilistic dementia—meaning that over the years the numerous blows to his head had caused his frozen movement, tremors, and near inability to speak.

It is important to define terms clearly. Developmental autism involves a range, or spectrum, of behaviors; hence the preferred term autism spectrum disorder (ASD). Individuals on the spectrum primarily have difficulty communicating and understanding how to interact with others. Secondary features include a tendency to regard people as objects, which further impedes the development of social relationships and adds to their isolation. Autistic individuals may have extremely narrow interests, an affinity for technology, and a penchant for repetitive self-stimulation such as spinning, rocking, and counting. These symptoms make it difficult to succeed socially, academically, and professionally. Some children with ASD never escape their self-referential world, which is why the specter of autism alarms many parents. Even the National Institutes of Health ominously calls autism "incurable" and cautions that affected individuals "may never live self-sufficiently."

Peer-reviewed research is only now beginning to explore the connection between screen exposure and virtual autism.[1] Screens are not the only variable that can explain the sudden rise in autism-like behavior, especially in boys. But there is a correlation, and screen media compete with youngsters' social and emotional development. For example, autistic individuals studiously avoid eye contact, a behavior also shared widely by heavy screen users. One reason autistic individuals bungle social interactions is that they don't process faces in the usual way. They fail to pick up subtleties as well as the nuances of overall body language (facial expression being only one aspect of body language). This leaves them with gaps in empathy and social insight.[2] Simon Baron-Cohen at Cambridge University, one of the world's leading autism experts, sees autistic individuals as suffering from a faulty "theory of mind." A theory of mind is what lets us attribute thoughts, desires, and intentions to others, predict their actions, explain them after the fact, and fathom their aims. Those on the autism spectrum don't "get" other people or even realize that others might have thoughts, feelings, and points of view that differ from their own.[3] As the Autism Spectrum Foundation points out, "Youth with ASD are often less mature relative to their chronologic age or physical development and their autism symptoms do not 'disappear' once they go online."[4]

As for screen dependency in adolescents who do not have ASD, child psychiatrist Victoria Dunckley notes, "the proliferation of the iPad and smartphones has produced more problems and setbacks in my practice than any other single factor." Regression of language or social behavior, or both, is not uncommon when families hand children a device that seems—at first—to quiet them down. Parents feel pressured by schools and friends to use flashy apps based on word of mouth but whose use is not validated by any evidence whatsoever. When parents do relent and bemoan that an iPad "is the only thing that works," they unwittingly reinforce via Pavlovian conditioning its malign influence, in the same way that giving in once to a dog at the dinner table reinforces the very begging behavior you are trying to stop.[5] In *Mind Change*, polymath neuroscientist Baroness Susan Greenfield drew parallels between developmental autism and autism-like behavior in young people who are "conditioned by an enthusiastic screen culture to avoid eye contact."[6]

According to Common Sense Media, over two-thirds of U.S. teenagers now prefer to communicate via text rather than face-to-face.[7] The result is young people who freak out at the sound of doorbells, ringing telephones, or knocking at the door. They expect friends to text "I'm here" upon arrival, and don't know how to react to an immediate physical presence.[8] Their preference for texting has been a boon to home security companies like Ring and Nest that manufacture WiFi-enabled doorbells with embedded cameras. These allow a mediated exchange with someone at the front door without ever having to open it and look them in the eye.

Texting and phone checking many times a day is a continual distraction, of course. A longitudinal study in the *Journal of the American Medical Association* tracked 2,500 teenagers over two years and found an association between frequent screen checking and symptoms of attention deficit disorder. In an accompanying editorial, Dr. Jenny Radesky, professor of developmental-behavioral pediatrics at Michigan Medicine, said we need to "dig deeper into how frequent split attention, instant gratification, and emotional arousal from media might be influencing teens' thinking processes."[9]

The first year of life is a peak period of neural plasticity. During the first three months the visual cortex is developing its forest of connections most rapidly, while the postnatal brain volume grows 1 percent every day, tripling by age two. This makes the influence of early experience particularly important. Enduring patterns of synaptic connections establish themselves via recurring sensory simulation and repeated movement—exactly what the infinite scroll offers via positive intermittent reinforcement. Critical time windows are periods when a specific type of input such as vision or hearing exerts its greatest effect on brain development. But a reversal of connections already laid down occurs just as easily when expected stimulation is absent or competing experiences such as prolonged screen exposure interfere with it.

Family physician Leonard Oestreicher, author of *The Pied Pipers of Autism* and the uncle of three autistic boys, cautions, "By the time we are 12 months old we have either become embedded in the complicated physical, social, and linguistic world around us, or we are on our way to autism, becoming more and more embedded in the illusory virtual world."[10] He raises the question of whether heavy screen exposure contributes not just to virtual autism but to developmental ASD itself. Among U.S. children, one in fifty-nine will develop autism. Studies in twins make it clear that genes account for 50–80 percent of the risk. The rest comes from environmental factors we do not yet completely understand.

Because human genes cannot change quickly, genetics alone cannot explain the steady rise of ASD, now thirty times more common than it was between 1950 and 1980.[11] If genes cannot change quickly, then we can ask what environmental factor did take off during that period. The answer: exposure to television. In 1957, 26 million teenagers watched *American Bandstand* at 3 p.m. Monday through Friday. By 1970 the mean age at which children began watching television fell to four years, and dedicated children's programming began in the morning; by 2006 the starting viewing age had plummeted to four months.[12]

Has screen exposure, now starting in infancy, contributed to the rise of ASD, and if so, what are the implications for virtual autism? The correlation is positive. A study in *JAMA Pediatrics* that followed 2,152 children noted that "on average, the group with ASD was the youngest to begin screen viewing (6 months of age) and had the most hours of media watching (4.6 hours/day), while the group with typical development was the oldest to begin screen viewing (12 months of age) and had the least hours of viewing (2 hours/day)."[13]

Oestreicher and ophthalmologist Karen Heffler at Drexel University College of Medicine and the Drexel Autism Institute published the "audiovisual model" suggesting how autism-like behavior might develop in otherwise normal kids. Because brains at every age adapt to whatever environment they find themselves in, the two doctors hypothesized that prolonged stimulation by screen-based sights and sounds competes with the laying down of brain pathways normally destined to support social cognition. This kind of competition is typical in that a focal gain in one capacity often exacts a loss somewhere else. Content coming from a screen may not mean anything to a young child, but the pacing and edits are rapid and abrupt, and continued stimulation fortifies basic sensory pathways at the expense of ones destined to participate in social intelligence. Hence the name "audiovisual model." This provocative idea warrants further study, which is already under way, in light of how much time young children spent indoors in front of screens during the pandemic.[14]

The model's predictions align with demographic data. According to the Kaiser Family Foundation, a decade ago 68 percent of children under age two viewed screen media for more than two hours a day, and 33 percent of households kept the TV on all the time. These numbers have since escalated, and we now have data showing that background television sound correlates with delayed language acquisition.[15] Compared to households in which the TV is usually off, constant background television reduces the number of words a child hears spoken by an adult as well as the child's own babblings: they don't get to hear themselves and the auditory feedback that normally helps shape emerging language. There are no known benefits of early screen exposure, whereas detrimental effects are measurable. In older children, more time spent glued to screens correlates with falling grade point averages, inattentiveness, impulsiveness, social isolation, moodiness, emotional immaturity, and delayed sleep. In one study of students in secondary school, grade point averages plotted against screen time per day showed a drop of 31 percent (zero to thirty minutes being the least time reported and greater than four hours the most).[16] While seemingly impressive, it, like similar studies, is difficult to assess because of wildly differing experimental methods: who is reporting the numbers, the parents or the child, what kind of content is being viewed, and is it comparable across different groups?

If mere behavior is the expression of virtual autism, then restricting screen exposure can hypothetically reverse it. Young children are awake for only ten to twelve hours a day, so lengthening screen exposure crowds out activities that normally drive cognitive, social, and emotional development—precisely the prediction of the audiovisual model. Time spent glued to screens correlates in older children with falling grade point averages, inattentiveness, impulsiveness, social isolation, moodiness, emotional immaturity, and greater time needed to fall asleep. These difficulties afflict young people with ASD, too. The more that either group stares at screens, the more they rob themselves of the chance to socially engage with actual people. Pediatric organizations worldwide have urged parents not to expose infants and toddlers to screen media at all, yet it remains common practice. According to psychologist Susan Linn at Harvard University, "Most parents who allow their youngest children screen time are under the impression that it had educational benefits."[17]

Table 16.1 summarizes recommendations from the American Academy of Child and Adolescent Psychiatry regarding age-appropriate use.

Imitation learning (mirroring) is important for developing a theory of mind, a capacity for empathy, and a secure sense of self, as MIT neuroscientist Rebecca Saxe reminds us.[18] We forget that youngsters faithfully

TABLE 16.1

0–18 MONTHS	• Avoid screens completely. • Hands-on activities with human engagement facilitate normal cognitive, motor, and social-emotional development. • Most time should be spent in hands-on activities without media in the child's environment.
18–24 MONTHS	• Most time should be spent in hands-on activities without media. • Very brief intervals. • Focus on high-quality educational programming. • Parent or caregiver watches with children and explains content.
2–5 YEARS	• Most time should be spent in hands-on activities without media. • <1 hr per day. • Still emphasize educational and age-appropriate programming. • Parent or caregiver still watches with children and explains content.
6–12 YEARS	• Consistent time limits. • Limit types of media. • Monitor sleep, physical activity, and behavioral health effects. • Screen-free zones: bedroom, dinner table. • Screen-free times: meals, bedtime, family interaction.
ADOLESCENTS	• <2 hr per day. • Media-free zones and times. • Ongoing education and communication. • Parental supervision and limit setting. • Parental modeling of healthy use. • Limit media use when doing homework.

Source: Gwynette et al., "Electronic Screen Media Use in Youth with Autism Spectrum Disorder," *Child and Adolescent Psychiatric Clinics of North America* 27, no. 2 (2018): 203–219. Reproduced with permission.

imitate the adults around them, who all too often are fixated on their own screens.[19] Exploring the link between parental cellphone preoccupation and the escalating incidence of autism-like symptoms, child neurologist Michael Davidovich proposed a framework centered on the parent-child eye contact that we know is essential to normal brain development.[20]

Eye contact is the principal means by which we all instigate communication. We build on it to read other people, their emotions, and their intentions. During the first months of life a baby grows more visually curious, fixating on objects and faces. I call these explorations its "visual apprenticeship with the world." A one-year-old normally responds to entreaties for attention. At eighteen months, a toddler's ability to initiate mutual attention predicts that language (sentences of two words or more) will appear at twenty-four months. Social behaviors such as cooperation, manipulation, group participation, social learning, and classroom learning soon follow. Functional MRI and diffusion tensor imaging have mapped the neural networks behind this kind of developing joint attention. A parent whose glance shifts from child to phone interrupts this coupling, with heartbreaking consequences, as the still face experiments illustrated. Perhaps, Dr. Davidovich speculates, chronic parental distraction disproportionately affects children who are vulnerable to developing ASD and virtual autism, whereas others are more resilient. "The parental use of cell phones does not render any benefits to an infant's development," he says, "while it may adversely affect the development of joint attention."[21]

The still face experiments illustrate how parental preoccupation precludes emotional bonding. The results made Davidovich suspect that digital screens may degrade a child's mirror learning. His team first measured how intently parents turned to their own cellphones during a one-hour assessment of their child's neurological and psychological development. If parents were not fully engaged during this relatively stressful session, he reasoned, then disengagement in day-to-day scenarios would likely be even greater and more problematic. He noticed that it was common to see mothers texting, taking pictures, or talking on their phone while nursing or pushing their babies in strollers. "The urge to use phones has become almost uncontrollable and often unconscious."

Prior to their child's assessment parents had agreed to take part in a study that observed their "parent-child interaction." But agreeing to participate did not prevent them from reaching multiple times for their phones during the ten minutes spent quietly with their child while the doctor prepared a summary of the evaluation. Some parents grabbed their phone four times over the course of the ten-minute waiting period, while two-thirds of them spent more than a minute each time sending texts, reading them, or talking. But the real kicker is this: parents whose children already had delayed language or motor skills resorted to their phones significantly more often than those whose children had congenital anomalies and thus served as a natural control group.

What does diminished eye contact portend? It is a core symptom of autism. The Marcus Autism Center at Emory University showed in a prospective, longitudinal study that infants later diagnosed with ASD had normal eye fixation that subsequently declined between ages two to six months (autism is usually not diagnosed before age eighteen to twenty-four months, often much later). This was the earliest objective record and laid to rest an earlier notion that a basic mechanism of social adaption, eye contact, is absent from the start in infants later diagnosed with ASD. Happily, the decline of visual fixation rather than its outright absence at birth suggests that there might be an interval during which intervention can preserve it while simultaneously nurturing social engagement.[22] The remedy could be as simple as not exposing young children to screens of any kind. Yet until we have more to go on, parents would be prudent to err on the side of caution, keeping smartphone and tablet use to a minimum in the presence of their young. Better yet, shut them off and put them away. Young children don't have the mental maturity to derive meaning from what they see and hear on a screen. Screen-based faces never return smiles or make eye contact either, rendering the child for all intents invisible and bereft of emotional feedback from these devices.

To be clear, Drs. Davidovich, Heffler, and Oestreicher show correlation, not causation, between the emergence of autism-like behavior and early exposure to digital media. Correlation is disconcerting nonetheless, for autism has no single cause.[23] Perhaps there is an early closure of neuroplasticity during a critical time window that leaves networks for socializing underdeveloped. The possibility is bolstered by a number of early intervention regimens that successfully reduce autism-like symptoms, such as Denver's Early Start approach that encourages social interaction while switching off TVs, electronic toys, and other screens that compete with face-to-face engagement with real people.[24]

VIRTUAL AUTISM AND UNDEVELOPED IMAGINATION

Follow-up studies of the neglected Romanian orphans provide a reminder of how, over millennia, our Stone Age brain evolved under predictable social conditions. When children are deprived of opportunities to socialize during early life, socialization never develops well. From the original orphanage cohort in Bucharest, Dr. Marius Zamfir at Spiru Haret University analyzed a subset of 110 children who met both the *Diagnostic and Statistical Manual of Mental Disorders, Fourth Edition,* and the ICD-10 criteria for ASD. His longitudinal survey confirmed the findings of Michael Rutter, the Romanian psychiatrist who first reported the high incidence of autistic symptoms in these neglected children, and Charles Nelson, the American neurologist who subsequently illuminated their stunted cognitive and emotional growth.[25]

Dr. Zamfir discovered that those who had been exposed to screen media for merely four hours a day had developed "a syndrome similar to autism spectrum disorder," which he attributed to "sensory-motor and socio-affective deprivation."[26] Developmental autism never improves spontaneously during early childhood, so it was unusual that this subset of children should show "dramatic improvements" after authorities removed screens from them. Improvements included more spontaneous speech, social interaction, and an escalating preference to be outdoors. Dr. Rutter originally took away televisions, then sent the children outside to play and encouraged them to talk among themselves and with staff, and to read to one another. Although their social communication rapidly improved, their autism-like behavior failed to reverse fully.

"After recovery," says Dr. Zamfir in a later analysis, "these children are very much like people with Asperger disorder [high functioning with normal language development]" in that they have persistent trouble understanding *"human emotions and relationships"* and correctly expressing emotion themselves (my emphasis). Compared to the two-way dialogue typically established with parents and grandparents during mutual play, he conjectures that "aggressive" screen stimuli that "move so quickly" overwhelm a child's capacity to grasp their meaning. They thus encourage a passive mindset antithetical to imagination and mental flexibility. He closes his argument by noting that mice subjected to virtual rather than naturalistic environments "showed that those exposed were hyperactive, heedless of danger and showed more liabilities to any new challenge." This should give pause when we encounter infants already infatuated with screens.[27]

What iPad babies get thrust before them is an abstract, disembodied experience cut off from firsthand sensory feedback of touch, time, what their own body feels like (proprioception), and what it feels like to imitate other people. It is time not learning nursery rhymes, time stolen from pattycake, peekaboo, and games like "grocery store" and "restaurant" that instill social skills in the young. It is time not hearing linguistically competent adults talk to them, read to them, and engage their imagination.

Without question, iPads seduce. For a parent trying to cope, iPads may appear to hold a child's attention and render them quiet and happy. But screens actually amputate them from embodied cognition, which is not an optional appendage but a fundamental aspect of rational thinking.[28]

Worry not about displaced sensation but displaced socialization. We know that deaf and blind children are at increased risk for developing autism—but only to the extent that their sensory deficit throws up a barrier to socialization. Once they begin social interactions, normal development proceeds apace. Deaf youngsters who learn sign language, for example, are not at higher risk for becoming autistic, as Helen Keller demonstrated. After the encephalitis she contracted at age two left her blind and deaf, she did become isolated and displayed the kind of disruptive behaviors typically seen in autism. Once her teacher, Annie Sullivan, was able to restore social communication through the sense of touch, Keller matured and went on to lead her remarkably accomplished life.

To summarize: virtual autism—the manifestation of autism-like behavior—correlates with heavy screen exposure, whereas the spectrum of developmental autism (ASD) results from as yet unidentified changes in brain structure during the first year of life. But screen exposure may play a role in the latter condition, too. The fact that affected individuals initially appear normal and then decline sharply does suggest it is external factors that nudge brain development onto its aberrant trajectory. The audiovisual model proposes that heavy exposure to screen media fortifies pathways devoted to sensation. These amped-up pathways then compete with the maturation of circuits usually destined to support socialization and emotional intelligence.

Despite explicit recommendations by pediatric organizations, many parents remain unaware of the association between screen exposure and virtual autism, or that encouraging their child to play with others and engage with them socially can have a protective effect. The best advice might be "unplug, don't drug," meaning eschew screens from birth to age two, and severely limit exposure for children twelve and under.[29] Take them outside. Talk to them. Engage together in arts and crafts. Play with them the games you yourself played as a child. Doing so will teach them more than any app ever could.

17

Social Learning: Kindergarten, Handwriting, and Dexterity

Friedrich Fröbel, an influential German educator, coined the word *Kindergarten* in 1852. It translates as "garden for children" and implies a realm in which they can grow up close to the natural world. Compared to typical American preschools, structured around adult schedules, routines, and lesson plans, German kindergartens to this day emphasize mutual cooperation, learning from nature, and self-discovery of one's inner feelings.[1]

Kindertagesstätte is a related word referring to preschool in general. From its shortened form, *kita,* we get *Kitafahrten,* or "daycare outings" *(fahren* means "to travel" or "wander"). These are the antithesis of hypervigilant helicopter parenting and slavish adherence to a preset curriculum. Instead of being chained to a seat or shuffled about in a group while clinging to a guide rope, three- to six-year-olds head for the woods for a few days accompanied by teachers, classmates, and their own wits. No books or smartphones come along because Germans don't learn to read and write until age six. Phone calls and texts are strictly verboten. "We'll contact you if we need to," is all the teachers tell parents. For their part, parents are often glad to have three or four days to themselves. Many adults look back at their own *Kitafahrt* excursions and fondly remember the life lessons learned about independence, resilience, and self-confidence.

A related German phenomenon is *Naturliebe,* meaning love of nature, wilderness, or countryside, and a national attitude underlies the emphasis on childhood self-reliance. By law, German kindergartens must strive to develop their pupils into independent, self-sufficient individuals. Families do their part to instill social skills, too: it is common to send young children to the bakery or on similar errands. The baker becomes part of the exchange, too. In America, "free-range" parents who encourage such behavior risk arrest for child endangerment and loss of custody to Child Protective Services (CPS). Montgomery Country actually charged Danielle and Alex Meitiv of Silver Spring, Maryland, with child neglect for allowing their children, ages ten and six, to walk home alone from a nearby park—in the daytime. The Meitivs practice a free-range style of parenting that encourages independence and exploration. Yet Montgomery County authorities threatened to forcibly remove the children unless the father signed "a safety plan pledging he would not leave his children unsupervised until the following Monday, when CPS would follow up." Three months later, when the children were out walking, they were again picked up and held in custody for more than five hours.[2] One size does not fit all situations, of course. In Michigan's Upper Peninsula rural families routinely let their kids wander in woods populated with bears, wolves, and rapidly flowing rivers; in Detroit such insouciance would be unlikely.

By contrast, *Kitafahrt* teachers don't panic when their charges wander out of sight. Experience tells them they will be fine. By nudging the very young to explore on their own, German teachers prepare them to cope with challenging circumstances. Youngsters on *Kitafahrten* outings sleep under the stars, roast sausages around a campfire, and invent their own games with sticks and stones and mud. As they explore the forest they choose their own playmates, settle their own disputes, learn how to ask for help, and begin to develop the lifelong gift of resilience.

France has a similar *classe verte*, or "green class," between the first and fifth grades that likewise encourages students to learn from hands-on experience such as pottery, weaving, and exploring the woods rather than relying solely on books. Finland's version of *classe verte* is kindergarten in the wild, with an emphasis on exploratory learning. "The work of a child is to play," says one Finnish educator. "When they are moving their brains work better." The American Academy of Pediatrics agrees in its most recent report: "The importance of playful learning for children cannot be overemphasized." Play-based building with natural materials, problem solving, and physical activity all help to develop fine motor skills.[3]

The Stone Age brain is at home in an environment where it can engage its disposition to socialize, especially during the early years of robust plasticity. The Meitiv family would have been glad to know, contrary to the authorities' accusations, that research supports letting kids roam. Moreover, distance traveled rather than the mere fact of being outside predicts that they will become better navigators as adults than peers who are more constrained.[4] When you are free to satisfy your curiosity about what lies over the hill, it pays off later in life.

Screen dependency has simultaneously wrought an epidemic of loneliness for millions whom social media once promised to unite. Realization that "social media" is an oxymoron became more widespread during COVID-19 lockdowns as Zoom chats and happy hours left users still hungry for genuine contact, even if it had to be six feet away.[5] The political economist William Davies distinguishes digital networks from social ones. In *Nervous States: How Feeling Took Over the World*, he asks, "How can we at once be so digitally connected and so lonely?"[6] The disconnect is particularly acute among adolescents and teens, a root cause being their inability to ask for help, offer help, or even recognize when they need it. "Arriving snowflakes are lonely," says *New York Times* commentator Frank Bruni of college freshmen, because "they don't know how to engage other people." Thanks to "their fixation on digital screens, communicating almost exclusively by text and avoiding face-to-face interactions," they have had no practice in making small talk or listening attentively to others so they can sustain an actual conversation. They realize "to their horror that they are quite unprepared to navigate the real world. The social world."[7]

And that's the crux, isn't it—the paltry development of social brain pathways in kids who have been enmeshed in screens their entire lives. What if America adopted something closer to the German approach, whereby primary school teachers would coach students on how to build, nurture, and maintain friendships? Older *Kitafahrt* students act as guides for less experienced ones because teachers know that assisting others builds character, confidence, and resilience. Everyone learns to give and take. An atmosphere in which students practice making and sustaining friendships helps in forging social bonds that can mitigate isolation, loneliness, and FOMO. In October 2018 Britain appointed its first minister of loneliness. Mandatory lessons in "relationships education" now begin in primary school and continue through the secondary years.[8]

Because social media inordinately amplify FOMO, says Brigham Young psychologist Julianne Holt-Lunstad, who studies loneliness, the United States "desperately" needs to follow Britain's lead. Naysayers complain that time shouldn't be taken away from existing school programs that address bullying, drug abuse, and suicide. But "addressing social isolation, loneliness, and social disconnection helps us to address those other issues, too," she points out.[9]

Rachel Wurzman at the University of Pennsylvania links loneliness to an increased risk of opioid addiction, now responsible for more than 115 deaths every day in the United States.[10] In her eighteen-minute TEDx talk she sees tackling loneliness as a potent counter to addiction. The driving forces behind social media obsession and addiction converge on an area in the brain's basal ganglia called the striatum that overlaps with networks supporting reward and reinforcement. The striatum participates in compulsive behaviors, and the presence or absence of social relationships strongly influences it. Isolating—which frequent screen use encourages—sensitizes the striatum and drives individuals to seek a reward—any reward—to assuage their loneliness. "We are so ravenous for our social neurochemistry to be rebalanced," says Dr. Wurzman, that "we're likely to seek relief from anywhere."[11] Isolation is a well-known contributor to relapse and overdosing, whereas deep, meaningful connections reduce compulsive behaviors and lessen the likelihood of such untoward outcomes.[12]

* * *

Building on Friedrich Fröbel's kindergarten is Dr. Rudolph Steiner's Waldorf Education. Though founded more than a century ago, Waldorf schools may be the best kept secret in education. More than one thousand such schools and two thousand kindergartens currently operate in sixty countries. The United States additionally has Waldorf-inspired charter and traditional public schools. Steiner's philosophy emphasizes nourishing the imagination and holistically integrating a pupil's intellectual, artistic, and practical development. In 1919 Emil Molt, owner of the Waldorf-Astoria cigarette factory in Stuttgart, asked Steiner to set up a new kind of classroom for his worker's children—hence the name Waldorf. In the aftermath of World War I, both men agreed that education had to be a "cultural deed." The first school was revolutionary for its time—coeducational and open to all social, religious, racial, and economic backgrounds. Students would be taught to think clearly, attend to their feelings, and act with purpose, a far cry from the emphases in today's classrooms.[13]

Showering material goods on children (especially distracting goods such as phones and tablets) at the expense of psychological and humanistic growth stunts or even arrests the development of clear thinking and communication. Mastering the social skills required to become an active participant in society is at least as important as learning long division. The imagination thrives when children actively listen to stories and conjure their own mental images instead of passively consuming so-called educational videos. At Waldorf schools, hands-on endeavors replace regimented classroom drills and standardized tests. Students engage in pottery, painting, knitting, or feltwork. Drawing, music, and creative movement are likewise regular activities.

Waldorf students are not exposed to digital technology in class or (hopefully) at home before the age of twelve. Faculty Chair Jennifer Page at the Washington, D.C., Waldorf School tells me, "It can be hard to persuade parents of the benefits of withholding smartphones, tablets, and television, especially for the very young."[14] Yet the school manages to convince them without a lot of fuss; some parents even pitch in to help other parents overcome their reservations.[15] Dr. Richard House from the *Nursery World* battle, himself a certified Waldorf-Steiner teacher, assures me that "late techno-starters are often more proficient and mature users than those introduced to technologies at a young age. Some Steiner graduates go on to highly successful careers in the high-tech field."[16] In Silicon Valley the Waldorf School of the Peninsula (WSP) concurs, boasting that many of its graduates "have gone on to successful careers in the computer industry." This school, which educates the offspring of tech titans, affirms Dr. House's point that early adoption of screen technology is "unnecessary, inappropriate, and harmful." Highlighting its hands-on "experiential curriculum," WSP's promotional literature says that "Waldorf educators believe it is far more important for students to interact with one another and their teachers, and work with real materials than to interface with electronic media or technology." The school agrees that socialization is crucial to brain development when it says, "The lure of electronic entertainment in our media-infused society influences the emotional and physical development of children and adolescents on many levels and can detract from their capacity to create a meaningful connection with others and the world around them."[17]

If the irony here is rich, the hypocrisy and felt guilt of Silicon Valley parents are even richer. Journalist Nellie Bowles, who reports on technology and internet culture, writes that Silicon Valley executives are "increasingly frantic about keeping their children away from the very screen technology they helped create."[18] Chris Anderson, CEO of 3D Robotics, founder of GeekDad.com and father of five, feels a twinge of cosmic retribution at having to enforce "a dozen rules" that shield his children from tech, which he likens—astutely—to crack cocaine flooding the growing brain's reward center. He says, "I didn't know what we were doing to their brains until I started to observe the symptoms and the consequences." Still, he excuses himself as blameless for the Pandora's box he helped open as "a force beyond our power to control."[19] Numerous Silicon Valley execs who send their children to WSP require their nannies to sign contracts promising not only to hide screens of any sort from their charges but also to refrain from using their own phones during work hours. Placement agencies say this is the new norm.[20]

According to observers like Bowles, affluent families, especially those around Silicon Valley, have become more attuned to the detrimental effects that screen technology has on their kids, whereas families of comparatively lower socioeconomic status are not as aware of the issue. Talk of a digital divide used to center on the haves and the have-nots: well-off families could afford multiple computers with fast internet connections, while poorer kids had to scrounge for computer time at the library to do their homework. Now the digital divide today is between highly informed parents who doubt the benevolent effects of screens and send their children to enlightened schools and parents who haven't yet heard the alarms from either media watchdogs or medical authorities. According to Common Sense Media, less-well-off kids now spend considerably more time in front of screens than their more affluent counterparts.

How did this unintended consequence come about? Recall that Tristan Harris was once an industry insider, a Google project manager whose concerns about user distraction pushed him to become Google's first "design ethicist." Yet after two years spent watching management systematically ignore his ideas, he quit Google to found Time Well Spent, which later became the Center for Humane Technology. In his Senate testimony on "Human Downgrading," he says:

> Today's tech platforms are caught in a race to the bottom of the brain stem [the reptilian brain] to extract human attention. It's a race we're all losing. The result: addiction, social isolation, outrage, misinformation, and political polarization—all part of one interconnected system called human downgrading that poses an existential threat to humanity.[21]

In this context an existential threat means the merging of individual identity into the mob's stream of consciousness. The threat lies in the profound way digital devices affect body, mind, and soul. It is a call to action to fight against digital incursions.

Schools across America remain in thrall to technology and so-called "personalized learning." A prime example would be the policy debacle in which the Los Angeles Unified School District opted to spend $1.3 billion putting 65,000 iPads into the hands of every child, teacher, and administrator in the school district. Each iPad came preloaded with a for-profit curriculum written by Pearson, a corporate publishing giant. All accounts conceded that it was a total waste and "spectacularly foolish." Pupils never got the hyped-up education they were promised, and the school district "denounced the [curriculum] material as utterly unusable." Even the FBI became involved because of irregular contracting and crippling technological issues that contributed to the disaster.[22] Technology often looks like an attractive solution, but just as often it fails to solve the problem it is supposed to address. Worse, parents didn't push back against Pearson or Apple to hold them accountable.

That has started to change. One community effort called Stand Together and Rethink Technology, or START, motivated parents in Kansas City, Missouri, to get together and do something. Its organizers felt stymied trying to "get their addicted kids off screens." The problem was a web-based curriculum from Summit Learning and developed by Facebook engineers. As one father complained, allowing computers to teach made the kids look "like zombies." When it became mandatory for students to be online most of the school day and the teacher's role was demoted to being "a mentor," a twelve-year-old with a seizure disorder began having multiple seizures a day (recall the discussion of reflex epilepsy in chapter 6). Her neurologist had recommended no more than thirty minutes of daily screen time to reduce the likelihood of seizures. But when parents started demanding solutions, they encountered stiff resistance. Nearly 80 percent asked that their child not be in classes that used the electronic curriculum, but the superintendent dismissed their concerns with bureaucratic babble about students becoming "self-directed learners" and getting greater "ownership of their learning activities." Parents pulled their kids out of the school.

Parents in Utah similarly fought back against state-funded preschools that were suddenly available only online, a bureaucratic scheme that robbed them of choice and ran the remaining brick-and-mortar preschools out of business. The top-down fiat monopolized the attention of 10,000 youngsters, edging out opportunities to socialize face-to-face with peers and responsive adults. Federal grants in 2019 expanded digital daycare to five neighboring states. The digital divide has also now become about who has the power to limit the intrusion of technology and who doesn't.[23]

Dr. Richard Freed practices child and adolescent psychology in Walnut Creek, California, just east of San Francisco Bay. He is also an advisory board member of the Children's Screen Time Action Network.[24] When he speaks to affluent audiences in San Francisco, Palo Alto, and Silicon Valley, he finds that most have read his book, *Wired Child*, and readily voice concerns about their offsprings' over-connected lives. By contrast, when he speaks only eight miles across the bay in less affluent Oakland or Antioch, he is typically "the first person to mention any of these risks about technology overuse." These parents work long hours for little pay, and he calls their knowledge gap around the dangers of tech "enormous"

Whether a school is affluent, such as Pine Crest in Fort Lauderdale, or impoverished, such as Anacostia High School in the nation's capital, a common concern is that today's prekindergarten and kindergarten students lack the motor skills that earlier generations had. "Few if any of them play with coloring books, cut out pictures, experiment with modeling clay, or even know what a jigsaw puzzle is," Pine Crest's admissions director told me. "They all have phones, and all they do is swipe and type."[25] The social worker at Anacostia said much the same thing: "Their fine motor skills simply aren't there. I can't imagine them doing anything like surgery let alone a trade like cake decorating or carpentry that requires a steady hand."[26]

THE EFFECTS OF CURSIVE WRITING

Students who type their classroom notes do not retain lesson material as well as those who write their notes by hand, particularly in cursive compared to printing. There is a well-established connection between the dominant writing hand and the brain, which is why everyone's signature is unique. In lower-grade classrooms, beautiful examples of each letter are typically pasted around the room. In learning to write, pupils practice copying these examples over and over in their own hand. Logically, since the example template is the same for each pupil, everyone's handwriting should look the same. Tech enthusiasts cannot explain why this is so. But neurology can.

Different brain areas are engaged when writing by hand compared to typing.[27] The neurology behind writing differs greatly from that of tapping a keyboard. Handwriting forces connections between areas for memory and learning. It requires a filter on what you are writing, whereas typing does not. Committing pen to paper forces you to think ahead to where the sentence is going. As the nineteenth century was ending, health issues forced Friedrich Nietzsche to give up penmanship for a typewriter, a modern invention at the time. Friends noted a change "from sustained argument and prolonged reflection" to a more abrupt "telegram style." Sensing the change himself, Nietzsche said, "Our writing tools work on our thoughts."[28]

Writing dates back five thousand years, typewriters one hundred fifty, the Blackberry thirty-eight, the smartphone less than twenty, and the tablet about twelve. The elimination of cursive handwriting from curriculums started in 2010, yet elementary and high school students continue to sign up for formal penmanship classes. At Fahrney's Pens, a ninety-year-old establishment in Washington, D.C., Elizabeth Bunn teaches third to sixth graders cursive handwriting and italic calligraphy. When the director of the Danbury Museum and Historical Society in Connecticut advertised "cursive camp," he was hoping to train future interns so that they could read documents from the city's past. Most of the society's collection is in cursive script and has never been transcribed. The response to the museum's ad was overwhelming. After five days, campers began to decipher historical documents. Ten-year-old twins David and Benjamin Miller were excited by their new ability to read a postcard addressed to another Miller in 1908. "I think we found a relative!" The problem of historical ignorance will only grow as cursive skills wither. Who is going to be able to read the National Archive's billions of handwritten documents?[29]

A person learning cursive makes each letter stroke slightly different every time they practice matching a new visual form to the hand-printed counterparts they already know. The brain activations of seven-year-olds who practiced cursive writing compared to those of controls who only watched others show that self-directed effort rather than passive visual feedback drives the learning. Self-generated actions generally have lasting effects and carry over to other skills.[30]

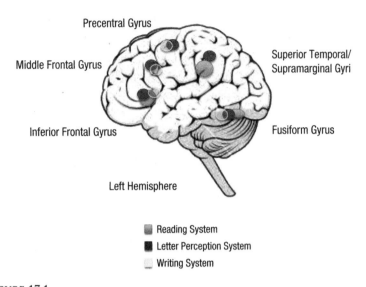

FIGURE 17.1
There is considerable overlap of visual-motor letter perception and letter production. The latter is a necessary precursor to reading. Results are taken from multiple studies.
Source: Karin James, *The Importance of Handwriting Experience on the Development of the Literate Brain*, SAGE Creative Commons. [See also color plate 9.]

The sequential movements used in handwriting activate brain regions involved in thinking, language, and working memory. These are part of a larger network of left hemisphere structures specifically involved in writing, including finger writing and imagined writing.[31] In evolutionary terms, writing is a very recent addition to human intellectual skill set, but the better you write, the better you read and comprehend (figure 17.1). Figure 17.2 illustrates in simplified form the circuitry involved in speaking a word you hear versus speaking a word that you read. Each task requires the participation of different regions of cortex.

The connecting strokes in cursive writing render it speedier and more efficient than the detached stop-and-start of printing, while from the brain's point of view cursive demands greater mental engagement and thus promotes a higher level of cognitive development. In a world that demands so much conformity from a child, developing a unique style, especially a signature, must feel gratifying. Success motivates a student's desire to learn and achieve. Unlike so-called "educational" and "brain training" apps, mastering cursive does have measurable crossover effects: its greater demands

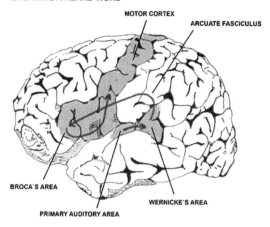

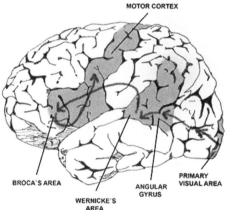

FIGURE 17.2

Top: The sensation from a heard word reaches the primary auditory cortex but is not understood until the input is transformed by Wernicke's area. To speak it, an auditory representation of it passes via the arcuate fasciculus to Broca's area. There it evokes a program for articulation that feeds into the face area of the motor cortex, which then drives the muscles of the tongue, lips, and larynx. *Bottom*: A word that is read first registers in primary visual cortex, and then transfers via the angular gyrus, which associates the visual form of the word to its corresponding sound pattern in Wernicke's area. Speaking the word then calls on the same motor pathways as above. [See also color plate 10.]

for concentration and hand-eye coordination enhance interconnections in brain wiring and creates the mental scaffolding that will be used later on for yet other skills.

Cursive measurably increases a note taker's attention span—a side benefit in the age of digital distractions. Three fingers may carry out the writing but the whole brain is at work, as a high-density EEG study illustrates.[32] Note takers who type have a propensity to transcribe lectures verbatim rather than listening attentively and recasting the material in their own words. The best way to retain new knowledge is to rewrite and organize your rough notes after class but before bedtime. Doing so in your own hand while the material is fresh lets you expand on what you originally wrote down. You will surprise yourself at how much more you remember this way, and not just until the next morning's exam but for the long term. Repetition sears the material into your mind and makes clear whether you understood the substance of a lecture or need remedial clarification and study.

Handwritten notes facilitate learning even better when you add graphic elements such as marginal drawings, arrows, stars, or other symbols.[33] These doodles give memory an extra hook for the same reason we all remember material better when read from a physical book compared to reading it from a screen. "Place memory" (the method of loci invented by the Romans), meaning where on the page you read something within the context of surrounding items, gives you another handle by which to recall it. By contrast, a screen-based Word document is just one more instance of the infinite scroll.

Unlike the world of the *Matrix* films, we cannot jack in and instantly download knowledge that would otherwise take years to attain. Learning takes time, as the relaxed, slow-tech approach of Waldorf education attests. A 2010 study by the renowned Gessell Institute for Human Development reconfirmed that the stages of childhood development have not changed in the sixty years the institute has been monitoring them. In fact, they haven't changed since the Stone Age. The world may have sped up, but our brains are hardly different from those of our distant ancestors. We still develop and learn at a natural pace that cannot be hurried.

And yet the unforgiving trend in American preschools, public and private, is to prod youngsters to be "school ready" through prescribed "seat work," as child educator Erika Christakis calls it in *The Importance of Being Little*. She laments that remote bureaucrats control the pace and content of what a child is supposed to learn rather than teachers close at hand who have taken the time to get to know their charges. Free exploration and self-directed play are now curtailed. Today's preschoolers face the pressure of premature expectations and fast-paced competition, which goads them to achieve external benchmarks at the expense of the self-discovery that naturally occurs in youngsters like those on *Kitafahrt* outings.[34]

This kind of premature pressure has led young learners to score worse on many academic measures compared to earlier students. As one father at the Waldorf School of the Peninsula says, "I have no interest in raising insecure overachievers." He understands firsthand how the common core and No Child Left Behind squeezed the time children need for free play, outdoor recreation, working together to figure out the rules of a game, or just daydreaming.

Do the self-unaware individuals we see staring down at their phones really represent the best and brightest, or are they the future know-nothings of America? We are going to find out.

18

War Games: Is the Only Winning Move Not to Play?

Trying to vanquish screen distractions is a hard battle. The fight against the machine may feel like the climax in the film *WarGames*, in which the U.S. nuclear launch capability is handed to an AI computer that runs simulations for the best attack plan for nuclear retaliation. In the film the computer mistakenly thinks the Soviet Union has launched missiles at the United States. While military authorities scramble to avoid mass destruction, the computer hunts for the perfect winning strategy by cycling through all possible scenarios. After finding not one in which the human race survives, the machine renders its conclusion:

A strange game. The only winning move is not to play.

With screen distractions, the winning move need not be so drastic. Taking strategic breaks and setting devices aside in the evening when your brain needs to wind down or powering it down for periods can be enough. The winning move asks you to question your priorities and do the important stuff first before giving in to scrolling and more. Small steps can make a big difference.

Indifferent forces of evolution have crafted the brain and its psychology in ways that guarantee we are distracted, and no amount of civilized progress or technology can oppose the inexorable biological forces that are otherwise aligned. It isn't a question of willpower outsmarting technology designed to ensnare your attention. If you want to reduce the frequency of distractions and reclaim your mental focus, then you have to acknowledge and work within the limits and vulnerabilities of the Stone Age brain as it operates in modern life. There are no shortcuts, and failures are inevitable. But you can always pick yourself up and try again.

The Pandora's box of the screen, with its infinite possibilities and branching points, will neither close nor go away. By design, screens are addictive. They hog our bandwidth of attention. Who knows what challenges future iterations will bring? Despite the considerable benefits a smartphone endows, it is still a narcotizing agent. The magic that puts the internet in our hands easily turns malevolent and, as the sorcerer's apprentice discovered, beyond our ability to control.

While it is technically true that we are more connected to one another than ever before, we bond on social media platforms now engineered to make outrage, affront, and indignation infectious. They exploit the principle that emotion, like yawning, is contagious. They have turned what used to be two-way conversations into public performances that escalate into shouting matches, sometimes violence. Cancel culture reeks of invective, a haughty sense of moral superiority, and a demand that everyone think like them. Online antagonists pride themselves in how nasty and intolerant they can be. Social media force us to heed the scandal or conflict of the moment instead of listening to and weighing other points of view. This background din makes listening difficult, and finding silence has become impossible without deliberate effort. Silence and stillness are essential nutrients for both mind and soul, yet millions are chronically starving for want of them.

The business model of today's media tries to persuade us that modern life is going to hell. Its sensational approach and relentless emphasis on "breaking" news plays on the brain's exquisite sensitivity to change. The result is an atmosphere of constant churn. The media see themselves as entitled to interrupt us at any time, believing their intrusion is more important than anything else we could be doing.

Ever-present screens that emphasize audiovisual sensation at the expense of thought and reflection leave young people poorly educated and with limited knowledge of the real world. Wealthy parents in Silicon Valley, where online life was invented, increasingly feel guilt and remorse for the zombification, inattention, and distractibility that their work has wrought—even as they shield their own offspring from the consequences of their creations. Today's digital divide is not between the haves and the have-nots but between those who can escape the malign effects of screen tech and the left-behind majority who cannot or do not even know they have been bewitched.

They are not powerless, though. They can resist the sirens of digital allure starting with easy adjustments such as turning down the display intensity of TVs, monitors, and mobile devices. Next, adjust the color settings and filters to minimize exposure to short-wavelength light blasting out at the wrong time of day. Small steps like these might lead to involving yourself in larger issues such as advocating Waldorf-like policies in your local kindergartens and grade schools. A bibliography at the end suggests different strategies.[1] Meanwhile, here are some things you can do for yourself:

- Rethink your internet, gaming, and screen media consumption. How much is based on need, how much on want, and how much on habit? Adjust accordingly. It will take deliberate, even hard, effort, but only you can make the changes. For example, out of habit I normally read the *Wall Street Journal,* the *Washington Post,* the *Atlantic,* and the *New York Times* each morning. Tallying up the time this took me, I cut down and now spend time with only one paper in rotation. If anything is truly newsworthy, I will find out eventually. "Breaking news" is merely the media's version of cocaine and equally as numbing.

- Try what the Dutch call *niksen,* the art of doing nothing and putting life on pause for a few minutes. Look out the window. Seek out and exploit daily opportunities for play, doodling, and engaging in unstructured activity that can take place beyond the screen. The payoff will be many times the investment compared to the technical tweaks you make to screen brightness and color filters.

- Immerse yourself more often in nature and natural sounds. It may help to make a mantra you recite to yourself. A short phrase of five to seven syllables works best. For example: Silence is essential; Silence is food for the soul; My life needs silence. Only three generations ago, Americans spent an average of ten hours per day outdoors. Now 90 percent of our time is spent inside where the quality of light works against our circadian clock, and most of the population live in metropolises that offer them little green space.

- Reading physical print in lieu of just a screenful at a time is a good start to pushing back against the intermittent reinforcement of screen dependence. We challenge our bodies with athletic feats and our wits with puzzles but have no exercises for strengthening mental discipline. Linear reading can be a restorative tonic for a mind degraded by interruptions and forced to flit from one thing to another. If you spend hours keeping your body fit, doesn't your brain deserve a few minutes of training? Just as you'd exercise a muscle, linear reading strengthens the ability to focus attention, sustain it, and resist intrusions. It builds emotional muscles, too, because stories force you to impute motives and intentions, anticipate a character's actions, and predict upcoming turns of plot. Research affirms that readers have greater empathy than nonreaders.

- Writing by hand rather than keyboard likewise hones your focus and capacity to concentrate. You will remember more by taking handwritten notes, thanks to the hand-brain connection. Keeping a diary is one way to reach your inner self. Even more adventurous is keeping a dream diary instead of reaching for your phone the second you wake up. Whereas a journal or conventional diary focuses on daytime, chronological affairs, a dream diary explores feelings and the creative realm of imagination. The appendix instructs how to keep one.

- Avoid reading, streaming, or watching TV while you eat. While doing so has the biggest payoff, it is hard for many to stick with. Keep at it, and don't berate yourself for failure. A Zen saying advises, "Eat when you eat, walk when you walk, die when you die," which means you should give yourself fully to whatever you are doing while you are doing it. Be mindful of times when you do multitask, and then stop to focus on one thing only until you finish it.

- Reignite your pleasure in reading or fire up a passion for it if you don't read regularly. Before turning to online streaming try picking up a book and consuming just a page or two. Small steps can lead to big changes. Magazines that feature essay articles and original fiction such as the *Atlantic*, the *New Yorker*, or the *Economist* offer another way to practice linear reading besides being enjoyable and informative to read.

- Consider your habit of walking, working, or even having sex while wearing earbuds (yes, people do this). How much of the real world are you missing when your ears are off elsewhere? When the Walkman became popular in 1980 it was a time of fewer distractions. Only a small number of people owned the device, and listening to cassette tapes didn't render them oblivious to their surroundings. No one walked into lampposts or lost themselves by staring at the rotating spools either. Reflect on why you think today's devices affect us so differently.

- To preclude oblivious absorption in sound and screen, look up and take in your immediate surroundings. Are you a part of them or apart from them? Who is responsible for that?
- Libraries are free. Visit yours to discover surprising offerings in your neighborhood. Frequenting a library may change you in multiple ways. It gets you out of the house and away from deeply ingrained screen habits. Engaging in person-to-person conversation with a librarian can remind you what making eye contact and small talk feels like. Reference librarians are a trove of knowledge.
- Yes, the brain adapts marvelously. But it adapts to bad stuff just as easily as to the good. Cumulative exposure plastically changes brain structure: scans show a reduction of brain volume in areas also involved in conventional addictions, bolstering the conclusion that we are, in fact, addicted to our phones.
- Tech companies and ordinary businesses hook you via positive intermittent reinforcement, making it hard to disengage from their products. They know there are only 1,440 minutes in a day and use manipulative gamification to hold you as long as possible.
- If something is "free," then you are almost certainly the product being sold.
- The average eight-year-old in the United States and the UK has spent more than a full year of twenty-four-hour days in recreational screen time.
- Selfies kill more people than sharks. People continue to text while driving despite repeated evidence that texting is one of the most distracting and lethal things we can do.
- Social media compete with physical face-to-face socialization, and texting in lieu of in-person interactions handicaps the development of social skills and causes those we already have to atrophy.
- Have what Oscar-winning producer Brian Grazer calls "curiosity conversations," talks with interesting people for no reason other than the possibility of learning something unexpected. Occasional "just because" lunches or coffee with people whose views diverge from your own can't help but expand your perspective. Meeting in person, undistracted and in full focus, is an opportunity to practice making eye contact and reading body language. Grazer prepares in advance and comes to each conversation curious and ready to ask questions. He'll bring a token gift "or some knowledge they would find useful or interesting."[2]

- If you can't meet in person, then call. Phone calls commit you to a sequence of beginning, middle, and end; they help you think ahead about what you want to say and achieve. It is a kind of mindfulness, and phone calls can fortify friendships. Maintaining friendships takes work, yet keeping those you already have is more rewarding than rustling up a thousand new virtual ones.
- "Attention span" is a colloquial term, whereas neuroscience speaks of sustained, selective, and alternating attention. Each has its own energy cost, and the tab is especially high when you force attention to shift back and forth. Linear reading is good practice for focusing and sustaining attention.
- In evolutionary terms, the psychology of FOMO is like the fear of abandonment in that it relates to survival in a group. Social media can be a security blanket, and feeling that you belong is highly salient and rewarding, which is why we are drawn to being included.
- Multitasking is a misnomer, and energy considerations easily refute its claims. It is really the repeated division and back-and-forth refocusing of attention. The only true multitaskers are mothers, simultaneous translators, and air traffic controllers—and the latter two get a break every forty-five minutes.
- Hearing is not the same as listening, and inattentional blindness, as in the famous "invisible gorilla" experiment, happens all too often to the distracted mind. The Na'vi in the film *Avatar* highlight the distinction when they say, "I *see* you." The brain devotes more cortex to facial recognition than to recognizing any other object, and the face uses forty-two muscles to broadcast an individual's emotional state and intent. Infants read faces years before they learn to speak. Yet people today are more often cut off and ignored rather than heard and understood. Good listeners engage one-on-one and never assume to already know what the other is going to say. They ask open-ended questions, such as "Tell me . . ." or "Tell me more. . . ."[3]
- Energy demand is always a limiting factor. An adult brain consumes 20 percent of daily ingested calories, a child's brain 50 percent, and the infant brain a whopping 60 percent.[4] An iPad monopolizes a youngster's limited bandwidth of attention, making it the worst possible

babysitter, and it isn't much better for babysitters themselves or parents, either. Brain circuits operate at low speeds compared to their electronic counterparts and are not fully myelinated until about age twenty-five. Myelination is essential to learning and social development. Blocking a youngster's developing central vision with bassinet, potty, and car seat screens interferes with this myelination.

- The difference with digital devices is that we treat them socially, and emotional attachment makes us regard them as an extended iSelf. Yet by monopolizing so much time and attention, smart devices unintentionally end up interfering with socialization. The epic prevalence of depression and loneliness has been repeatedly linked to the saturation of smart screens.

- Digital devices are like secondhand smoke in that their mere presence degrades attention and efficiency, raises anxiety, and makes the mind drift. The most peripheral parts of the visual field have robust projections to the emotional limbic brain, which is why things caught in the corner of the eye grab our attention and seem salient and meaningful. The visual push of a notification is enough to distract us many times throughout the day. Try putting your phone face down or turning it off if possible. Discover the thrill of being unavailable and unreachable—if only for a spontaneous hour or for scheduled office hours. Why does doing so feel like a guilty pleasure?

- Next time you are in a waiting room or at the airport, try to not look at the TV monitors and advertisements that surround you. The effort should convince you how distractible you are. Our brain's impulse to socialize hasn't changed from what it has always been; it's just that the powerful distractors of smart devices overwhelm the salience of everything else.

- If you always take pictures with your phone, then how much of what is before you do you honestly see? Are you alive or living life secondhand through a view screen?

- Two separate networks lie behind reward and pleasure. Dopamine is the main reward molecule but is involved in wanting, too. Its network is extensive, easy to engage, and hard to satiate, meaning that winning rewards and satisfying wants doesn't last (hence the hedonic treadmill).

The smaller pleasure and liking network, based on opioids, is harder to activate and its effects are briefer. It is easy to yank these two systems in opposite directions, as illustrated by the "lusting while loathing" experiment. Unfortunately, addictions that successfully commandeer the dopamine network can alter it permanently for the worse.

- Any reinforced behavior, such as compulsive phone checking, can sustain itself for a long time without the original reinforcement needing to be repeated. This means that breaking the cycle will take sustained effort. As with the begging dog at the table, a single instance of giving in makes the conditioned behavior infinitely worse and harder to break.
- The ability to connect dots turns factoids into knowledge. But the internet encourages shallow learning, while offloading everything to external agents such as Google leaves us with no dots to connect and a paltry base of shared common knowledge.
- Go beyond the binary logic of whether screens are good or bad to ask, "Are digital screen media causing social and emotional disruptions in my life?" The "still face" experiment is a vivid demonstration of how easily they can. Screens compete fiercely with natural processes. Will you let them?
- People fuss endlessly over what they put in their bodies, but not over what they allow in through the senses. Try a sensory fast, and reflect on what it means for silence to be an essential nutrient.
- Carving out "disconnect time" when you can't be beckoned can be a godsend. Choose the times when you have to be available, and then establish space outside them when someone else can cover for you and you can restore yourself. No one is indispensable; you don't always have to be reachable immediately. So maintain perspective: *Homo sapiens* has been around for 3,000,000 years, smartphones for less than twenty.
- From an energy perspective, sleep is decidedly not passive. Depriving your brain of expected REM episodes degrades memory and learning for up to three days.
- Compared to an adult, the teenage circadian clock is twice as sensitive to short-wavelength light. But even adults suffer unintended consequences from late-night screen viewing. Their EEGs show decreased delta and theta brain waves, leaving them sleepier the next morning and mentally foggy.

- The consequences of inescapably viewing bright LED screens outside the home are worrisome, and not yet fully understood. For now, if it's ultra-bright, look away.
- Peer-reviewed literature on virtual autism—meaning autism-like symptoms induced by heavy screen exposure—is only now emerging. The causation model blames an excess of external audiovisual input for derailing normal development via neuroplasticity. Many parents may not know about the association between virtual autism and heavy screen exposure, so the prudent course may be to unplug, not drug. And please, spend some time outside.

19

CODA: LESSONS FROM THE LOCKDOWN YEARS

What did we learn from what felt like interminable lockdowns? Long before the pandemic, screen distractions from smartphones, tablets, fitness trackers, and televisions were already taking their toll on mental life. Many people claimed to be addicted, even though saying so had become a cliché. The COVID-19 pandemic highlighted the pernicious effects of screens as social skills atrophied and we abandoned our connections with nature. Relying on digital intermediaries for two-plus years revealed just how hungry people were for physical connection. For a touch. For a hug. At a minimum, to shake hands.

Is a screen the last thing you see at night or the first thing you clutch in the morning? Despite a slew of news reports about atrophied attention spans, sleep deprivation, social alienation, blurred vision, and other maladies blamed on heavy screen exposure, too many of us still couldn't switch off.

Experience with earlier coronaviruses led physicians to anticipate a rise in neurological and psychiatric illness following the surge of COVID-19 and its variants. The early months indeed saw an increase in stroke and brain hemorrhage, along with upticks in rates of depression, anxiety, panic attacks, substance abuse, and insomnia. The largest study to date established a direct link between infection and neurological aftereffects. *The Lancet Psychiatry* concluded that one in three survivors receive a substantial neurological or psychiatric diagnosis within six months of becoming infected, subsequently rending them more susceptible than others to the downsides of screen distractions.

The *Lancet*'s observational study of over 230,000 health records, culled from a database of more than 81 million, examined fourteen specific neurological and mental health diagnoses. After controlling for age, sex, ethnicity, and preexisting conditions, it found a 44 percent greater incidence in these patients than in those who had endured influenza or other respiratory infections.[1] Four conditions it found exacerbated by screens during lockdowns were loneliness, anxiety, reactionary outrage, and a mindset of exhausted numbness.

Hours once devoted to commuting to work and frequenting restaurants, cinemas, bars, and gyms suddenly became free time. These unexpected dividend hours were a natural opportunity to stop, sit still, reflect, and rediscover intimacy and the pleasures of deep conversation with a small circle of actual friends compared to the banal chatter of a shallow online hoard.

But many of us did not pause. We spent much of our newfound hours binging on news and video streaming. Gaming shot up 75 percent, according to Verizon.[2] We Zoomed (an activity so ubiquitous it became a verb) until we could no longer think or see straight. In December 2019, Zoom hosted fewer than ten million meeting participants; four months later, after the start of the pandemic, that figure had increased thirtyfold, to 300 million. Soon the world was swept up in mob mentality and the streams of misinformation promoted by social media. What ill effects did the pandemic accelerate, and what future trends might these portend?

Let's begin with loneliness, which affected the broadest range of those trapped in lockdowns. Social media and video calling services initially promised to keep us connected while we were cooped up, away from the workplace, cut off from colleagues, and separated from family and friends. But paradoxically, technology made loneliness worse by inflaming emotions and bringing out the worst in people. Young mothers had a particularly hard time. Suddenly they had to pull off the unfamiliar demands of working from home while managing a household that might include a home-officed spouse taking up space and wanting attention, plus children in need of full-time supervision and the piled-on stress of having to monitor their screen-based schooling.

When personal needs are not being met, it is normal to feel lonely in a crowd. Needs were definitely not met for young mothers forced to turn to Facebook and Instagram for the support they used to receive for free at the playground or the dog park. According to columnist Julie Jargon, who writes about tech's influence on family life, discussions about best parenting practices began to heat up once they moved online. "Haughty, judgmental chatter" left users feeling shamed and dismissed at a time when many were already weary and on edge from prolonged isolation. One parent confided that "even asking a question in those groups opens the door to feeling inadequate."[3]

Having a baby during the pandemic instilled unexpected feelings of disconnection. Before COVID-19 happened, 20 percent of new mothers experienced postpartum depression; after the onset of the pandemic, it rose to 36 percent.[4] Pandemic loneliness plagued young adults, too. A 2021 Harvard University study examined how the pandemic "deepened an epidemic of loneliness" when people who experience social disruption "have nothing but technology to fall back on."[5] Social media are of course a highly edited theater of self-display, which can be hell for those already feeling lonely. Its deceptive posturing invites false comparisons to the ostensible happiness of others and intensifies feelings of rejection, envy, and FOMO-based anxiety.

According to the Institute for Family Studies, Generation Z, born between 1997 and 2012, is "the loneliest generation on record," and the struggle of life online is the primary reason. Lead author Jean Twenge, who coined the term "iGen" to describe the first generation to grow up enmeshed in smartphone technology, sees the hours spent on social media and the resulting plunge in person-to-person socializing as the main cause for the high rates of anxiety, depression, and self-injury among this age group.[6] An uptick in self-cutting and suicidal rumination illustrates how anxious and fragile these young people are. For those aged twelve to seventeen, the months of pandemic isolation were a nerve-wracking tableau of emergency department visits—up 31 percent.[7] In the United States, EDs are traditionally the first point of contact for young people suffering acute blows to self-esteem and their sense of worth. But neither EDs nor families were equipped to handle the surge of adolescents in crisis. Psychiatric beds were already in short supply, and kids having a meltdown did not carry a conventional psychiatric diagnosis such as schizophrenia, meaning that insurance was unlikely to pay for hospitalization (an exception was eating disorders, which also rose markedly). Many hospitals, especially those outside major cities, didn't even have an existing pediatric mental health policy. EDs dispensed some meds and sent teens and children home without resources for follow-up support.

Months of remote screen-based schooling and enforced social isolation turned even the most outgoing student moody, withdrawn, and reclusive. One mother of three blamed the unpredictable disruption of school life and upending of routines for changing her extroverted thirteen-year-old "in profound ways I would never have anticipated."[8] Young brains cannot wait for social policy to meet them where they are: they are developing vigorously (hence teenage drama), do not fully mature until age twenty-five, and cannot be expected to sustain focused attention for the long hours that remote learning demands. Hours of screen viewing left many students bored, burned out, and falling behind, and the pandemic disproved the buoyant promises of what screen devices and an internet connection could accomplish, especially when young people are given free rein to apportion their own use of technology. The most brilliant teachers can do only so much during screen sessions to engage their students. A huge number of pupils disappeared from school rosters and went truant, while those with special needs were left behind. The latter "cannot access the services properly over the computer," said one frustrated mother. "It literally doesn't work."[9]

The downside costs of tech saturation sharply shifted the perspective of those who had heretofore championed its spread as an unalloyed good. We normally expect students to master academic skills; the pandemic reminded us that they should also be leaning social skills that will serve them a lifetime. To partly address the glaring absence of socialization during the lockdowns, libraries across the country hosted drawing and calligraphy classes. These sessions simultaneously gave students a break from their screens.

Media reports in 2021 bolstered earlier findings by the CDC and scholarly researchers that loneliness and anxiety take a heavy toll, especially on the youngest. During lockdowns, students with initiative created study pods with their friends. But bickering ensued. Cliques formed, and the pods soon fell apart. Breakdown of daily routine compounded pandemic-related stresses: loss of friends, loss of school, loss of sports and group activities. These cumulative losses prolonged their isolation and amplified the drumbeat of mounting deaths broadcast nightly on screens big and small. The pandemic saw rises in parental neglect, substance abuse, mental illness, and divorce. Thousands lost a parent or grandparent to COVID-19. "All of the people I look up to," said one thirteen-year-old, "they are all, like, breaking down."[10] Distressed or emotionally unavailable parents are ineffective (compare this with the "still face" experiments using iPhones described in chapter 13). Pileups of stressful events had even worse effects on those with preexisting conditions such as autism, learning disabilities, anorexia, and mood disorders.

According to a different Harvard study still in progress, structure and routine are especially important when times get tough.[11] Youngsters aged seven to fifteen who maintained structured routines showed fewer symptoms of anxiety and depression than their more scattered peers. Prior to the pandemic, 20 to 30 percent in that age group exhibited troublesome behavior; during lockdown the proportion rose to 66 percent. These are prime years for the neuroplasticity that shapes developing brains (discussed in chapters 5 and 9). We already know from pediatric PTSD that brain changes induced by stress can lead to emotional, cognitive, and academic issues that persist well into adulthood. Children with PTSD have deleterious alterations in the amygdala (involved in fear and negative emotions), the hippocampus (memory), and the prefrontal cortex (flexibility, choice, and resilience). In the same way that associative learning influenced Pavlov's dogs, adverse incidents such as divorce, parental neglect, drug use, or a death condition the young to become overly sensitive to triggers such as isolation, separation from friends, or FOMO that remind them of the original shock.[12] This we know, thanks to pediatric brain imaging at Stanford's Early Life Stress Research together with the measurement of cortisol—the body's primary stress hormone. Cortisol concentrations in hair samples serve as a chronological biomarker of distressing experiences.[13] Compared to PTSD from accidents and natural disasters, damage inflicted by the pandemic may be worse because of its lengthy and uncertain duration.

What might the long-term consequences of excessive early stress be? The question is pertinent given the elevated levels of self-injury, eating disorders, and cratering mental health in the young. After a year spent away from the classroom, previously high-performing students watched their grades tank. Teachers and parents, already harried by the demands of remote learning, had no time to think about ways in which pupils might build resilience when faced with one challenge after another. With hardly any travel or party photos posted to social media during quarantines, did older students still feel they were missing out on fun that peers might be having without them? Studies show FOMO to be replaced by assessments as to whether online concerts, happy hours, and group games were worth it. Whatever remote setup was used, distractions continued. YouTube, video games, and TikTok proved to be far more compelling than schoolwork, beckoning users to ditch their virtual classrooms.[14]

According to the CDC's ongoing Household Pulse Survey, adults also suffered from screen distractions and inattention.[15] Industry monitors reported weekly podcast listening up 17 percent compared to pre-COVID-19 years. While escape from people normally in your orbit might have been welcomed initially, the unwitting end result was reduced intimacy and further social isolation.[16] When newcomers flooded the invitation-only Clubhouse app that places multiple users in two-way chat rooms, the quality of debate quickly deteriorated. Nonetheless, the *Economist* predicted that "group chats are likely to remain a feature of social media after the pandemic ends."[17]

Meanwhile, a year of extra time to spend staring at screens illustrated how easy it is to stir up and succumb to reactionary outrage. For global audiences eager to take umbrage, social media flawlessly exploits grievances. Gen-Zers' short attention spans give them an equally short fuse, says Professor Twenge, because "complex ideas require sustained attention. The idea that you're going to be patient and sit down to read a book for two hours and do nothing else is kind of mind-blowing to an iGener." To confirm her point, she notes that the percentage of high school students who read books of any kind has dropped from 60 percent to 15 percent since the 1980s.

It is astonishingly easy for Twitter and the sprawling social media sphere to short-circuit common sense and persuade broad swaths of all ages that their values are under attack. In my own 2021 Twitter poll, 61 percent of respondents said "the constant outrage" bothered them the most, followed by "loneliness" at 28 percent and "screen addiction" at 11 percent, with "FOMO" earning zero votes. Gerald Seib, executive editor of the *Wall Street Journal,* said, "The national mood seems to be one of outrage," while the Center for Strategic and International Studies called our present climate "the age of mass protests."[18] People who disagree with a point of view are not merely mistaken or have a different perspective; they are dismissed as asinine, idiots, and evil. In a polarized climate, debate is replaced by denouncements by those eager to vent. Even President Biden got into the fray, blaming "Neanderthal thinking" for the decision of some governors to lift mask mandates (even though the governors' decision was later proven correct). India, Australia, Italy, Spain, Brazil, South Africa, and the UK likewise saw angry mobs react to lockdown restrictions imposed during the pandemic.

News broadcasts illustrated how easily ideological protests can turn into physical ones, fueled by individuals and institutions with vested interests in stoking political anger: the rampages of Seattle's "summer of love," the Black Lives Matter demonstrations, the riots following the death of George Floyd, the antilockdown protests, the racist violence against Asian Americans, and the televised-live storming of the U.S. Capitol. All illustrated how easily invective spreads to sweep up people into a mob mentality. Yet affected individuals became strangely detached from the inevitable consequences: participants who stormed the Capitol gladly posted incriminating selfies and angled to give press interviews. They never imagined anyone would recognize their online marauding or report them to the FBI. Likewise, organizers of Seattle's "autonomous zone" and a sympathetic media blithely minimized the property destruction the riots inflicted on the local police station and surrounding residences (the city settled out of court in 2023 for $3.65 million).

The way online media shape belief systems that are increasingly out of touch with reality has been widely reported and discussed. The January 6 Capitol protestors zealously believed that the election was stolen despite recounts, reverifications, and eighty-six lost court cases. YouTube claims to have tweaked its recommendation engine in favor of established journalism outlets over "borderline content and harmful misinformation."[19] Yet continued exposure to false and unvetted opinion still leads people away from reality to become wedded to conspiracy theories. For various reasons people do not turn to established authorities or institutions that have deep and broad knowledge about the topic at hand. Instead, they open themselves to any voice that feels entitled to spout an opinion, regardless of qualifications or motivation. For a thoughtful analysis of why people adopt beliefs that are patently untrue and resist all evidence, see Dan Ariely's *Misbelief*.[20]

For example, large numbers of people believe COVID-19-related tweets by celebrities more than those of medical authorities.[21] When the internet first opened to the public in 1993, the dominant top-level domains were dot-edu(cation), dot-mil(itary), dot-gov(ernment), and dot-org(anization). The dot-com(mercial) domain accounted for a tiny fraction of early traffic, while opinion and discussion were relegated to the dot-alt(ernative) domain. The latter included debates about aliens, conspiracies, and everything else. While dot-alt users took a stance of "anything goes," all sides insisted on debatable evidence rather than wild and unsupportable claims.

The dot-com domain now dwarfs all others many times over. Once the gates opened it became hard to know whether you were listening to a trusted authority or the indignant rantings of a twelve-year-old.

All this chaos left many in a state of exhausted numbness. During lockdowns everyone had to deal with issues ranging from noise, interruptions, and the stress of multitasking to isolation, uncertainty, and the disappearance of comforting routines. These interfered with the ability to focus. The brain can spend energy trying to filter out distractions, but at a high cost, particularly for those already sleep deprived or overwhelmed. One couldn't be vigilant about the pandemic or the latest political kerfuffle and still get things done as efficiently as before. Only a few months into 2020, "Zoom fatigue" became a discussion topic.[22] Most complaints were about sound or video dropping out because of poor connectivity. But video callers speak about 15 percent louder than they do in person, need effort to focus and shift attention while looking on-screen at a dozen or more participants, and must constantly gauge whether they are coming across as engaged. All this requires extra emotional effort compared to face-to-face conversation. As a result, virtual meetings went from novelty to a source of exhaustion.

The cognitive load of online meetings eats up your capacity to think, too. When face-to-face we process a slew of signals without having to consciously think about them: facial expression, gesture, posture, vocal tone and rhythm, and the distance between speakers. We read body language and make emotional judgments about whether others are credible or not. This is easy to do in person, whereas video chats force us to work to glean the same cues. This consumes a lot of energy. Recall that compared to electronic devices, the human brain operates at ridiculously slow speeds of about 120 bits (approximately 15 bytes) per second.[23] Listening to one person takes about 60 bits per second of brainpower, or half our available bandwidth. Trying to follow two people speaking at once is fairly impossible for the same reason multitaskers fare poorly: attempting to handle two or more simultaneous tasks quickly maxes out our fixed operating bandwidth.

As attention flags, we fatigue. Yet it is the audio gaps, not the video, that makes Zoom sessions draining. All languages have clear rules for conversation that assure no overlap but no long silences. Online meetings disrupt that convention because the separate sound and video streams are chopped into tiny digital packets and sent via different pathways to the recipient's end where they are electronically reassembled. When some packets arrive late the software must decide whether to wait to reassemble them—causing a delay—or stitch together whatever packets are available, giving rise to stuttering audio.

Video conferencing platforms have opted to deliver audio that arrives quickly but is low in quality. Platforms aim for a lag time of less than 150 milliseconds. Yet that is long enough to violate the no-overlap/no-gap convention to which speakers are accustomed. A round-trip signal can take up to 300 milliseconds before one gets a reply, a pause that makes speakers seem less convincing and trustworthy. Repeatedly having to sort out talking over one another and who goes first is also tiresome and draining to everyone on the call.

Speaking by telephone takes comparatively less effort because we can concentrate on a single voice. We can pace about during the call if we want to, which helps thinking. The simple act of walking is well proven to aid problem solving and creativity. An office provides an abundance of locations for conversation, and every physical location in which we engage coworkers carries implicit meaning that colors decisions and the way we think. Working from home merges a life of previously demarcated areas into one amorphous mass shaded with domestic associations. This, along with the temptation to multitask, degrades your ability to think.

The mental strain of having to look at oneself over hours of Zoom meetings results in what Stanford University psychologists call "mirror anxiety," while "Zoom dysmorphia" describes a user's anxiety about dark circles, wrinkles, or bad hair. From a sample of 10,322 subjects, 14 percent of women felt "very" or "extremely" fatigued after Zoom calls, compared to only 6 percent of men. The researchers devised a Zoom and Exhaustion Fatigue scale to assess how serious the problem felt across five dimensions of fatigue: general, social, emotional, visual, and motivational (readers can test how susceptible they are at bit.ly/332zRaS). In addition to mirror anxiety, more women than men felt trapped: they took fewer breaks and felt obligated to hover within the camera's frame. Established research tells us that looking in a mirror raises self-consciousness and self-criticism about one's appearance. In what sounds like a good idea, researchers suggest making some meetings audio-only as a way to "reduce the psychological costs, especially given that these costs are born unequally across society."[24]

Videoconferencing has not surprisingly contributed to the "Zoom boom" spike in cosmetic surgery consultations, up 64 percent from the prepandemic baseline, according to the American Society of Plastic Surgeons.[25] Jeremy Bailenson, founding director of Stanford's Virtual Human Interaction Lab, likens our sudden immersion in videoconferencing to spending an entire workday holding a mirror in our hand. "Zoom users are seeing reflections of themselves at a frequency and duration that hasn't been seen before in the history of media and likely the history of people."[26]

The relationships we forge in communal workspaces are often highly meaningful. In *Someone to Talk To,* Harvard sociologist Mario Small notes that we are more likely to discuss important matters with people who are readily available than with family members or friends with whom we are emotionally close. Work is where many engage in the bulk of their social interactions, making it feel natural to turn to colleagues when the need arises. "Access equals [having] confidants," he says.[27] When people work from home, these casual interactions either do not happen or else must be scheduled—thus interfering with the timeliness of what we feel the need to discuss. And because Zooming, email, and texting feel qualitatively different from face-to-face talk, we also weigh them as less meaningful.

As the pandemic barreled on, productivity burnout set in. The added demands on our attention proved to be unsustainable. With too much to handle we got less and less done. Some suspected that their memory was failing. In the writer's version of Groundhog Day, the usually prolific Susan Orlean tweeted, "Good morning everyone but especially to the sentence I just rewrote for the tenth time."[28] Malaise, ennui, and lethargy crept in—*non voglia,* as the Italians say: no will, no volition, not feeling like it. We had to stop and ask ourselves what day it was. We labored to remember what movie we had streamed only last night or what we ate for dinner. And my personal favorite: What did I come into this room for? Responding to a *New York Times* questionnaire, a college student attending a year of remote classes said, "I'm so burnt out that even this form is way, way too long."[29] She had become at the mercy of every digital distraction.

What can we learn from the lockdown years?

- The neurologic and cognitive sequelae from COVID-19 make it harder to combat and cope with the effects of screen overload. So do loneliness, isolation, and boredom.
- Socialization is a fundamental human need (a primary reinforcer, discussed in chapter 11). Technology has never been a good substitute for personal connections and never will be.

- Mind and body are fragile. Both need downtime and rebel if not given it.
- The bandwidth the brain has to work with is limited; heavy demands on attention are not sustainable.
- We may want to rethink our relationship with screen technology and ask how happy it makes our short life on this planet. What really matters and what is simply distraction?
- One can quickly become starved for intimacy. Music can stand in for touch, as Yo-Yo Ma discovered during his fifteen-minute post-vaccination wait in a school gym. When he played a little impromptu Bach under the basketball hoops, the whole crowd turned quiet. "Our skin is our largest organ," he said. "The sound of music moves air molecules. So when air floats across your skin and touches the hairs of your skin, that's touch."[30]
- One lesson from being cooped up is how heavily we rely on the physical presence of others. U.S. Secretary of Transportation Pete Buttigieg said, "How little of what we have to tell each other is communicated in words." Swapping constant travel for home confinement and the steady presence of his spouse, he discovered how much is communicated by a look. "A short word or facial expression became the equivalent of a whole discussion."[31]
- Google is betting on our return to hybrid work—some days remotely, others at the office—with a $7 billion investment in postpandemic office designs. How will we handle such a new transition, and in what way will we work differently? Will work still be dominated by screens?[32]
- Take a walk. Look up at the trees. Soak in the silence. Remember to breathe. The way to be more productive may not be to use more technology but to do less, and only what really matters.

Acknowledgments

Special thanks to Andrew Lampack of the Peter Lampack Agency for representing me over the decades in print, radio, television, and international lecture, concert, and museum appearances; to Philip Laughlin, my longtime editor at the MIT Press, for steadily championing my work; to the Virginia Center for the Creative Arts, where many of my books have been written amid Appalachian solitude and where an attentive kitchen crew understands that you can't make art on an empty stomach; to the media department at George Washington University and Laura Abate, director of the Himmelfarb Library; and to the DC Commission on the Arts and Humanities, which for multiple years has made me an Artist Fellow. Gaby Cardoso created and improved many of the illustrations. Tom Estlack and Tommy Keefer have designed and maintained my website for many years.

Thanks, too, to advance readers who offered thoughtful comments and challenges: Barbara Bellman, Carolyn Christov-Bakargiev, Joe Clement, Sean Day, Adam Garfinkle, Susan Greenfield, Armin Haracic, Alexander Howard, Charles Karelis, David Keplinger, Ellen Lupton, Marcus Lutyens, Matt Miles, Roger Rosenblatt, Richard C. Shah, George Siscoe, Gabriella Warren-Smith, Harry Whitaker, Maryanne Wolf, and four anonymous reviewers arranged by the press.

Lastly, I am beyond grateful for the support and collaboration of my late husband of twenty-six years, Stephen P. Gorman. Childhood hearing loss endowed him with a capacious visual memory, an enormous help as he navigated us through the unlabeled streets of Tokyo or Hong Kong or the forests of Australia and the reefs off Bali, but also in critiquing my writing as no one else could. He was enormously well read and equally well traveled. Everyone knew him as warm, open, and loquacious. His only fault is that I could never get him to understand that interrupting a writer at work is like waking a dreamer from sleep—you can't go back to the dream or the train of thought. But love abides all, and such endearing foibles are what make me smile still.

Appendix: Keeping a Dream Diary

How does a dream diary differ from a regular journal? A regular diary commits to what I call mere chronicling: "I did this, then this, and then that. . . ." An ordinary diary is essentially a period log of events that happen around a given time. You can look back and see who came to Thanksgiving dinner or your Guy Fawkes bonfire, who did what and when and where, who snubbed you or complimented you, and so on. An ordinary diary stays on the surface as it chronicles quotidian events. It can cover either a day-to-day interlude or a burgeoning time during which matters of import happen. "My years at college," for example.

What a regular diary doesn't do is get to the meaning of those events. And this is where a dream diary excels. Dream diaries, like those based on depth psychology propounded by the likes of Carl Jung or Ira Progoff, delve into meaning. They ask what the images, narratives, and symbols of your personal nighttime cinema have to do with ongoing life, specifically the inner self that is connected to the values, beliefs, and attitudes that give life meaning.

Everybody dreams, even those who claim never to remember theirs. EEG recordings confirm that we spend a quarter of each night doing so. That amounts to about two hours of dreamtime each evening, so it should be possible to tap into at least a part of that. Many people find dreaming inherently interesting with its tantalizing air of mystery—why are dreams so strange, what do they mean, what purpose do they serve? When Leonardo da Vinci asked, "Why does the eye see a thing more clearly in dreams than the mind while awake?" he was calling attention to a dream's vivid, often surreal aspect. He underscored the intuition that dreams relate to something important even though their meaning is not immediately obvious.

Just because dreams are rooted in the brain's physiology doesn't mean they are "simply random electrical impulses." They draw on intellect, memory, and imagination. But what are dreams pointing to? Fortunately, techniques exist to bring the nonanalytical way of thinking we engage in during sleep into the light of day, where we can make objective what is inherently symbolic.

The first step toward finding out what dreams are trying to tell us is a factual recapitulation. But how do you remember your dreams? We have all had the experience of waking from an episode fraught with sensation and insight only to have it fade within minutes once awake. To reduce the chance of this happening, it helps to remind yourself as you hit the pillow that you wish to awaken from your dreams and remember them. It may sound ridiculous, but the proper mental attitude can make this happen. An Englishwoman named Mary Arnold-Foster first wrote about autosuggestion in her 1921 book, *Studies in Dreams*. Before falling asleep, she told herself to notice her subsequent dreams. And she did. She found she could influence the content of her dreams, too, by autosuggestion. Mary was especially fond of flying.

* * *

As to getting material down, the first problem is writing in the dark so you'll be able to read back your notes in the morning. A bedside pad or medium-size spiral notebook of about five by seven inches is good for this, and you'll also need a pen that doesn't dry out. In *A Life in the Dark*, the famous movie critic Pauline Kael wrote about how to take legible notes in the dark. Get accustomed to short, scratch pad entries, she advised. Write big and leave space to elaborate later, taking as many pages as you need. Jot down just the gist of the dream, not every little detail. In the morning you copy your bedside notes legibly into a larger notebook. Here is where you will flesh out the details and note any strong feelings in the dream sequence you haven't already mentioned.

Having prompted yourself at bedtime, you wake up a few hours later and realize you have been dreaming. What do you do? First:

- Don't move a muscle. As soon as you realize you are having a dream, freeze. Stay completely still and resist the urge to move. During REM sleep, your voluntary muscles were paralyzed, so moving now will erase the dream from your working memory.

- Do not open your eyes either. That counts as a movement that will also hasten dream erasure.
- Recapitulate backward. Continue to lie still, eyes closed, and retell the dream story to yourself in reverse. Let's say you are driving alone, scared, down a deserted street at night in the rain. What happened before that? Oh, I was in my old sports car with two strangers and my father; he was trying to get the radio to work. And what happened before that, and so on. As you rewind backward you may land on a feeling that signals the central point of the dream. When you hit an emotional jolt you have usually hit upon a well of meaning.
- Having reviewed the plot, you can now open your eyes and sit up. You have successfully transferred the dream narrative into short-term memory.

Turn next to your bedside pad and jot down key points, not the full narrative. As you do this you will remember more details as you transfer them this time from short-term to long-term memory. The sooner you write things down the more detail you will recall. Clarify any moods, feelings, or attitudes. Always jot down any dialogue or written signage. At first, try to write down every dream, no matter how fragmentary or simple it is. With practice you will become increasingly adept at this and can choose to write down only those dreams you find compelling. Follow these steps, and you will soon be able to remember four or more dreams a night.

*   *   *

Recall that REM periods grow longer as the night wears on. This is why we are better able to remember dreams from the latter part of the night. The first REM period lasts about five to ten minutes; the fourth and fifth may extend to thirty minutes. By contrast, the most restful sleep of stages III and IV peter out as the night wears on. You might try setting an alarm to wake you up shortly after bedtime to see what your dreaming mind has come up with. At this point you don't have to be concerned about writing anything down, just getting accustomed to being more aware of the times when you are dreaming. Setting an alarm for four and one-half, six, or seven and one-half hours after sleep onset gives you a good chance of awakening during the third, fourth, or fifth REM cycle.

Nathaniel Kleitman discovered rapid eye movement sleep in 1953. He was the first to notice that sleepers had periods when their eyeballs darted back and forth under their lids. The EEG during these times looked like that of a person who was alert rather than showing the slow delta waves characteristic of someone deeply asleep. When Kleitman woke sleepers up and asked "What's going on?," they turned out to be having vivid dreams. The moment marked the discovery that REM was correlated with dreaming.

We have since learned that we can dream during every sleep stage from I to IV, but that REM dreams are the longest and most vivid. During stage I or II, the lighter end of slow-wave sleep, most people don't even realize they are asleep. If awakened, subjects may insist that they were lying awake, thinking quietly. Yet their EEG brain waves confirm they are sound asleep. When awakened and asked what they were dreaming they say things like "I was looking around in a drawer," or "I was thinking about getting my car fixed." Slow-wave sleep dreams tend to be mundane, repetitive, and have little emotional affect. After being awakened from an REM dream, sleepers quickly sink back into slow-wave sleep, traveling through stages II, III, and IV and then back up again for another REM session. The pattern repeats itself four to six times a night, depending on how long you stay asleep. Obviously sleep is not a single state of passive inertia but an oscillation between energetic extremes, with peaks during REM and troughs during stage IV slow-wave sleep.

Starting in the 1970s, Allan Hobson at Harvard showed why dreams are not only bizarre and unpredictable but also highly emotional. A dream has a narrative structure. It is an irrational yet sequential story wherein the impossible routinely happens. He proposed a DUI framework of discontinuity, uncertainty, and incongruity. For example:

Dreams change course abruptly without warning.

Identities are unstable and uncertain. In our notes we write, "some people," "I think it was my sister," "I was in what appeared to be a courtroom."

Identities change abruptly: "The man turned into my mother."

Incongruous or improbable features and actions are common: "I had to walk along a chain under water." There are shifts in time and location: "Suddenly, I was in a dark cave."

Strong emotions are common such as fear, anxiety, frustration, or sexual arousal and engagement.

Take these elements in stride. When recording a dream, you want just the facts, the plot, the characters. Mention discontinuities, changing identities, affect, and how you felt as each one changed. Is your dream in color? Conventional wisdom says dreaming in color signifies heightened emotion, yet many people say they never dream in color. When you interrogate sleepers during the night rather than in the morning, however—during or close to a REM period—85 percent of individuals do report dreaming in color. It seems nearly everyone dreams in color but many don't remember that they do.

In a little while you will probably see patterns and recurrent themes emerge in what you are dreaming. Let this guide your insights. It is useless and likely misleading to turn to external guidebooks for what a particular image means. Dreams come from the dreamer's head and nowhere else, so no one else can possibly opine on the meaning of a particular dream. It may take patience and digging, but eventually insight will strike. A powerful tool for exploring this issue further is the "dream enlargement" section of Ira Progoff's Intensive Journal. Progoff was one of the better-known practitioners of depth psychology, a reaction against Freud that emphasized not analyzing the psyche but evoking it to gain perspective on one's inner life. Progoff, like his mentor Carl Jung, saw dreams not as otherworldly but as part of the whole integrated person.[1]

Notes

Your Stone Age Brain, 3 BCE versus Today

1. Mark Dyble et al., "Sex Equality Can Explain the Unique Social Structure of Hunter-Gatherer Bands," *Science* 348, no. 6236 (2015): 796–798, https://doi.org/10.1126/science.aaa5139.

2. "The Future of Dating Is Fluid," Tinder Newsroom, March 24, 2021, tinderpressroom.com/futureofdating.

3. Kelly Servick, "In Two-Person MRI, Brains Socialize at Close Range," *Science* 367, no. 6474 (2020): 133; Ville Renvall et al., "Imaging Real-Time Tactile Interaction with Two-Person Dual-Coil fMRI," *Frontiers in Psychiatry* 11 (2020): 279; Elizabeth Redcay and Leonhard Schilbach, "Using Second-Person Neuroscience to Elucidate the Mechanisms of Social Interaction," *Nature Reviews Neuroscience* 20, no. 8 (2019): 495–505.

4. Ann Gibbons, "The Calorie Counter," *Science* 375, no. 6582 (2022): 710–713, https://science.org/doi/pdf/10.1126/science.ada1185; Richard E. Cytowic, What Percentage of Your Brain Do You Use?, TED-Ed: Lessons Worth Sharing, video, 5:15, January 24, 2014, https://ed.ted.com/lessons/what-percentage-of-your-brain-do-you-use-richard-e-cytowic.

5. Some readers may ask what happens when the brain runs out of energy. How do people survive a long fast or starvation? The answer is that the brain never runs out of energy when survival is at stake: it will convert fat stores into ketone bodies and then muscle tissue into glucose (gluconeogenesis).

Chapter 1

1. Parent's letter to the author, *Psychology Today*, March 26, 2019.

2. Maryanne Wolf, *Reader, Come Home: The Reading Brain in a Digital World* (New York: HarperCollins, 2018).

3. Ben Sasse, *Them: Why We Hate Each Other and How to Heal* (New York: St. Martin's Press, 2018); Daniel J. Levitin, "Why the Modern World Is Bad for Your Brain," *The Guardian*, January 18, 2015, https://www.theguardian.com/science/2015/jan/18/modern-world-bad-for-brain-daniel-j-levitin-organized-mind-information-overload.

4. Thomas S. Kuhn, *The Structure of Scientific Revolutions*, 4th ed. (Chicago: University of Chicago Press, 2012).

5. To be fair, leeches are making a comeback in certain surgical applications, especially with the very recent discovery of a new species of leech on the eastern U.S. coast that has potentially useful medicinal qualities. See Anna J. Phillips et al., "Phylogenetic Position and Description of a New Species of Medicinal Leech from the Eastern United States," *Journal of Parasitology* 105, no. 4 (219): 587, https://doi.org/10.1645/18-119.

6. World Health Organization, *Guidelines on Physical Activity, Sedentary Behavior and Sleep for Children under 5 Years of Age* (Geneva: World Health Organization, 2019), https://who.int/iris/handle/10665/311664; Adolescent Brain Cognitive Development Study (web page), National Institutes of Health, 2019, https://abcdstudy.org; "Media and Children," Patient Care, American Academy of Pediatrics, updated June 4, 2021, aap.org/en-us/about-the-aap/aap-press-room/news-features-and-safety-tips/Pages/Children-and-Media-Tips.aspx.

7. Adrian F. Ward et al., "Brain Drain: The Mere Presence of One's Own Smartphone Reduces Available Cognitive Capacity," *Journal of the Association for Consumer Research* 2, no. 2 (April 3, 2017): 140–154, https://doi.org/10.1086/691462; Richard E. Cytowic, "A Link Between Screen Exposure and Autism-Like Symptoms," *The Fallible Mind* (blog), *Psychology Today* (online), December 11, 2020, https://www.psychologytoday.com/us/blog/the-fallible-mind/202012/link-between-screen-exposure-and-autism-symptoms; Susan Greenfield, *Mind Change: How Digital Technologies Are Leaving Their Mark on Our Brains* (New York: Random House, 2015).

8. DeVry University Achievement Academy, email communication with the author, April 2, 2019.

9. Unsigned review of *Stone-Age Brains in the Screen Age*, by Richard E. Cytowic, 2024. Such sentiments are broadly representative of academic criticism.

10. Richard E. Cytowic, "Your Brain on Screens," *American Interest*, June 9, 2015, 53–61, the-american-interest.com/2015/06/09/your-brain-on-screens.

11. Adam Garfinkle, "The Erosion of Deep Literacy," *National Affairs*, Spring 2020, nationalaffairs.com/publications/detail/the-erosion-of-deep-literacy.

12. Michael Crichton, "Michael Crichton / Reflections of a New Designer," interview by Kathy Yakai, *Compute!*, February 1985, 44–45, archive.org/stream/1985-02-compute-magazine/Compute_Issue_057_1985_Feb#page/n45/mode/2up.

13. John Bohannon, "Credit Card Study Blows Holes in Anonymity," *Science* 347, no. 6221 (2015): 468, https://doi.org/10.1126/science.347.6221.468; Yves-Alexandre de Montjoye et al., "Unique in the Shopping Mall: On the Reidentifiability of Credit Card Metadata," *Science* 347, no. 6221 (2015): 536, https://doi.org/10.1126/science.1256297.

14. "Introducing Meta's Next-Gen AI Supercomputer," Meta Newsroom, updated January 24, 2022, https://about.fb.com/news/2022/01/introducing-metas-next-gen-ai-supercomputer.

15. Leaders, "Self-Driving Cars Offer Huge Benefits—but Have a Dark Side," *Economist*, March 1, 2018, https://www.economist.com/leaders/2018/03/01/self-driving-cars-offer-huge-benefits-but-have-a-dark-side. See also Laurel Wamsley, "Uber Ends Its Controversial Post-Ride Tracking of Users' Location," *The Two-Way* (blog), NPR, August 29, 2017, npr.org/sections/thetwo-way/2017/08/29/547113818/uber-ends-its-controversial-post-ride-tracking-of-users-location.

16. Andrew Van Dam, "What We Buy Can Be Used to Predict Our Politics, Race or Education—Sometimes with More Than 90 Percent Accuracy," *Washington Post*, July 9, 2018, https://washingtonpost.com/business/2018/07/10/rich-people-prefer-grey-poupon-white-people-own-pets-data-behind-cultural-divide; Marianne Bertrand and Emir Kamenica, "Coming Apart? Cultural Distances in the United States over Time," NBER Working Paper 24771 (Cambridge, MA: National Bureau of Economic Research, June 2018), https://nber.org/papers/w24771; Jennifer Valentino-DeVries et al., "Your Apps Know Where You Were Last Night, and They're Not Keeping It Secret," *New York Times*, December 10, 2018, nytimes.com/interactive/2018/12/10/business/location-data-privacy-apps.html.

17. Zenyep Tufekci, "Think You're Discreet Online? Think Again," *New York Times*, April 21, 2019, nytimes.com/2019/04/21/opinion/computational-inference.html.

18. Shoshana Zuboff, *The Age of Surveillance Capitalism: The Fight for a Human Future at the New Frontier of Power* (New York: PublicAffairs, 2020); *The Social Dilemma*, directed by Jeff Orlowski, Netflix, 2020.

19. Alex Hern, "Netflix's Biggest Competitor? Sleep," *Guardian* (U.S. edition), April 18, 2017, https://www.theguardian.com/technology/2017/apr/18/netflix-competitor-sleep-uber-facebook; Rina Raphael, "Netflix CEO Reed Hastings: Sleep Is Our Competition," *Fast Company*, June 17, 2017, https://fastcompany.com/40491939/netflix-ceo-reed-hastings-sleep-is-our-competition.

20. Yasuhiko Kubota et al., "TV Viewing and Incident Venous Thromboembolism: The Atherosclerotic Risk in Communities Study," *Journal of Thrombosis and Thrombolysis* 45, no. 3 (2018): 353–359, https://doi.org/10.1007/s11239-018-1620-7; Jenna Birch, How Binge-Watching Is Hazardous to Your Health, *Washington Post*, June 3, 2019, https://www.washingtonpost.com/lifestyle/wellness/how-binge-watching-is

-hazardous-to-your-health/2019/05/31/03b0d70a-8220-11e9-bce7-40b4105f7ca0
_story.html.

21. Joe Keeley, "Binge-Watching vs. Binge-Racing: What It Is and Why You Should Try It," MUO, June 4, 2020, makeuseof.com/tag/binge-racing-next-netflix-trend-try.

22. Netflix Media Center, "Ready, Set, Binge: More Than 8 Million Viewers 'Binge Race' Their Favorite Series," press release, Netflix, October 17, 2017, doi: media.netflix.com/en/press-releases/ready-set-binge-more-than-8-million-viewers-binge-race-their-favorite-series.

23. David Pogue, "Asimov's Predictions from 1964: A Brief Report Card," *Scientific American*, March 1, 2014, https://www.scientificamerican.com/article/asimov-predictions-from-1964-brief-report-card/#:~:text=He%20foresaw%20underground%20and%20underwater,on%20jets%20of%20compressed%20air.

24. Soocial, "23 Binge-Watching Statistics You Should Know in 2022"; "23 Binge-Watching Statistics You Should Know (2023)," https://www.soocial.com/binge-watching-statistics.

25. Kristine Phillips, "Mark Zuckerberg Apologizes 'For the Ways My Work Was Used to Divide People,'" *Washington Post*, October 1, 2017, https://washingtonpost.com/news/the-intersect/wp/2017/10/01/mark-zuckerberg-apologizes-for-the-ways-my-work-was-used-to-divide-people.

26. Hayley Tsukayama, "Experts Grade Apple's and Google's New Tools to Fight Smartphone Addiction," *Washington Post*, June 7, 2018, https://washingtonpost.com/news/the-switch/wp/2018/06/07/experts-grade-apples-and-googles-new-tools-to-fight-smartphone-addiction.

27. Aric Sigman, "Screen Dependency Disorders: A New Challenge for Child Neurology," *Journal of the International Child Neurology Association* 17, no. 119 (2017), https://jicna.org/index.php/journal/article/view/67.

28. Russell Jago et al., "Cross-Sectional Associations between the Screen-Time of Parents and Young Children: Differences by Parent and Child Gender and Day of the Week," *International Journal of Behavioral Nutrition and Physical Activity* 11, no. 1 (2014): 54.

29. Greenfield, *Mind Change*.

30. Steve Sharman, Michael R. F. Aitken, and Luke Clark, "Dual Effects of 'Losses Disguised as Wins' and Near-Misses in a Slot Machine Game," *International Gambling Studies* 15, no. 2 (2015): 212–223, https://doi.org/10.1080/14459795.2015.1020959; Steve Sharman and Luke Clark, "Mixed Emotions to Near-Miss Outcomes: A Psychophysiological Study with Facial Electromyography," *Journal of Gambling Studies* 32, no. 3 (2016): 823–834, http://doi.org/10.1007/s10899-015-9578-2.

31. Chanel J. Larche, Natalia Musielak, and Mike J. Dixon, "The Candy Crush Sweet Tooth: How 'Near-Misses' in Candy Crush Increase Frustration, and the Urge to Continue Gameplay," *Journal of Gambling Studies* 33, no. 2 (2017): 599–615; Dana Smith, "This Is What Candy Crush Saga Does to Your Brain," *Guardian* (U.S. edition), April 1, 2014, https://www.theguardian.com/science/blog/2014/apr/01/candy-crush-saga-app-brain.

32. Nicholas Carr, *The Shallows: What the Internet Is Doing to Our Brains* (New York: W. W. Norton, 2011), 116.

33. John S. Hutton et al., "Associations between Screen-Based Media Use and Brain White Matter Integrity in Preschool-Aged Children," *JAMA Pediatrics* 174, no. 1 (2019): e193869, https://doi.org/10.1001/jamapediatrics.2019.3869.

34. Kate Julian, "The Sex Recession," *Atlantic*, December 2018, 78–94; Elizabeth Bernstein, "Dating 101, for the Romantically Challenged Gen Z," *Wall Street Journal*, March 11, 2019, https://www.wsj.com/articles/dating-101-for-the-romantically-challenged-gen-z-11552318023.

35. Conor Friedersdorf, "How Does Hookup Culture Affect Sexual Assault on Campus?," *The Atlantic*, June 28, 2016, https://www.theatlantic.com/politics/archive/2016/06/how-does-hookup-culture-affect-sexual-assault-on-campus/489098.

36. B. F. Skinner, *Beyond Freedom and Dignity* (New York: Knopf, 1971).

Chapter 2

1. David A. Sbarra, Julia L. Briskin, and Richard B. Slatcher, "Smartphones and Close Relationships: The Case for an Evolutionary Mismatch," *Perspectives on Psychological Science* 14, no. 4 (2019), https://doi.org/10.1177/1745691619826535.

2. John Kelly, "People Keep Falling into the Grand Canyon. After a Visit, I Can See Why," *Washington Post*, May 5, 2019, https://www.washingtonpost.com/local/people-keep-falling-into-the-grand-canyon-after-a-visit-i-can-see-why/2019/05/05/352b24b0-6f39-11e9-8be0-ca575670e91c_story.html.

3. Leo Benedictus, "Look at Me: Why Attention-Seeking Is the Defining Need of Our Times," *Guardian* (UK edition), February 5, 2018, https://www.theguardian.com/society/2018/feb/05/crimes-of-attention-stalkers-killers-jihadists-longing.

4. World 5 List, "12 Shockingly Bizarre Selfie Deaths," video, 12:05, January 6, 2017, https://web.archive.org/web/20170810152559/https://www.youtube.com/watch?v=2JvXQmL26Zo; Pramod Madhav, "Chennai: Bakery Fire Injures 48 Onlookers Standing Too Close, Taking Selfies," *India Today*, July 17, 2017, https://www.indiatoday.in/india/story/chennai-bakery-fire-onlookers-injured-selfie-craze-1024652-2017-07-16.

5. Ishita Mishra, "Selfie in Front of Running Train Costs Three College-Goers Their Life," *Times of India*, January 27, 2015, https://timesofindia.indiatimes.com/india/Selfie-in-front-of-running-train-costs-three-college-goers-their-life/articleshow/46025185.cms; "Snuff Selfie: 21yo Russian Woman Shoots Self in Head Posing for Photo," *Russia Today*, May 23, 2015, https://www.rt.com/news/261445-selfie-sticks-deadly-fatal; Jim Goad, "Snuffies: 18 of the Most Horrifying Selfie-Related Tragedies," *Thought Catalog*, February 10, 2016, https://thoughtcatalog.com/jim-goad/2016/02/snuffies-18-of-the-most-horrifying-selfie-related-tragedies; "Pilot Taking 'Selfies' before Fatal Colorado Crash: Report," Reuters, February 3, 2015, reuters.com/article/us-usa-crash-colorado/pilot-taking-selfies-before-fatal-colorado-crash-report-idUSKBN0L720720150203; Isobel Markham, "Polish Couple Killed in Cliff Fall after Posing for 'Selfies,'" *Telegraph*, video, August 11, 2014, https://web.archive.org/web/20140811220204/http://www.telegraph.co.uk/news/worldnews/europe/poland/11024800/Polish-couple-killed-in-cliff-fall-after-posing-for-selfies.html.

6. Cleve R. Wootson Jr., "YouTube Daredevils Filmed Dangerous Stunts for Clicks—Then Died Going over a Waterfall," *Washington Post*, July 7, 2018, washingtonpost.com/news/the-intersect/wp/2018/07/07/youtube-daredevils-filmed-dangerous-stunts-for-clicks-then-died-going-over-a-waterfall.

7. Cara Buckley, "New Oscar Rules Aim to Avoid Another Envelope Mix-Up," *New York Times*, January 22, 2018, nytimes.com/2018/01/22/movies/oscar-rules-change-mixup.html; Stefan van der Stigchel, "Dangers of Divided Attention," *American Scientist*, January–February 2021, 46–53.

8. Van der Stigchel, "Dangers of Divided Attention"; Jeffrey H. Kunzekoff and Scott Titsworth, "The Impact of Mobile Phone Usage on Student Learning," *Communication Education* 62, no. 3 (2013): 233–252; Eyal Ophir, Clifford Nass, and Anthony D. Wagner, "Cognitive Control in Media Multitaskers," *Proceedings of the National Academy of Sciences* 106, no. 37 (2009): 15583–15587.

9. Zlatan Krizan and Anne D. Herlache, "The Narcissism Spectrum Model: A Synthetic View of Narcissistic Personality," *Personality and Social Psychology Review* 22, no. 1 (2017), https://doi.org/10.1177/1088868316685018; Alison Miller, "The 50 Shades of Social Rejection: The Role of Rejection-Sensitivity in Everyday Exclusion Experiences" (PowerPoint presentation, Symposium on Undergraduate Research and Creative Expression, Iowa State University, Ames, April 14, 2015), https://dr.lib.iastate.edu/entities/publication/743c6e8b-12a9-49f5-9862-83e433df4d1a.

10. *The Social Dilemma*, directed by Jeff Orlowski, Netflix, 2020.

11. Susruthi Rajanala, Mayra B. C. Maymone, and Neelam A. Vashi, "Selfies—Living in the Era of Filtered Photographs," *JAMA Facial Plastic Surgery* 20, no. 6 (2018): 443–444, https://doi.org/10.1001/jamafacial.2018.0486; "Body Dysmorphic Disorder,"

Cleveland Clinic, updated January 11, 2023, https://my.clevelandclinic.org/health/diseases/9888-body-dysmorphic-disorder.

12. Daria Hamrah, "What Do My Cosmetic Surgery Patients Want? To Look Better in Selfies," *Washington Post*, March 13, 2019, https://www.washingtonpost.com/outlook/my-cosmetic-surgery-patients-want-one-thing-to-look-better-in-selfies/2019/03/13/f8114f44-4429-11e9-9726-50f151ab44b9_story.html.

13. American Academy of Facial Plastic and Reconstructive Surgery, "As Americans Return to the Office, AAFPRS Unveils Aesthetic Trends from Annual Facial Plastic Surgery Survey," press release, February 13, 2023, https://aafprs.org/Media/Press_Releases/New-Trends-in-Facial-Plastic-Surgery.aspx.

14. Daria Hamrah, "What Do My Cosmetic Surgery Patients Want? To Look Better in Selfies," *Washington Post*, March 13, 2019, https://wapo.st/2TQPr6A.

15. Peter Holley, "In Silicon Valley, Some Men Say Cosmetic Procedures Are Essential to a Career," *Washington Post*, January 9, 2020, https://washingtonpost.com/technology/2020/01/09/silicon-valley-some-men-say-cosmetic-procedures-are-essential-career.

16. Holley, "In Silicon Valley."

17. Roberto Bibas et al., "Cryoliposis for Body Sculpting," in *Lasers, Lights, and Other Technologies*, ed. Maria Claudia Almeida Issa and Bhertha Tamura (Cham, CH: Springer, 2018); Jörg Faulhaber et al., "Effective Noninvasive Body Contouring by Using a Combination of Cryolipolysis, Injection Lipolysis, and Shock Waves," *Journal of Cosmetic Dermatology* 18, no. 4 (2019): 1014–1019, https://doi.org/10.1111/jocd.12953.

18. Jack Kelly, "Google Settles Age Discrimination Lawsuit, Highlighting the Proliferation of Ageism In Hiring," *Forbes*, July 23, 2019, https://www.forbes.com/sites/jackkelly/2019/07/23/google-settles-age-discrimination-lawsuit-highlighting-the-proliferation-of-ageism-in-hiring/?sh=21807c9e5c67.

19. Jesse Fox and Margaret C. Rooney, "The Dark Triad and Trait Self-Objectification as Predictors of Men's Use and Self-Presentation Behaviors on Social Networking Sites," *Personality and Individual Differences* 76 (April 2015): 161–165; Raymond L. M. Lee, "Diagnosing the Selfie: Parody or Pathology?," *Third Text* 30, no. 3–4 (2017): 264–273, https://dx.doi.org/10.1080/09528822.2017.1315901.

20. Allyson Chiu, "'Never Approach Animals': Video Shows 9-Year-Old Girl Tossed in the Air by Charging Bison at Yellowstone," *Washington Post*, July 24, 2019, https://washingtonpost.com/nation/2019/07/24/bison-charges-flips-year-old-yellowstone-national-park-video; Jessica Mendoza, "Enough with the Bison Selfies at Yellowstone, Says National Park Service," *Christian Science Monitor*, July 26, 2015,

https://csmonitor.com/Technology/2015/0726/Enough-with-the-bison-selfies-at-Yellowstone-says-National-Park-Service.

21. Lindsey Bever, "A Woman Was Trying to Take a Selfie with a Jaguar When It Attacked Her, Authorities Say," *Washington Post*, March 10, 2019, https://www.washingtonpost.com/science/2019/03/10/woman-was-trying-take-selfie-with-jaguar-when-it-attacked-her-authorities-say.

22. Roman Polotovsky et al., "Head and Neck Injuries Associated with Cell Phone Use," *JAMA Otolaryngology–Head and Neck Surgery* 146, no. 2 (2020): 122–127, https://doi.org/10.1001/jamaoto.2019.3678.

23. National Safety Council, "Distracted Walking Injuries on the Rise; 52 Percent Occur at Home" (2015), doi: nsc.org/Connect/NSCNewsReleases/Lists/Posts/Post.aspx. See also "Watch Where You're Going!" (May 1, 2017), https://www.nsc.org/community-safety/safety-topics/pedestrian-safety.

24. Max Benwell, "Now Liverpool Is Getting a Fast Walking Lane, Here's What Else Our Cities Need," *Independent*, November 5, 2015, https://independent.co.uk/voices/now-liverpool-is-getting-a-fast-walking-lane-heres-what-else-our-cities-need-a6722946.html; Rob Pegoraro, "Cellphone Talkers Get Their Own Sidewalk Lane in D.C.," *Yahoo Finance*, July 17, 2014, https://finance.yahoo.com/news/cellphone-talkers-get-their-own-sidewalk-lane-in-d-c-92080566744.html.

25. Tom Dingus and Mindy Buchanan-King, *Survive the Drive*, 2nd ed. (Blacksburg: Virginia Tech Transportation Institute, 2020), https://publishing.vt.edu/site/books/m/10.21061/survive-the-drive; Michael Laris, "Thomas Dingus Knows How to Keep Americans Alive on the Road, but Many Are Not Listening," interview, *Washington Post*, May 22, 2021, https://washingtonpost.com/transportation/2021/05/22/thomas-dingus-knows-how-keep-americans-alive-road-many-are-not-listening.

26. Hisashi Murakami et al., "Mutual Anticipation Can Contribute to Self-Organization in Human Crowds," *Science Advances* 7, no. 12 (2021), https://doi.org/10.1126/sciadv.abe7758.

27. Olga Mecking, *Niksen: Embracing the Dutch Art of Doing Nothing* (Boston: Houghton Mifflin Harcourt, 2021).

28. Sherry Turkle, *Reclaiming Conversation: The Power of Talk in a Digital Age* (New York: Penguin, 2015).

29. Eran Zaidel, "Interhemispheric Transfer in the Split Brain: Long-Term Status Following Complete Cerebral Commissurotomy," in *The Asymmetrical Brain*, ed. Kenneth Hugdahl and Richard J. Davidson (Cambridge, MA: MIT Press, 1994), 491–532; Richard E. Cytowic, "Disconnection Syndromes," in *The Neurological Side of Neuropsychology* (Cambridge, MA: MIT Press, 1996), 265–281. The former is a classic review paper.

Chapter 3

1. Hans Selye, *The Stress of Life*, rev. ed. (New York: McGraw-Hill, 1978); Hans Selye, "The Evolution of the Stress Concept," *American Scientist* 61, no. 6 (November–December 1973): 692–699.

2. Tom Lamont, "Can We Escape from Information Overload?," *1843 Magazine*, April 29, 2020, https://www.economist.com/1843/2020/04/29/can-we-escape-from-information-overload. See also the website of the Information Overload Research Group at iorgforum.org.

3. Gloria Mark, "How to Restore Our Dwindling Attention Spans," *Wall Street Journal*, January 6, 2023, https://www.wsj.com/articles/how-to-restore-our-attention-spans-11673031247.

4. Ian C. Fiebelkorn and Sabine Kastner, "A Rhythmic Theory of Attention," *Trends in Cognitive Science* 23, no. 2 (2019): 87–101; Randolph F. Helfrich et al., "Neural Mechanisms of Sustained Attention Are Rhythmic," *Neuron* 99, no. 4 (2018): 854–865; Jordana Cepelewicz and Quanta Magazine, "What a New Theory of Attention Says about Consciousness," *Atlantic*, September 29, 2019, https://www.theatlantic.com/science/archive/2019/09/how-brain-helps-you-pay-attention/598846.

5. Richard E. Cytowic, "Dopamine Fasting to 'Reset' Your Brain: Pros, Cons and Results," *The Fallible Mind* (blog), *Psychology Today* (online), August 25, 2020, https://www.psychologytoday.com/us/blog/the-fallible-mind/202008/dopamine-fasting-reset-your-brain-pros-cons-results.

6. Mitch Daniels, "Purdue President Daniels Sends Graduates Off with Advice for a Lonely Generation," speech, Purdue University, West Lafayette, IN, May 13, 2020, https://www.purdue.edu/newsroom/releases/2020/Q2/purdue-president-daniels-sends-graduates-off-with-advice-for-a-lonely-generation.html.

7. Jun Sawada et al., "TrueNorth Ecosystem for Brain-Inspired Computing: Scalable Systems, Software, and Applications," in *Proceedings of the International Conference for High Performance Computing, Networking, Storage and Analysis* (Piscatawy, NJ: IEEE Press, 2016).

8. Anne-Dominique Gindrat et al., "Use-Dependent Cortical Processing from Fingertips in Touchscreen Phone Users," *Current Biology* 25, no. 1 (2015): 109–116, https://www.sciencedirect.com/science/article/pii/S0960982214014870; Myriam Balerna and Arko Ghosh, "The Details of Past Actions on a Smartphone Touchscreen Are Reflected by Intrinsic Sensorimotor Dynamics," *npj Digital Medicine* 1, no. 4 (2018), https://www.nature.com/articles/s41746-017-0011-3.

9. Lauren L. Orefice, "Outside-In: Rethinking the Etiology of Autism Spectrum Disorders," *Science* 366, no. 6461 (2019), https://doi.org/10.1126/science.aaz3880; Lauren L. Orefice, "Peripheral Somatosensory Neuron Dysfunction: Emerging Roles

in Autism Spectrum Disorders," *Neuroscience* 445:120–129, https://doi.org/10.1016/j.neuroscience.2020.01.039.

10. D. A. Sbarra, J. L. Briskin, and R. B. Slatcher, "Smartphones and Close Relationships: The Case for Evolutionary Mismatch," *Perspectives on Psychological Science* 14, no. 4 (July 2019), https://pubmed.ncbi.nlm.nih.gov/31002764.

11. Aric Sigman, "Screen Dependency Disorders: A New Challenge for Child Neurology," *Journal of the International Child Neurology Association* 17, no. 119 (2017), https://jicna.org/index.php/journal/article/view/67.

12. Kate Conger, "Driver Charged in Uber's Fatal 2018 Autonomous Car Crash," *New York Times*, September 15, 2020, nytimes.com/2020/09/15/technology/uber-autonomous-crash-driver-charged.html.

13. National Transportation Safety Board, Collision between Vehicle Controlled by Developmental Automated Driving System and Pedestrian, investigation no. HWY18MH010, 2018, https://www.ntsb.gov/investigations/Pages/HWY18MH010.aspx.

14. Eliot Brown and Tim Higgins, "Operator in Fatal Self-Driving Uber Crash Was Streaming 'The Voice,' Police Say," *Wall Street Journal*, June 22, 2018, https://www.wsj.com/articles/operator-in-fatal-self-driving-uber-crash-was-streaming-the-voice-police-say-1529700387; Michael Laris, "Pedestrian in Self-Driving Uber Crash Probably Would Have Lived If Braking Feature Hadn't Been Shut Off, NTSB Documents Show," *Washington Post*, November 5, 2019, https://www.washingtonpost.com/local/trafficandcommuting/pedestrian-in-self-driving-uber-collision-probably-would-have-lived-if-braking-feature-hadnt-been-shut-off-ntsb-finds/2019/11/05/7ec83b9c-ffeb-11e9-9518-1e76abc088b6_story.html.

15. Michael Laris, "'A Terrible Idea': Multitasking State Senator Drives While Videoconferencing," *Washington Post*, May 6, 2021, https://www.washingtonpost.com/transportation/2021/05/06/andrew-brenner-video; David Tuller, "Using Hands-Free Cellphones When You Drive Is Not as Safe as You Think," *Washington Post*, March 22, 2021, https://www.washingtonpost.com/health/hands-free-cellphone-driving-risk/2021/03/22/780d7ae8-8112-11eb-ac37-4383f7709abe_story.html.

Chapter 4

1. Kathy Kleiman, *Proving Ground: The Untold Story of the Six WomenWwho Programmed the World's First Modern Computer* (Boston: Grand Central Publishing, 2022).

2. Richard E. Cytowic, "Synesthesia's Challenge to Brain-Inspired Computing" (speech, IBM Research Cognitive Systems Colloquium, "Brain-Inspired Computing," IBM Research–Almaden, San Jose, CA, November 12, 2014), video, 17:24, June 11, 2015, https://www.youtube.com/watch?v=tAtmNYBObkw; Dharmendra S. Modha,

"Brain-Inspired Computing: A Decade-Long Journey" (speech, IBM Research Cognitive Systems Colloquium, "Brain-Inspired Computing," IBM Research–Almaden, San Jose, CA, November 12, 2014), video, 20:27, December 16, 2014, https://www.youtube.com/watch?v=qE4kQh_30bA; Paul A. Merolla et al., "A Million Spiking-Neuron Integrated Circuit with a Scalable Communication Network and Interface," *Science* 345, no. 6197 (2014): 668–673.

3. The binary computer analogy is a particularly poor one. Rather than either "on" or "off" states, some neurons can selectively release more than one transmitter; excitatory and inhibitory receptors can and do exist for the same transmitter; at least fourteen brain serotonin receptors exist, as do a number for dopamine. We once thought that the synapse was the whole story; now peptides, hormones, and other molecules are known to act as signal carriers in what is called "volume transmission." If you think of the hard wiring of axons and synapses being like a train going down a track, then volume transmission is the train leaving the track. Far from being binary, brain circuitry is multiplex, parallel, and massively recursive.

4. Liqun Luo, "The Brain Achieves Its Computational Power through a Massively Parallel Architecture," in *Think Tank: Forty Neuroscientists Explore the Biological Roots of Human Experience*, ed. David J. Linden (New Haven, CT: Yale University Press, 2018), 82–87; Mihaly Csikszentmihalyi and Jeanne Nakamura, "Effortless Attention in Everyday Life: A Systematic Phenomenology," in *Effortless Attention: A New Perspective in the Cognitive Science of Attention and Action*, ed. Brian Burya (Cambridge, MA: MIT Press, 2010); Daniel J. Levitin, *The Organized Mind: Thinking Straight in the Age of Information Overload* (New York: Dutton, 2014); Terrence J. Sejnowski, personal communication with author at Karles conference, U.S. Navy Labs, October 11, 2018.

5. David Marchese, "Cal Newport Says the Digital Workplace Is a Disaster," interview, *New York Times Magazine*, January 29, 2023, 11–13.

6. Clifford Nass, "Stopping Multitasking Is Mostly a Losing Battle," interview, *Frontline*, PBS, December 1, 2009, http://www.pbs.org/wgbh/pages/frontline/digitalnation/interviews/nass.html.

7. Clifford I. Nass and Byron Reeves, *The Media Equation: How People Treat Computers, Television, and New Media Like Real People and Places* (Cambridge: Cambridge University Press, 1996).

8. Melina R. Uncapher and Anthony D. Wagner, "Minds and Brains of Media Multitaskers: Current Findings and Future Directions," *Proceedings of the National Academy of Sciences* 115, no. 40 (2018): 9889–9896, https://doi.org/10.1073/pnas.1611612115; Kevin P. Madore et al., "Memory Failure Predicted by Attention Lapsing and Media Multitasking," *Nature* 587, no. 7832 (2020): 87–91, https://doi.org/10.1038/s41586-020-2870-z; Pornchada Srisinghasongkram et al., "Effect of Early Screen Media Multitasking on Behavioural Problems in School-Age Children," *European Child and Adolescent Psychiatry* 30, no. 5 (2021): 1–17.

9. Cristopher P. Bonafide et al., "Association between Mobile Telephone Interruptions and Medication Administration Errors in a Pediatric Intensive Care Unit," *JAMA Pediatrics* 174, no. 2 (2020): 162–169, https://doi.org/10.1001/jamapediatrics.2019.5001.

10. See, for example, Charles Krauthammer on precisely these points, and how "a heretofore-unheard-of profession has been invented—the 'scribe' who just enters the data so the doctor can actually do doctoring." Krauthammer, "Why Doctors Quit, Chapter 2," *Washington Post*, June 4, 2015, https://www.washingtonpost.com/opinions/why-doctors-quit-chapter-2/2015/06/04/1b2de91c-0ade-11e5-9e39-0db921c47b93_story.html.

11. Richard E. Cytowic, "Lessons in the Art of Dying," *Washingtonian*, March 11, 2014, https://washingtonian.com/articles/health/lessons-in-the-art-of-dying.

12. Richard E. Cytowic, "What Percentage of Your Brain do you Use?," TED-Ed: Lessons Worth Sharing, video, 5:15, January 24, 2014, https://ed.ted.com/lessons/what-percentage-of-your-brain-do-you-use-richard-e-cytowic.

13. Roger Shattuck, *The Forbidden Experiment: The Story of the Wild Boy of Aveyron* (New York: Farrar, Straus and Giroux, 1980).

14. Susan Curtiss et al., "The Linguistic Development of Genie," *Language* 50, no. 3 (1974): 528–554; Susan Curtiss, "Genie: A Psycholinguistic Study of a Modern-Day 'Wild Child,'" *Perspectives in Neurolinguistics and Psycholinguistics* (New York: Academic Press, 1977); Leonard L. LaPointe, "Feral Children," *Journal of Medical Speech-Language Pathology* 13, no. 1 (March 2005), https://go.gale.com/ps/anonymous?id=GALE%7CA131004083&sid; Victoria Fromkin et al., "The Development of Language in Genie: A Case of Language Acquisition beyond the 'Critical Period,'" *Brain and Language* 1, no. 1 (January 1974): 81–107.

15. Miguel Nicolelis, *The True Creator of Everything: How the Human Brain Shaped the Universe as We Know It* (New Haven, CT: Yale University Press, 2019).

16. Simon B. Laughlin and Terrence J. Sejnowski, "Communication in Neuronal Networks," *Science* 301, no. 5641 (2003): 1870–1874, https://doi.org/10.1126/science.1089662.

17. I distinguish maintenance of the brain's physical structure as a steady baseline cost, while operational costs cover perception, thinking, emotion, and action.

18. Robert G. Shulman et al., "Energetic Basis of Brain Activity: Implications for Neuroimaging," *Trends in Neuroscience* 27, no. 8 (August 2004), 489–495.

19. Laughlin and Sejnowski, "Communication in Neural Networks."

20. Kim Armstrong, "Interoception: How We Understand Our Body's Inner Sensations," *Observer*, September 25, 2019, https://www.psychologicalscience.org/observer/interoception-how-we-understand-our-bodys-inner-sensations.

Chapter 5

1. Clifford I. Nass and Byron Reeves, *The Media Equation: How People Treat Computers, Television, and New Media Like Real People and Places* (Cambridge: Cambridge University Press, 1996); Clifford Nass, Jonathan Steuer, and Ellen R. Tauber, "Computers Are Social Actors," in *Proceedings of the SIGCHI Conference on Human Factors in Computing Systems* (Boston: ACM 1994), 72–78.

2. Gloria Mark, "How to Restore Our Dwindling Attention Spans," *Wall Street Journal*, January 6, 2023, https://www.wsj.com/articles/how-to-restore-our-attention-spans-11673031247; Sherry Turkle, *Alone Together: Why We Expect More from Technology and Less from Each Other* (New York: Basic Books, 2012); Jonathan Haidt and Tobias Rose-Stockwell, "The Dark Psychology of Social Networks," *Atlantic*, December 2019, https://theatlantic.com/magazine/archive/2019/12/social-media-democracy/600763.

3. Giosuè Baggio, *Neurolinguistics* (Cambridge, MA: MIT Press, 2022).

4. Miguel Nicolelis, *The True Creator of Everything: How the Human Brain Shaped the Universe as We Know It* (New Haven, CT: Yale University Press, 2019). I am referring to myelination especially in what used to be called the tertiary association areas, and more recently in terms of topological organization are called heteromodal isocortex.

5. Impulses along small unmyelinated nerve fibers with a diameter of 0.2–1.5 μm travel at about 1 m/sec (2.2 mph), whereas large, myelinated fibers conduct the same impulse at 120 m/sec (460 mph).

6. Patrick Long and Gabriel Corfas, "To Learn Is to Myelinate," *Science* 346, no. 6207 (2014): 298–299; Teal Burrell, "Brain Boosting: It's Not Just Grey Matter That Matters," *New Scientist*, February 18, 2015, 30–33; Alison C. Lloyd and Beth Stevens, "Editorial Overview: Glial Biology," *Current Opinion in Neurobiology* 47: iv–vi.

7. Matthias Brand, "Can Internet Use Become Addictive?," *Science* 376, no. 6595 (2022): 798–799, https://doi.org/10.1126/science.abn4189.

8. Julie Jargon, "Teens Smuggle Burner Phones to Defy Parents," *Wall Street Journal*, May 14, 2019, https://www.wsj.com/articles/teens-smuggle-burner-phones-to-defy-parents-11557826201.

9. Adrian F. Ward et al., "Brain Drain: The Mere Presence of One's Own Smartphone Reduces Available Cognitive Capacity," *Journal of the Association for Consumer Research* 2, no. 2 (April 3, 2017): 140–154, https://doi.org/10.1086/691462; Faria Sana, Tina Weston, and Nicholas J. Cepeda, "Laptop Multitasking Hinders Classroom Learning for Both Users and Nearby Peers," *Computers and Education*, 62 (2013): 24–31; Clay Shirky, "Why I Just Asked My Students to Put Their Laptops Away," *Medium*, September 8, 2014, medium.com/@cshirky/why-i-just-asked-my-students-to-put-their-laptops-away-7f5f7c50f368.

10. Kristen Duke et al., "Having Your Smartphone Nearby Takes a Toll on Your Thinking," *Harvard Business Review*, March 20, 2018, https://hbr.org/2018/03/having-your-smartphone-nearby-takes-a-toll-on-your-thinking; Ryan J. Dwyer, Kostadin Kushlev, and Elizabeth W. Dunn, "Smartphone Use Undermines Enjoyment of Face-to-Face Social Interactions," *Journal of Experimental Social Psychology*, 78 (2018): 233–239, https://doi.org/10.1016/j.jesp.2017.10.007.

11. David A. Sbarra, Julia L. Briskin, and Richard B. Slatcher, "Smartphones and Close Relationships: The Case for an Evolutionary Mismatch," *Perspectives on Psychological Science* 14, no. 4 (2019), https://doi.org/10.1177/1745691619826535.

12. Linda S. Pagani, Marie Josée Harbec, and Tracie A. Barnett, "Prospective Associations between Television in the Preschool Bedroom and Later Bio-Psycho-Social Risks," *Pediatric Research* 85, no. 7 (June 2019): 967–973, https://doi.org/10.1038/s41390-018-0265-8.

13. Alixandra Barasch et al., "Photographic Memory: The Effects of Volitional Photo Taking on Memory for Visual and Auditory Aspects of an Experience," *Psychological Science* 28, no. 8 (2017): 1056–1066, https://doi.org/10.1177/095679761769486.

14. Peter Sokol-Hessner and Robb B. Rutledge, "The Psychological and Neural Basis of Loss Aversion," *Current Directions in Psychological Science* 28, no. 1 (2018): 20–27, https://doi.org/10.1177/0963721418806510.

15. Various methods of speed reading work to a degree because they fully focus one's attention on reading and filter out distractions. And disclosure: my sister and I took Evelyn Wood's course in the 1960s. We stopped using her methods shortly thereafter.

16. Jonathan Haber, *Critical Thinking* (Cambridge, MA: MIT Press, 2020).

17. Christopher Mims, "Say No to the Distraction-Industrial Complex," *Wall Street Journal*, June 29, 2014, https://online.wsj.com/articles/say-no-to-the-distraction-industrial-complex-1404081926.

18. Matthew B. Crawford, *The World beyond Your Head: On Becoming an Individual in an Age of Distraction* (New York: Farrar, Straus and Giroux, 2015); James Miller, "Intensifying Mediatization: Everyware Media," in *Mediatized Worlds* (New York: Palgrave Macmillan, 2014), 107–122.

19. Marshall McLuhan, *Understanding Media: The Extensions of Man* (Cambridge, MA: MIT Press, 1994).

20. McLuhan, *Understanding Media*.

21. Gregory S. Berns et al., "Short- and Long-Term Effects of a Novel on Connectivity in the Brain," *Brain Connectivity* 3, no. 6 (2013): 590–600; Paula Leverage et al., eds., *Theory of Mind and Literature* (West Lafayette, IN: Purdue University Press, 2011);

David Comer Kidd and Emanuele Castano, "Reading Literary Fiction Improves Theory of Mind," *Science* 342, no. 6156 (2013): 377–380, https://doi.org/10.1126/science.1239918.

22. "William Styron, The Art of Fiction," interview by Peter Matthiessen and George Plimpton, *Paris Review* 5 (Spring 1954), https://theparisreview.org/interviews/5114/the-art-of-fiction-no-5-william-styron.

23. Glucose is the brain's sole source of food except in times of fasting or starvation. Then it burns ketones, a water-soluble fatty acid produced in the liver by cannibalizing fat and muscle tissue that helped our ancestors survive times of food scarcity.

24. Eric Schmidt, quoted at BrainyQuote.com, 2024, https://www.brainyquote.com/quotes/eric_schmidt_557874, accessed January 14, 2024.

25. Lizzie Widdicombe, "The Zuckerberg Bump," *New Yorker*, January 12, 2015, https://www.newyorker.com/magazine/2015/01/19/zuckerberg-bump.

26. Adam Fisher, "Sam Bankman-Fried Has a Savior Complex—and Maybe You Should Too," Sequoia Capital, September 22, 2022, archive.ph/Yg64e#selection-1695.320-1707.251.

27. Thomas Chatterton Williams, "The People Who Don't Read Books," *Atlantic*, January 25, 2023, https://www.theatlantic.com/ideas/archive/2023/01/kanye-west-sam-bankman-fried-books-reading/672823.

Chapter 6

1. "Borg Collective," *The Borg Collective Star Trek Wiki*, last updated August 21, 2020, https://borgcollective.fandom.com/wiki/Borg_Collective; Anthony Rotolo, "Are The Borg the Ultimate Social Network?," StarTrek.com, July 11, 2011, https://www.startrek.com/article/is-the-borg-the-ultimate-social-network.

2. I am not conflating visual and auditory noise with screen use per se. For example, I can quietly read an article on my iPad while enjoying the silence of my den.

3. Cal Newport, *A World without Email: Reimagining Work in an Age of Communication Overload* (New York: Penguin, 2021).

4. Richard Cohen, "The Evil Coming into Your Home via Cable and Internet," *Washington Post*, October 10, 2017, https://washingtonpost.com/opinions/the-evil-entering-your-home-via-cable-and-internet/2017/10/09/ebf1aa2e-ad03-11e7-be94-fabb0f1e9ffb_story.html.

5. Eliot Hearst, "Psychology and Nothing," *American Scientist* 79, no. 5 (September–October 1991): 432–443; Harold I. Lief, "Silence as Intervention in Psychotherapy," *American Journal of Psychoanalysis* 22, no. 1 (1962): 80–83.

6. Alan Fletcher, *The Art of Looking Sideways* (London: Phaidon, 2001).

7. Leonard Koren, *Wabi-Sabi for Artists, Designers, Poets and Philosophers* (Berkeley, CA: Stone Bridge Press, 1994); Leonard Koren, *Wabi-Sabi: Further Thoughts* (Point Reyes, CA: Imperfect Publishing, 2015).

8. Koren, *Wabi-Sabi for Artists, Designers, Poets and Philosophers*; Koren, *Wabi-Sabi: Further Thoughts*; Stephen Treffinger, "Could Japanese Design Be the Cure for Your Tech Addiction?," *Wall Street Journal*, April 25, 2015, https://www.wsj.com/articles/could-japanese-design-be-the-cure-for-your-tech-addiction-1429897425.

9. Anna Lucia Spear King et al., "Nomophobia: Dependency on Virtual Environments or Social Phobia?," *Computers in Human Behavior* 29, no. 1 (2013): 140–144.

10. David Sax, *The Revenge of Analog: Real Things and Why They Matter* (New York: PublicAffairs, 2016).

11. Charles Spence, "The Multisensory Experience of Handling and Reading Books," *Multisensory Research* 33, no. 8 (2020): 902–928, https://doi.org/10.1163/22134808-bja10015.

12. Susan Greenfield et al., "Technology, the Brain, and Audience Expectation: Vying for Attention in 'Generation Elsewhere'" (debate, Phillips Collection, Washington, DC, October 27, 2016).

13. Erin C. Westgate, Timothy D. Wilson, and Daniel T. Gilbert, "With a Little Help for Our Thoughts: Making It Easier to Think for Pleasure," *Emotion* 17, no. 5 (2017): 828–839.

14. Thích Nhất Hạnh, *Silence: The Power of Quiet in a World Full of Noise* (New York: Random House, 2015).

15. "Scientific Review: The Benefits of Forest Bathing," Guides and Programs, Association of Nature and Forest Therapy, https://www.natureandforesttherapy.earth/about/the-science.

16. Qing Li, "Forest Bathing Enhances Human Natural Killer Activity and Expression of Anti-cancer Proteins," *International Journal of Immunopathology and Pharmacology* 20, no. 2 (suppl. 2, April–June 2007): 3–8; Allison Aubrey, "Forest Bathing: A Retreat to Nature Can Boost Immunity And Mood," *Morning Edition*, NPR, July 17, 2017, npr.org/sections/health-shots/2017/07/17/536676954/forest-bathing-a-retreat-to-nature-can-boost-immunity-and-mood.

17. Rahawa Haile, "'Forest Bathing': How Microdosing on Nature Can Help with Stress," *Atlantic*, June 30, 2017, https://www.theatlantic.com/health/archive/2017/06/forest-bathing/532068; Byeongsang Oh et al., "Health and Well-being Benefits of Spending Time in Forests: Systematic Review," *Environmental Health and Preventative Medicine* 22, no. 71 (2017). See also the *Washington Post*, May 7, 2017.

18. Mary Helen Immordino-Yang, Joanna A. Christodoulou, and Vanessa Singh, "Rest Is Not Idleness: Implications of the Brain's Default Mode for Human Development and Education," *Perspectives on Psychological Science* 7, no. 4 (2012): 352–364, https://doi.org/10.1177/1745691612447308.

19. Marcia Davis, "Luxury 2.0," April 27, 2017, https://www.washingtonpost.com/sf/style/2017/04/27/luxury-2-0.

20. Gregory N. Bratman et al., "Nature Experience Reduces Rumination and Subgenual Prefrontal Cortex Activation," *Psychological and Cognitive Sciences* 112, no. 28 (2015): 8567–8572, https://doi.org/10.1073/pnas.1510459112.

21. Gillian Flaccus and Hillel Italie, "Ursula K. Le Guin, Best-Selling Science Fiction Author, Dies at 88," *Chicago Tribune*, January 23, 2018, https://www.chicagotribune.com/entertainment/ct-ursula-k-le-guin-dies-20180123-story.html.

22. Ye Wen et al., "Medical Empirical Research on Forest Bathing (Shinrin-Yoku): A Systematic Review," *Environmental Health and Preventative Medicine* 24, no. 70 (2019): https://doi.org/10.1186/s12199-019-0822-8; Sandrine Mathias et al., "Forest Bathing: A Narrative Review of the Effects on Health for Outdoor and Environmental Education Use in Canada," *Journal of Outdoor and Environmental Education* 23, no. 3 (2020): 309–321; Margaret M. Hansen, Reo Jones, and Kirsten Tocchini, "Shinrin-Yoku (Forest Bathing) and Nature Therapy: A State-of-the-Art Review," *International Journal of Environmental Research and Public Health* 14, no. 8 (2017): 851; Aaron J. Schwartz et al., "Visitors to Urban Greenspace Have Higher Sentiment and Lower Negativity on Twitter," *People and Nature* 1, no. 4 (December 2019): 476–485, https://doi.org/10.1002/pan3.10045.

23. Virginia Woolf, *Moments of Being: Unpublished Autobiographical Writings*, edited by Jeanne Schulkind (New York: Harcourt Brace Jovanovich, 1976).

24. Alana G. DeLoach, Jeff P. Carter, and Jonas Braasch, "Tuning the Cognitive Environment: Sound Masking with 'Natural' Sounds in Open-Plan Offices," *Journal of the Acoustical Society of America* 137, no. 4 (April 2015): 2291.

25. Joseph W. Newbold et al., "Using Nature-Based Soundscapes to Support Task Performance and Mood," in *Proceedings of the 2017 CHI Conference Extended Abstracts on Human Factors in Computing Systems* (New York: ACM, 2017), 2802–2809.

26. Robert S. Fisher et al., "Photic- and Pattern-Induced Seizures: A Review for the Epilepsy Foundation of America Working Group," *Epilepsia* 46, no. 9 (September 2005): 1426–1441; Sanae Yoshimoto et al., "Visual Discomfort and Flicker," *Vision Research* 138 (2017): 18–28.

27. Brad Lehman and Arnold J. Wilkins, "Designing to Mitigate Effects of Flicker in LED Lighting: Reducing Risks to Health and Safety," *IEEE Power Electronic Magazine* 1, no. 3 (September 2014): 18–26; Jay N. Borger, Reto Huber, and Arko Ghosh,

"Capturing Sleep-Wake Cycles by Using Day-to-Day Smartphone Touchscreen Interactions," *npj Digital Medicine* 2, no. 73 (2019).

28. Sarah M. Haig, Nicholas R. Cooper, and Arnold J. Wilkins, "Cortical Excitability and the Shape of the Haemodynamic Response," *NeuroImage*, 111 (2015): 379–384; Oliver Penacchio and Arnold J. Wilkins, "Visual Discomfort and the Spatial Distribution of Fourier Energy," *Vision Research* 108 (2015): 1–7; Arnold J. Wilkins, "A Physiological Basis for Visual Discomfort: Application in Lighting Design," *Lighting Research and Technology* 48, no. 1 (2015): 44–54; Arnold J. Wilkins, Jie Huang, and Yue Cao, "Visual Stress Theory and Its Application to Reading and Reading Tests," *Journal of Research in Reading* 27, no. 2 (2004): 152–162.

29. An Trong Dinh Le et al., "Discomfort from Urban Scenes: Metabolic Consequences," *Landscape and Urban Planning* 160 (2017): 61–68.

30. Arnold J. Wilkins, "Disturbing Vision," November 5, 2014, TEDx video, 13:15, youtu.be/GBOzv9HgoWM.

31. Jochem O. Klompmaker et al., "Associations of Greenness, Parks, and Blue Space with Neurodegenerative Disease Hospitalizations among Older US Adults," *JAMA Network Open* 5, no. 12 (2022): e2247664. See also the Park Prescription website at https://parkrx.org, and Robert Zarr, "'Why I Prescribe Nature'—In D.C., Pioneering Pediatricians Offer New Hope and Health through Park Rx," *New Nature Movement* (blog), Children and Nature Network, November 5, 2013, https://web.archive.org/web/20190311220418/https://www.childrenandnature.org/2013/11/05/why-i-prescribe-nature-in-d-c-pioneering-pediatricians-and-park-rx-offer-new-hope-and-health. Park Prescription programs coordinate with managers of public lands nationwide, and include Walk with a Doc, Rx2Move, and ParkRxAmerica.

32. "Tango Classes, Ukulele Lessons: The Rise of 'Social Prescriptions,'" *Economist*, February 15, 2018, https://www.economist.com/britain/2018/02/15/tango-classes-ukulele-lessons-the-rise-of-social-prescriptions.

33. Cathy Free, "This Town's Solution to Loneliness? The 'Chat Bench,'" *Washington Post*, July 17, 2019, https://www.washingtonpost.com/lifestyle/2019/07/17/this-towns-solution-loneliness-chat-bench.

34. Rachel T. Buxton et al., "Noise Pollution Is Pervasive in U.S. Protected Areas," *Science* 356, no. 3667 (2017): 531–533.

35. "Protected Areas," GAP Analysis Project, USGS, https://www.google.com/search?client=safari&rls=en&q=Protected+Areas+%7C+U.S.+Geological+Survey+(usgs.gov)&ie=UTF-8&oe=UTF-8.

36. Christopher J. Preston, *The Synthetic Age: Outdesigning Evolution, Resurrecting Species, and Reengineering Our World* (Cambridge, MA: MIT Press, 2018).

37. David Owen, "Is Noise Pollution the Next Big Public-Health Crisis?," *New Yorker*, May 13, 2019, https://www.newyorker.com/magazine/2019/05/13/is-noise-pollution-the-next-big-public-health-crisis; Bianca Bosker, "Why Everything Is Getting Louder," *Atlantic*, November 2019, https://www.theatlantic.com/magazine/archive/2019/11/the-end-of-silence/598366.

38. M. Simpson and R. Bruce, *Noise in America: The Extent of the Noise Problem*, Environmental Protection Agency report 550/9-81-101 (Washington, DC: EPA, September 1981); Scott Creel et al., "Snowmobile Activity and Glucocorticoid Stress Responses in Wolves and Elk," *Conservation Biology* 16, no. 3 (June 2002): 809–814.

39. M. Jackman, "How Far Can Sound Travel," Sound, *Britannica* (online), https://www.britannica.com/science/sound-physics/Impedance.

40. Erling Kagge, *Silence in the Age of Noise*, trans. Becky L. Crook (New York: Pantheon, 2017).

Chapter 7

1. Iain Mathieson et al., "Genome-Wide Patterns of Selection in 230 Ancient Eurasians," *Nature* 528, no. 7583 (2015): 499–503, https://doi.org/10.1038/nature16152.

2. Steven Pinker, email to the author, May 31, 2019; Gary Marcus, email to the author, May 24, 2019.

3. David A. Sbarra, Julia L. Briskin, and Richard B. Slatcher, "Smartphones and Close Relationships: The Case for an Evolutionary Mismatch," *Perspectives on Psychological Science* 14, no. 4 (2019), https://10.1177/1745691619826535.

4. Tristan Harris, "How a Handful of Tech Companies Control Billions of Minds Every Day," TED video, 16:52, April 28, 2017, https://www.tristanharris.com/#about.

5. Gabe Zichermann, "Addictive Tech Giants Cannot Be Trusted to Self-Regulate," *Medium*, January 16, 2018, https://gzicherm.medium.com/addictive-tech-giants-cannot-be-trusted-to-self-regulate-96e9a0bbdfaf; Gabe Zichermann, personal communication withthe author, Phillips Museum, May 23, 2019. See also "Dopamine" (website), Wayback Machine, https://web.archive.org/web/20210801084105/http://dopa.mn.

6. Haley Sweetland Edwards, "You're Addicted to Your Smartphone. This Company Thinks It Can Change That," *Time*, April 13, 2018, https://time.com/5237434/youre-addicted-to-your-smartphone-this-company-thinks-it-can-change-that.

7. Ray Downs, "Experts: Tech Companies Use 'Persuasive Technologies' That Harm Children," *United Press International*, August 8, 2018, https://www.upi.com/Top_News/US/2018/08/08/Experts-Tech-companies-use-persuasive-technologies-that-harm-children/1521533778564; Behavior Design Lab, Stanford University, https://behav

iordesign.stanford.edu; Game Center, American University, https://www.american.edu/centers/gamecenter.

8. "Our Letter to the APA," Screen Time Action Network, Fairplay, screentimenetwork.org/apa.

9. Richard Freed, "The Tech Industry's War on Kids," *Medium*, March 12, 2018, https://medium.com/@richardnfreed/the-tech-industrys-psychological-war-on-kids-c452870464ce.

10. Jeff Horwitz et al., "The Facebook Files," *Wall Street Journal*, September 13–December 29, 2021, https://www.wsj.com/articles/the-facebook-files-11631713039.

11. American Psychological Association, "Stress in America: The State of Our Nation," press release, November 1, 2017, https://apa.org/news/press/releases/stress/2017/state-nation.pdf; Mark Aguiar et al., "Leisure Luxuries and the Labor Supply of Young Men," NBER Working Paper 23552 (Cambridge, MA: National Bureau of Economic Research, June 21, 2017), https://www.nber.org/papers/w23552.

12. Robert L. Isaacson, *The Limbic System* (New York: Plenum Press, 1982); Russel T. Geary, ed., *The Limbic System: Anatomy, Functions and Disorders* (Hauppage, NY: Nova Science Publishers, 2014).

13. Steven Pinker, *Rationality: What It Is, Why It Seems Scarce, Why It Matters* (New York: Viking, 2021), 1.

14. Antonio Damasio has written many popular accounts of this. See, for example, Damasio, *Descartes' Error: Emotion, Reason, and the Human Brain* (New York: Putnam, 1994), and Damasio, *The Feeling of What Happens: Body and Emotion in the Making of Consciousness* (New York: Harcourt, 1999).

15. Joseph Weizenbaum, *Computer Power and Human Reason: From Judgment to Calculation* (San Francisco: W. H. Freeman, 1976).

16. Richard M. Held and Alan Hein, "Movement-Produced Stimulation in the Development of Visually Guided Behavior," *Journal of Comparative and Physiological Psychology* 56, no. 5 (November 1963): 872–876.

17. Patricia K. Kuhl, "Is Speech Learning 'Gated' by the Social Brain?," *Developmental Science* 10, no. 1 (January 2007): 110–120, https://doi.org/10.1111/j.1467-7687.2007.00572.x; Nairán Ramírez-Esparza, Adrián García-Sierra, and Patricia K. Kuhl, "Look Who's Talking: Speech Style and Social Context in Language Input to Infants Are Linked to Concurrent and Future Speech Development," *Developmental Science* 17, no. 6 (November 2014): 880–891, https://doi.org/10.1111/desc.12172; Patricia K. Kuhl, "Early Language Learning and the Social Brain," *Cold Spring Harbor Symposia on Quantitative Biology* 79 (2014): 211–220; Patricia K. Kuhl, TEDx video, 10:01, October 10, 2010, https://www.ted.com/talks/patricia_kuhl_the_linguistic_genius_of_babies.

Chapter 8

1. Guillermo Gallego et al., "Event-Based Vision: A Survey," *IEEE Transactions on Pattern Analysis and Machine Intelligence* 44, no. 1 (January 2022): 154–180.

2. George O. Abell, *Exploration of the Universe*, 2nd ed. (New York: Holt, Rinehart and Winston, 1969), chap. 2; Misty E. Vermaat et al., *Discovering Computers, Enhanced Edition* (Boston: Cengage Learning, 2014).

3. Vermaat et al., *Discovering Computers, Enhanced Edition*; Maria A. Contin et al., "Light Pollution: The Possible Consequences of Excessive Illumination on Retina," *Eye* 30, no. 2 (2015): 255–263, https://doi.org/10.1038/eye.2015.221; Javier Vincente-Tejedor et al., "Removal of the Blue Component of Light Significantly Decreases Retinal Damage after High Intensity Exposure," *PLoS ONE* 13, no. 3 (2018): e0194218, https://doi.org/10.1371/journal.pone.0194218.

4. Adrian F. Ward et al., "Brain Drain: The Mere Presence of One's Own Smartphone Reduces Available Cognitive Capacity," *Journal of the Association for Consumer Research* 2, no. 2 (April 3, 2017): 140–154, https://doi.org/10.1086/691462.

5. Alexandre Marois et al., "Eyes Have Ears: Indexing the Orienting Response to Sound Using Pupillometry," *International Journal of Psychophysiology* 123 (2018): 152–162.

6. Sean C. L. Deoni et al., "The COVID-19 Pandemic and Early Child Cognitive Development: A Comparison of Development in Children Born during the Pandemic and Historical References," preprint, submitted August 16, 2022, https://doi.org/10.1101/2021.08.10.21261846.

7. Shari Lewis, personal communication to author with Shari Lewis in Washington, DC, approx..1993.

8. Michael A. Persinger, "Religious and Mystical Experiences as Artifacts of Temporal Lobe Function: A General Hypothesis," *Perceptual and Motor Skills* 57, no. 3, part 2 (1983): 1255–1262; Michael A. Persinger and Katherine Makarec, "The Feeling of a Presence and Verbal Meaningfulness in Context of Temporal Lobe Function: Factor Analytic Verification of the Muses?," *Brain and Cognition* 20, no. 2 (1992): 217–226.

9. Pierre Gloor, "Role of the Human Limbic System in Perception, Memory, and Affect: Lessons from Temporal Lobe Epilepsy," in *The Limbic System: Functional Organization and Clinical Disorders*, ed. Benjamin K. Doane and Kenneth F. Livingston (New York: Raven Press, 1986), 159–169.

10. Kate Conger, "Google Removes 'Don't Be Evil' Clause from Its Code of Conduct," *Gizmodo*, May 18, 2018, https://gizmodo.com/google-removes-nearly-all-mentions-of-dont-be-evil-from-1826153393.

11. Marcus Aurelius, *Meditations: With Selected Correspondence*, trans. Robin Hard, with an introduction and notes by Christopher Gill (New York: Oxford University Press, 2011).

12. Russell B. Clayton, Glenn Leshner, and Anthony Almond, "The Extended iSelf: The Impact of iPhone Separation on Cognition, Emotion, and Physiology," *Journal of Computer-Mediated Communication* 20, no. 2 (March 2015): 119–135.

13. Nicholas Epley, "A Mind Like Mine: The Exceptionally Ordinary Underpinnings of Anthropomorphism," *Journal of the Association for Computer Research* 3, no. 4 (October 2018): 591–598, https://doi.org/10.1086/699516; Jessica E. Bodford, Virginia S. Y. Kwan, and David S. Sobota, "Fatal Attractions: Attachment to Smartphones Predicts Anthropomorphic Beliefs and Dangerous Behaviors," *Cyberpsychology, Behavior, and Social Networking* 20, no. 5 (2017): 320–326, https://doi.org/10.1089/cyber.2016.0500.

14. Larry Rosen and Keith N. Hampton, "Is Technology Making People Less Sociable?," *Wall Street Journal*, May 10, 2015, https://www.wsj.com/articles/is-technology-making-people-less-sociable-1431093491; Larry D. Rosen, Nancy Cheever, and L. Mark Carrier, eds., *The Wiley Handbook of Psychology, Technology, and Society* (Hoboken, NJ: Wiley-Blackwell, 2015).

15. Nancy A. Cheever et al., "Out of Sight Is Not Out of Mind: The Impact of Restricting Wireless Mobile Device Use on Anxiety Levels among Low, Moderate and High Users," *Computers in Human Behavior* 37 (2014): 290–297.

16. Joseph Reagle, "Following the Joneses: FOMO and Conspicuous Sociality," *First Monday* 20, no. 10 (October 2015), https://doi.org/10.5210/fm.v20i10.6064.

17. Reagle, "Following the Joneses."

Chapter 9

1. Anthony Damasio, *Self Comes to Mind: Constructing the Conscious Brain* (New York: Pantheon Books, 2010); Michael S. A. Graziano, and Taylor W. Webb, "From Sponge to Human: The Evolution of Consciousness," in *Evolution of Nervous Systems*, ed. Jon H. Kaas, 2nd ed. (Oxford: Elsevier, 2017), 547–554.

2. Zoltan Torey, *The Conscious Mind* (Cambridge, MA: MIT Press, 2014).

3. Richard E. Cytowic, "A Link between Screen Exposure and Autism-Like Symptoms," *The Fallible Mind* (blog) *Psychology Today* (online), December 11, 2020, https://www.psychologytoday.com/us/blog/the-fallible-mind/202012/link-between-screen-exposure-and-autism-symptoms; Susan Greenfield, *Mind Change: How Digital Technologies Are Leaving Their Mark on Our Brains* (New York: Random House, 2015); Karen Frankel Heffler and Leonard M. Oestreicher, "Causation Model of Autism: Audiovisual Brain Specialization in Infancy Competes with Social Brain Networks,"

Medical Hypotheses 91 (June 2016): 144–122, https://doi.org/10.1016/j.mehy.2015.06.019.

4. David Brin, "Neoteny and Two-Way Sexual Selection in Human Evolution: Paleo-Anthropological Speculation on the Origins of Secondary-Sexual Traits, Male Nurturing, and the Child as a Sexual Image," *Journal of Social and Evolutionary Systems* 18, no. 3 (1995): 257–276, https://doi.org/10.1016/1061-7361(95)90006-3. See also David Brin, "Neoteny and Two-Way Sexual Selection in Human Evolution," Essays on Science and Technology, David Brin (website), http://davidbrin.com/nonfiction/neoteny1.html.

5. Herman Pontzer, "Economy and Endurance in Human Evolution," *Current Biology* 27, no. 12 (2017): R613–R621, https://doi.org/10.1016/j.cub.2017.05.031.

6. Ann Gibbons, "The Calorie Counter," *Science* 375, no. 6582 (February 18, 2022): 710–713, https://science.org/doi/pdf/10.1126/science.ada1185. To be more precise, the issue isn't whether keeping a trait is costly but whether it is too costly and thus detrimental to reproduction. Retained traits without purpose in contemporary men are functioning breasts and the propensity to have more than twenty types of hernias.

7. Ville Renvall et al., "Imaging Real-Time Tactile Interaction with Two-Person Dual-Coil fMRI," preprint, submitted January 24, 2020, https://doi.org/10.1101/861252.

8. Sophie Curtis, "'iPotty' Named Worst Toy of the Year," *Telegraph*, December 20, 2013, https://www.telegraph.co.uk/technology/news/10531063/iPotty-named-worst-toy-of-the-year.html.

9. Amy Bastian, "Children's Brains Are Different," in *Think Tank: Forty Neuroscientists Explore the Biological Roots of Human Experience*, ed. David J. Linden (New Haven, CT: Yale University Press, 2018), 40–44; Robert Needleman et al., "Effectiveness of a Primary Care Intervention to Support Reading Aloud: A Multicenter Evaluation," *Ambulatory Pediatrics* 5, no. 4 (July–August 2005): 209–215; Dominic W. Massaro, "Reading Aloud to Children: Benefits and Implications for Acquiring Literacy before Schooling Begins," *American Journal of Psychology* 130, no. 1 (Spring 2017): 63–72; John S. Hutton et al., "Home Reading Environment and Brain Activation in Preschool Children Listening to Stories," *Pediatrics* 136, no. 3 (2015): 466–478.

10. Thomas R. Verny, *The Embodied Mind: Understanding the Mysteries of Cellular Memory, Consciousness, and Our Bodies* (New York: Pegasus, 2021).

11. David A. Sbarra, Julia L. Briskin, and Richard B. Slatcher, "Smartphones and Close Relationships: The Case for an Evolutionary Mismatch," *Perspectives on Psychological Science* 14, no. 4 (2019), https://10.1177/1745691619826535.

12. Matt Miles and Joe Clement, *Screen Schooled: Two Veteran Teachers Expose How Technology Overuse Is Making Our Kids Dumber* (Chicago Review Press, 2017); Matt

Miles, email to the author, July 12, 2018; James Silver, Andre Simons, and Sarah Craun, *A Study of the Pre-Attack Behaviors of Active Shooters in the United States between 2000 and 2013*, report prepared at the request of the Federal Bureau of Investigation, U.S. Department of Justice (Washington, DC, June 2018).

13. Joël Billieux et al., "Problematic Involvement in Online Games: A Cluster Analytic Approach," *Computers in Human Behavior* 43 (2015): 242–250.

14. Joan L. Luby et al., "Preschool Is a Sensitive Period for the Influence of Maternal Support on the Trajectory of Hippocampal Development," *Proceedings of the National Academy of Sciences* 113, no. 20 (2016): 5742–5747; John Bowlby, *A Secure Base: Parent-Child Attachment and Healthy Human Development* (New York: Basic Books, 2008); Louis Cozolino, *The Neuroscience of Human Relationships: Attachment and the Developing Social Brain*, 2nd ed. (New York: W. W. Norton, 2014); T. Berry Brazelton and Bertrand G. Cramer, *The Earliest Relationship: Parents, Infants and the Drama of Early Attachment* (London: Routledge, 2018); Susan Hart, *Brain, Attachment, Personality: An Introduction to Neuroaffective Development* (London: Karnac, 2008); Peter Lovenheim, *The Attachment Effect: Exploring the Powerful Ways Our Earliest Bond Shapes Our Relationships and Lives* (New York: TarcherPerigree, 2018).

15. Torsten N. Wiesel, "The Postnatal Development of the Visual Cortex and the Influence of Environment" (Stockholm: Nobel Foundation, 1981); David H. Hubel and Torsten N. Wiesel, *Brain and Visual Perception: The Story of a 25-Year Collaboration* (Oxford: Oxford University Press, 2004); Samuel Wang, "The Brain Develops through Conversation with Itself," in *Think Tank: Forty Neuroscientists Explore the Biological Roots of Human Experience*, ed. David J. Linden (New Haven, CT: Yale University Press: 2018).

16. Eric I. Knudsen, "Sensitive Periods in the Development of the Brain and Behavior," *Journal of Cognitive Neuroscience* 16, no. 8 (2004): 1412–1425; Sharon E. Fox, Pat Levitt, and Charles A. Nelson III, "How the Timing and Quality of Early Experiences Influence the Development of Brain Architecture," *Child Development* 81, no. 1 (January–February 2010): 28–40; Eric I. Knudsen, "Early Experience and Sensitive Periods," in *Fundamental Neuroscience*, ed. Larry Squire et al. (San Diego: Academic Press, 2014), 479–495; Samuel S.-H. Wang, Alexander D. Kloth, and Aleksandra Badura, "The Cerebellum, Sensitive Periods, and Autism," *Neuron* 83, no. 3 (2014): 518–532.

17. Wang, "The Brain Develops through Conversation with Itself."

18. Charles A. Nelson, Nathan A. Fox, and Charles H. Zeanah, *Romania's Abandoned Children: Deprivation, Brain Development, and the Struggle for Recovery* (Cambridge, MA: Harvard University Press, 2014); Melissa Fay Greene, "30 Years Ago, Romania Deprived Thousands of Babies of Human Contact," *Atlantic*, July–August 2020, https://www.theatlantic.com/magazine/archive/2020/07/can-an-unloved-child-learn-to-love/612253.

19. Eliot Marshall, "Child Neglect Erodes the Brain," *Science*, January 26, 2015, https://www.science.org/content/article/childhood-neglect-erodes-brain.

20. Nelson et al., "Romania's Abandoned Children"; Eliot Marshall, "An Experiment in Zero Parenting," *Science* 345, no. 6198 (2014): 752–754, https://doi.org/10.1126/science.345.6198.752; Real Battle Ministries, "Virtual Autism—Autism Recovery Claims in Romania," June 23, 2017, video, 2:18, youtu.be/ZRR3pQ_i42M; Charles A. Nelson, "Romanian Orphans Reveal Clues to Origins of Autism," *Spectrum News*, April 25, 2017, https://www.spectrumnews.org/opinion/viewpoint/romanian-orphans-reveal-clues-origins-autism.

21. William Wan, "What Separation from Parents Does to Children: 'The Effect Is Catastrophic,'" *Washington Post*, June 18, 2018, https://www.washingtonpost.com/national/health-science/what-separation-from-parents-does-to-children-the-effect-is-catastrophic/2018/06/18/c00c30ec-732c-11e8-805c-4b67019fcfe4_story.html.

22. Matt Richtel, "Attached to Technology and Paying a Price," *New York Times*, June 6, 2010, https://www.nytimes.com/2010/06/07/technology/07brain.html.

23. Matt Bluemink, "Socrates, Memory and the Internet," *Philosophy Now*, October/November 2017, https://philosophynow.org/issues/122/Socrates_Memory_and_The_Internet.

24. David Foster Wallace, "David Foster Wallace on Life and Work," *Wall Street Journal*, September 19, 2008, wsj.com/articles/SB122178211966454607.

25. Rebecca Tan, "Young Adults Use Their Phones 6 Hours a Day but Barely Touch Their News Apps, a Study Shows. Here's Why," September 5, 2019, https://www.washingtonpost.com/arts-entertainment/2019/09/05/young-adults-use-their-phones-hours-day-barely-touch-their-news-apps-study-shows-heres-why.

26. Jennifer Aaker and Andy Smith, *The Dragonfly Effect: Quick, Effective, and Powerful Ways to Use Social Media to Drive Social Change* (San Francisco, CA: Jossey-Bass, 2010).

27. Carrie James et al., "Digital Life and Youth Well-Being, Social Connectedness, Empathy, and Narcissism," *Pediatrics* 140, suppl. 2 (2017): S71–S75.

28. Joe Clement and Matt Miles, *Screen Schooled: Two Veteran Teachers Expose How Technology Overuse Is Making Our Kids Dumber* (Chicago: Chicago Review Press, 2017).

29. James et al., "Digital Life and Youth Well-Being"; Linda Wilbrecht, "Your Twelve-Year-Old Isn't Just Sprouting New Hair but Is Also Forming (and Being Formed by) New Neural Connections," in *Think Tank*, ed. David J. Linden (New Haven, CT: Yale University Press, 2018), 45–51.

30. Min Fu and Yi Zuo, "Experience-Dependent Structural Plasticity in the Cortex," *Trends in Neuroscience* 34, no. 4 (April 2011): 177–187.

Chapter 10

1. See, for example, FMI Applications, Fatigue Management International, https://fmiapplications.com, and "Fatigue Risk Management System," *Circadian Technologies*, accessed January 26, 2022, https://circadian.com.

2. Kirsten E. G. Sanders et al., "Targeted Memory Reactivation during Sleep Improves Next-Day Problem Solving," *Psychological Science* 30, no. 11 (2019), https://doi.org/10.1177/0956797619873344.

3. Matthew Walker, *Why We Sleep: Unlocking the Power of Sleep and Dreams* (New York: Scribner, 2018).

4. A. Roger Ekirch, *At Day's Close: A History of Nighttime* (Sydney: Hatchette, 2013); A. Roger Ekirch, "Segmented Sleep in Preindustrial Societies," *Sleep* 39, no. 3 (March 2016): 715–716; Emerson M. Wickwire, review of *The Sleep Revolution: Transforming Your Life, One Night at a Time* by Arianna Huffington, *Journal of Clinical Sleep Medicine* 13, no. 1 (January 2017): 145–146, https://doi.org/10.5664/jcsm.6412.

5. Thomas A. Wehr, "In Short Photoperiods, Sleep Is Biphasic," *Journal of Sleep Research* 1, no. 2 (June 1992): 103–107, https://doi.org/10.1111/j.1365-2869.1992.tb00019.x; Thomas A. Wehr and Frederick K. Goodwin, eds., *Circadian Rhythms in Psychiatry*, vol. 2 of *Psychobiology and Psychopathology* (Pacific Grove, CA: Boxwood Press, 1983); Daniel Aeschbach et al., "A Longer Biological Night in Long Sleepers Than in Short Sleepers," *Journal of Clinical Endocrinology and Metabolism* 88, no. 1 (January 2003): 26–30.

6. Sandra J. Kuhlman, L. Michon Craig, and Jeanne F. Duffy, "Introduction to Chronobiology," *Cold Spring Harbor Perspectives in Biology* 10 (2018): a033613, https://doi.org/10.1101/cshperspect.a033613; Anna Wirz-Justice, "Chronobiology Comes of Age," *Acta Psychiatrica Scandinavica* 136, no. 6 (December 2017): 531–533, https://onlinelibrary.wiley.com/doi/full/10.1111/acps.12828.

7. Thomas A. Wehr, "Photoperiodism in Humans and Other Primates: Evidence and Implications," *Journal of Biological Rhythms* 16, no. 4 (August 2001): 348–364.

8. Richard E. Cytowic, "Sleep: The Clean-Up Crew of a Dirty Mind," *The Fallible Mind* (blog), *Psychology Today* (online), February 19, 2014, https://www.psychologytoday.com/us/blog/the-fallible-mind/201402/sleep-the-clean-crew-dirty-mind.

9. Rolf Sinclair, "Astronomical Phenomena North of the Arctic Circle . . . and How People Respond to Them," *Mediterranean Archaeology and Archaeometry* 18, no. 4 (2018): 363–369, https://doi.org/10.5281/zenodo.1478005.

10. Rolf Sinclair, "Moonlight and Circadian Rhythms," *Science* 235, no. 4785 (1987): 145, https://doi.org/10.1126/science.3798101.

11. Guy Bordin, *Beyond Darkness and Sleep. The Inuit Night in North Baffin Island* (Bristol, CT: Peeters, 2015).

12. Wendy Noble and Tara L. Spires-Jones, "Sleep Well to Slow Alzheimer's Progression?," *Science* 363, no. 6429 (2019): 813–814; Jerrah K. Holth, "The Sleep-Wake Cycle Regulates Brain Interstitial Fluid Tau in Mice and CSF Tau in Humans," *Science* 363, no. 6429 (2019): 880–884.

13. Pam Belluck, "Why Didn't She Get Alzheimer's? The Answer Could Hold a Key to Fighting the Disease," *New York Times*, November 4, 2019, https://www.nytimes.com/2019/11/04/health/alzheimers-treatment-genetics.html.

14. Torbjørn Elvsåshagen et al., "Widespread Changes in White Matter Microstructure after a Day of Waking and Sleep Deprivation," *PLoS One* 10, no. 5 (2015): e0127351.

15. Irene Voldsbekk et al., "Sleep and Sleep Deprivation Differentially Alter White Matter Microstructure: A Mixed Model Design Utilising Advanced Diffusion Modelling," *NeuroImage*, 226 (2021): 117540, https://doi.org/10.1016/j.neuroimage.2020.117540; Desana Kocevska et al., "The Prospective Association of Objectively Measured Sleep and Cerebral White Matter Microstructure in Middle-Aged and Older Persons," *Sleep* 42, no. 10 (October 2019), https://doi.org/10.1093/sleep/zsz140.

16. Erin L. Boespflug and Jeffrey J. Iliff, "The Emerging Relationship between Interstitial Fluid-Cerebrospinal Fluid Exchange, Amyloid-β, and Sleep," *Biological Psychiatry* 83, no. 4 (2018): 328–336 https://dx.doi.org/10.1016%2Fj.biopsych.2017.11.031; Jonathan Cedernaes, "Candidate Mechanisms Underlying the Association between Sleep-Wake Disruptions and Alzheimer's Disease," *Sleep Medicine Reviews* 31 (2017): 102–111.

17. Kate E. Sprecher et al., "Poor Sleep Is Associated with CSF Biomarkers of Amyloid Pathology in Cognitively Normal Adults," *Neurology* 89, no. 5 (August 2017): 445–453, https://doi.org/10.1212/WNL.0000000000004171; Kirusanthy Kaneshwaran et al., "Sleep Fragmentation, Microglial Aging, and Cognitive Impairment in Adults with and without Alzheimer's Dementia," *Science Advances* 5, no. 12 (December 2019), https://doi.org/10.1126/sciadv.aax7331; Séverine Sabia et al., "Association of Sleep Duration in Middle and Old Age with Incidence of Dementia," *Nature Communications* 12, no. 2289 (2021), https://doi.org/10.1038/s41467-021-22354-2.

18. Ehsan Shokri-Kojori et al., "β-Amyloid Accumulation in the Human Brain after One Night of Sleep Deprivation," *Proceedings of the National Academy of Sciences* 115, no. 17 (2018): 4483–4488, https://doi.org/10.1073/pnas.1721694115.

19. Michele Bellesi, "Sleep Loss Promotes Astrocytic Phagocytosis and Microglial Activation in Mouse Cerebral Cortex," *Journal of Neuroscience* 37, no. 21 (2017): 5263–5273, https://doi.org/10.1523/jneurosci.3981-16.2017.

20. Lisa Marshall et al., "Brain Rhythms during Sleep and Memory Consolidation: Neurobiological Insights," *Physiology* (Bethesda, MD) 35, no. 1 (January 2020): 4–15; Björn Rasch, "Sleep and Plasticity: Do We Consolidate Memories Separately in Each Hemisphere?," *Dispatch* 30, no. 8 (2020): R349–R351, https://doi.org/10.1016/j.cub.2020.02.079.

21. Walker, *Why We Sleep*.

22. Stanislas Dehaene, *How We Learn: Why Brains Learn Better Than Any Machine . . . for Now* (New York: Viking, 2020); Gideon P. Dunster et al., "Sleepmore in Seattle: Later School Start Times Are Associated with More Sleep and Better Performance in High School Students," *Science Advances* 4, no. 12 (December 2018): eaau6200; Valentina Alfonsi et al., "Later School Start Time: The Impact of Sleep on Academic Performance and Health in the Adolescent Population," *International Journal of Environmental Research and Public Health* 17, no. 7 (2020): 2574, https://doi.org/10.3390/ijerph17072574.

23. Kevin B. Koronowski and Paolo Sassone-Corsi, "Communicating Clocks Shape Circadian Homeostasis," *Science* 371, no. 6530 (2021), https://doi.org/10.1126/science.abd0951.

24. Walker, *Why We Sleep*; Josephine Arendt, "Approaches to the Pharmacological Management of Jet Lag," *Drugs* 78, no. 14 (2018): 1419–1431, https://doi.org/10.1007/s40265-018-0973-8; Deetja Iggena, York Winter, and Barbara Steiner, "Melatonin Restores Hippocampal Neural Precursor Cell Proliferation and Prevents Cognitive Deficits Induced by Jet Lag Simulation in Adult Mice," *Journal of Pineal Research* 62, no. 4 (May 2017): e12397, https://doi.org/10.1111/jpi.12397.

25. Stephanie J. Crowley et al., "Increased Sensitivity of the Circadian System to Light in Early/Mid-Puberty," *Journal of Clinical Endocrinology and Metabolism* 100, no. 11 (November 2015): 4067–4073, https://doi.org/10.1210/jc.2015-2775; Stacy Lu, "Enlightened Minds," *Monitor on Psychology*, January 2016, 61–65.

26. Olakunle J. Onaolapo and Adejoke Yetunde Onaolapo, "Melatonin, Adolescence, and the Brain: An Insight into the Period-Specific Influences of a Multifunctional Signaling Molecule," *Birth Defects Research* 109, no. 20 (2017): 1659–1671; Henry Nicholls, *Sleepyhead: The Neuroscience of a Good Night's Rest* (New York: Basic Books, 2018); Yumna Saeed, Phyllis C. Zee, and Sabra M. Abbott, "Clinical Neurophysiology of Circadian Rhythm Sleep-Wake Disorders," *Handbook of Clinical Neurology* 161 (2019): 369–380.

27. Michael B. Robb, *The New Normal: Parents, Teens, Screens, and Sleep in the United States* (San Francisco, CA: Common Sense Media, 2019).

28. Lisa A. Ostrin, Kaleb S. Abbott, and Hope M. Queener, "Attenuation of Short Wavelengths Alters Sleep and the ipRGC Pupil Response," *Ophthalmic and*

Physiological Optics 37, no. 4 (July 2017): 440–450, https://doi.org/10.1111/opo.12385.

29. Lisa A. Ostrin, "Ocular and Systemic Melatonin and the Influence of Light Exposure," *Clinical and Experimental Optometry* 102, no. 2 (2019): 99–108.

30. "Frequently Asked Questions," *Measuring the Effects of Light from Electronic Devices on Sleep: An Online Research Study,* Division of Sleep and Circadian Disorders, Brigham and Women's Hospital, Boston, accessed January 12, 2019, https://web.archive.org/web/20190112024043/flux.partners.org/ (site discontinued).

31. Search Amazon or sites such as lowbluelights.com. For Re-Timer, see "Discover," Re-Timer, https://www.re-timer.com/discover.

32. Joëlle Coutaz and James L. Crowley, "A First-Person Experience with End-User Development for Smart Homes," *IEEE Pervasive Computing* 15, no. 2 (April 2016): 26–39, https://doi.org/10.1109/MPRV.2016.24; Ronda Kaysen, "Light Bulbs That Help You Sleep," *New York Times*, February 10, 2017, https://nytimes.com/2017/02/10/realestate/light-bulbs-that-help-you-sleep.html. See also the website of Philips Hue at https://www.philips-hue.com/en-us.

33. Nathaniel P. Hoyle et al., "Circadian Actin Dynamics Drive Rhythmic Fibroblast Mobilization during Wound Healing," *Science Translational Medicine* 9, no. 415 (2017), https://doi.org/10.1126/scitranslmed.aal2774.

34. Lameese D. Akacem, Kenneth P. Wright Jr., and Monique K. LeBourgeois, "Sensitivity of the Circadian System to Evening Bright Light in Preschool-Age Children," *Physiological Reports* 6, no. 5 (March 2018), https://doi.org/10.14814/phy2.13617.

35. Perri Klass, "To Help Children Sleep, Go Dark," *New York Times*, March 5, 2018, nytimes.com/2018/03/05/well/family/children-sleep-light-melatonin.html.

36. Judith A. Owens and Maile Moore, "Insomnia in Infants and Young Children," *Pediatric Annals* 46, no. 9 (2017): e321–e326; Kiran Maski and Judith Owens, "Pediatric Sleep Disorders," *Continuum* (Minneapolis, MN) 24, no. 1 (2018): 210–227.

37. Kenneth P. Wright Jr. et al., "Entrainment of the Human Circadian Clock to the Natural Light-Dark Cycle," *Current Biology* 23, no. 16 (2013): 1554–1558, https://doi.org/10.1016/j.cub.2013.06.039.

38. Mariana G. Figueiro, "Light, Sleep and Circadian Rhythms in Older Adults with Alzheimer's Disease and Related Dementias," *Neurodegenerative Disease Management* 7, no. 2 (2017): 119–145.

39. Mark S. Rea et al., Luminous roof for NICU incubators for regulating circadian rhythms in infants and for providing high visibility of infant anatomy for healthcare staff, U.S. Patent 10,646,685 B2, filed December 2, 2014, and issued May 12, 2020.

40. Healthe (website), accessed August 17, 2020 (website now defunct).

41. Lu, "Enlightened Minds"; Ruth M. Munn et al., "Health Consequences of Electric Lighting Practices in the Modern World: A Report on the National Toxicology Program's Workshop on Shift Work at Night, Artificial Light at Night, and Circadian Disruption," *Science of the Total Environment* 607–608 (2017): 1073–1084, https://doi.org/10.1016/j.scitotenv.2017.07.056.

42. Mariana G. Figueiro and David Pedler, "Red Light: A Novel, Non-pharmacological Intervention to Promote Alertness in Shift Workers," *Journal of Safety Research* 74 (2020): 169–177, https://doi.org/10.1016/j.jsr.2020.06.003; Arne Lowden and Göran Kecklund, "Considerations on How to Light the Night-Shift," *Lighting Research and Technology* 53, no. 5 (August 2021): 437–452.

CHAPTER 11

1. "Warning: Gaming Diaper Required!?," Capital Medical Supply (blog), October 31, 2017, https://web.archive.org/web/20210115032035/capitalmedicalsupply.ca/blogs/capital-medical-supply/gamming-diapers; Fengshu Liu, *Urban Youth in China: Modernity, the Internet and the Self*, Routledge Research in Information Technology and Society 10 (New York: Routledge, 2011).

2. Patricia Wallace, "Internet Addiction Disorder and Youth," *EMBO Reports* 15, no. 1 (January 2014): 12–16.

3. American Psychological Association, "Stress in America: The State of Our Nation," press release, November 1, 2017, https://apa.org/news/press/releases/stress/2017/state-nation.pdf; Douglas Heaven, "Death by Video Game: A Power Like No Other," *New Scientist*, August 19, 2015, https://www.newscientist.com/article/mg22730350-700-death-by-video-game-a-power-like-no-other.

4. Byron Reeves et al., "Screenomics: A Framework to Capture and Analyze Personal Life Experiences and the Ways That Technology Shapes Them," *Human–Computer Interaction* 36, no. 2 (2021): 150–201, https://doi.org/10.1080/07370024.2019.1578652.

5. "Network" and "circuit" are roughly synonymous and convey the same idea of a circumscribed electrochemical pathway in the nervous system. A circuit may imply a smaller, more localized loop; a network can be more widespread and connect distant nodes to more central ones.

6. Shazia R. Chaudhry and William Gossman, "Biochemistry, Endorphin," in *StatPearls* (Treasure Island, FL: StatPearls, 2020).

7. Kenneth Blum and Rajendra D. Badgaiyan, "Reward Deficiency Syndrome (RDS): Entering the Genomics and Neuroscience Era of Addiction Medicine," *Journal of Reward Deficiency Syndrome and Addiction Science* 1, no. 1 (2015): 1–2.

8. Kenneth Blum et al., "'Liking' and 'Wanting' Linked to Reward Deficiency Syndrome (RDS): Hypothesizing Differential Responsivity in Brain Reward Circuitry," *Current Pharmaceutical Design* 18, no. 1 (2012): 113–118; Edward Justin Modestino et al., "Reward Deficiency Syndrome: Attentional/Arousal Subtypes, Limitations of Current Diagnostic Nosology, and Future Research," *Journal of Reward Deficiency Syndrome and Addiction Science* 1, no. 1 (2015): 6–9, https://jaddictionscience.com/2015/02/06/reward-deficiency-syndrome-attentionalarousal-subtypes-limitations-of-current-diagnostic-nosology-and-future-research; Kenneth Blum et al., "Clinically Combating Reward Deficiency Syndrome (RDS) with Dopamine Agonist Therapy as a Paradigm Shift: Dopamine for Dinner?," *Molecular Neurobiology* 52 (2015): 1862–1869, https://doi.org/10.1007/s12035-015-9110-9.

9. Robert Kubey and Mihaly Csikszentmihalyi, "Television Addiction Is No Mere Metaphor," *Scientific American* 286, no. 2 (February 2002): 74–80.

10. Eric Weiner, *The Geography of Bliss: One Grump's Search for the Happiest Places in the World* (New York: Twelve, 2008).

11. John F. Helliwell, Richard Layard, and Jeffrey D. Sachs, eds., *World Happiness Report 2019* (New York: Sustainable Development Solutions Network, 2019).

12. Philip Brickman and Donald T. Campbell, "Hedonic Relativism and Planning the Good Society," in *Adaptation-Level Theory*, ed. Mortimer H. Appley (New York: Academic Press, 1971), 287–305.

13. Joan Marques, "Leadership and Ambition," in *Leadership Today: Practices for Personal and Professional Performance*, ed. Joan Marques and Satinder Dhiman (Cham, CH: Springer, 2016), 353–370.

14. Lisa Feldman Barrett, *How Emotions Are Made: The Secret Life of the Brain* (Boston: Houghton Mifflin Harcourt, 2017). Note, however, that Barrett refutes Paul Ekman's idea of universal emotion.

15. Jeff Horwitz et al., "The Facebook Files," *Wall Street Journal*, September 13–December 29, 2021, https://www.wsj.com/articles/the-facebook-files-11631713039; Jeff Horwitz, "The Facebook Whistleblower, Frances Haugen, Says She Wants to Fix the Company, Not Harm It," *Wall Street Journal*, October 3, 2021, https://www.wsj.com/articles/facebook-whistleblower-frances-haugen-says-she-wants-to-fix-the-company-not-harm-it-11633304122.

16. Jiddu Krishnamurti, as quoted on the website of the Krishnamurti Foundation America (https://www.kfa.org).The philosopher's writings are vast.

17. Boulos-Paul Bejjani et al., "Transient Acute Depression Induced by High-Frequency Deep-Brain Stimulation," *New England Journal of Medicine* 340, no. 19 (1999): 1476–1480.

18. Myriam Balerna and Arko Ghosh, "The Details of Past Actions on a Smartphone Touchscreen Are Reflected by Intrinsic Sensorimotor Dynamics," *npi Digital Medicine* 1, no. 4 (2018), https://doi.org/10.1038/s41746-017-0011-3.

19. Jay N. Borger, Reto Huber, and Arko Ghosh, "Capturing Sleep-Wake Cycles by Using Day-to-Day Smartphone Touchscreen Interactions," *npj Digital Medicine* 2, no. 73 (2019); Arko Ghosh, "For Participants," CODELAB, https://www.arkoghosh.com/for-participants.html.

20. James Olds, "'Reward' from Brain Stimulation in the Rat," *Science* 122, no. 3175 (1955): 878.

21. Robert G. Heath, "Pleasure and Brain Activity in Man: Deep and Surface Electroencephalograms during Orgasm," *Journal of Nervous and Mental Disease* 154, no. 1 (1972): 3–18; Robert G. Heath, "Electrical Self-Stimulation of the Brain in Man," *American Journal of Psychiatry*, 120:571–577.

22. Gregory Berns, *How Dogs Love Us: A Neuroscientist and His Adopted Dog Decode the Canine Brain* (Boston: New Harvest, 2013).

23. Valery Grinevich and Inga D. Neumann, "Brain Oxytocin: How Puzzle Stones from Animal Studies Translate into Psychiatry," *Molecular Psychiatry* 26 (2021): 265–279, https://doi.org/10.1038/s41380-020-0802-9; Joel E. Dimsdale, *Dark Persuasion: A History of Brainwashing from Pavlov to Social Media* (New Haven, CT: Yale University Press, 2021).

24. David G. Schwartz, "How Casinos Use Math to Make Money When You Play the Slots," *Forbes*, June 4, 2018, https://forbes.com/sites/davidschwartz/2018/06/04/how-casinos-use-math-to-make-money-when-you-play-the-slots.

25. Ethan Kross et al., "Facebook Use Predicts Declines in Subjective Well-Being in Young Adults," *PLoS One* 8, no. 8 (2013): e69841.

26. Eric Weiner, personal communication with author, January 6, 2014.

27. Daniel T. Gilbert and Timothy D. Wilson, "Miswanting: Some Problems in the Forecasting of Future Affective States," in *Feeling and Thinking: The Role of Affect in Social Cognition*, ed. Joseph P. Forgas (New York: Cambridge University Press, 2000), 179.

28. Weiner, *The Geography of Bliss*; Philip Brickman, Dan Coates, and Ronnie Janoff-Bulman, "Lottery Winners and Accident Victims: Is Happiness Relative?," *Journal of Personality and Social Psychology* 36, no. 8 (1978): 917–927.

29. Gregory Burns, *Iconoclast: A Neuroscientist Reveals how to Think Differently* (Boston: Harvard Business School Press, 2010).

30. Susan Cain, *Quiet: The Power of Introverts in a World That Can't Stop Talking* (New York: Broadway Paperbacks, 2013), 92, 177.

Chapter 12

1. Leroy E. Burney, "Smoking and Lung Cancer: A Statement of the Public Health Service," *JAMA* 171, no. 13 (1959): 1829–1837, https://doi.org/10.1001/jama.1959.73010310005016. See also "History of the Surgeon General's Reports on Smoking and Health," in *Smoking and Tobacco Use* (Atlanta, GA Centers for Disease Control and Prevention), accessed January 15, 2022, https://web.archive.org/web/20220115201716/cdc.gov/tobacco/data_statistics/sgr/history/index.htm (page discontinued), for a summary of the Surgeon General's 1964 report, as well as Luther L. Terry et al., *Smoking and Health: Report of the Advisory Committee to the Surgeon General of the Public Health Service*, Public Health Service Publication 1103 (Washington, DC, 1964) for the original.

2. David Eagleman, *The Brain: The Story of You* (New York: Pantheon Books, 2015); David Eagleman, *Livewired: The Inside Story of the Ever-Changing Brain* (New York: Pantheon Books, 2020).

3. Christian Lüscher, Vincent Pascoli, and Meaghan Creed, "Optogenetic Dissection of Neural Circuitry: From Synaptic Causalities to Blue Prints for Novel Treatments of Behavioral Diseases," *Current Opinion in Neurobiology* 35 (2015): 95–100; Ana-Clara Bobadilla et al., "Corticostriatal Plasticity, Neuronal Ensembles, and Regulation of Drug-Seeking Behavior," *Progress in Brain Research* 235 (2017): 93–112.

4. Xiaosi Gu, "Incubation of Craving: A Bayesian Account," *Neuropsychopharmacology* 43, no. 12 (2018): 2337–2339, https://doi.org/10.1038/s41386-018-0108-7; Xuan Li, Daniele Caprioli, and Nathan J. Marchant, "Recent Updates on Incubation of Drug Craving: A Mini-Review," *Addiction Biology* 20, no. 5 (2015): 872–876.

5. See Myriam Balerna and Arko Ghosh, "The Details of Past Actions on a Smartphone Touchscreen Are Reflected by Intrinsic Sensorimotor Dynamics," *npj Digital Medicine* 1, no. 4 (2018): https://doi.org/10.1038/s41746-017-0011-3.

6. Julie Kauer, "Life Experiences and Addictive Drugs Change Your Brain in Similar Ways," in *Think Tank*, ed. David J. Linden (New Haven, CT: Yale University Press, 2018), 66–72.

7. Jaron Lainier, *Ten Arguments for Deleting Your Social Media Accounts Right Now* (New York: Henry Holt, 2018), 15.

8. Ab Litt, Uzma Khan, and Baba Shiv, "Lusting While Loathing: Parallel Counterdriving of Wanting and Liking," *Psychological Science* 21, no. 1 (2009): https://doi.org/10.1177/0956797609355633.

9. Peter Johansson et al., "Failure to Detect Mismatches between Intention and Outcome in a Simple Decision Task," *Science* 310, no. 5745 (2005): 116–119; Thomas Strandberg, "The Malleability of Political Attitudes: Choice Blindness, Confabulation and Attitude Change" (PhD diss., Lund University, 2020).

10. Amy Fleming, "The Science of Craving," *Intelligent Life*, May/June 2015, https://web.archive.org/web/20151002063556/https://moreintelligentlife.com/content/features/wanting-versus-liking (site discontinued); Kent C. Berridge, "From Prediction Error to Incentive Salience: Mesolimbic Computation of Reward Motivation," in "Beyond Simple Reinforcement Learning," ed. John P. O'Doherty, special issue, *European Journal of Neuroscience* 35, no. 7 (April 2012): https://doi.org/10.1111/j.1460-9568.2012.07990.x.

11. Kent C. Berridge, email to author, October 28, 2017.

12. Aric Sigman, "Screen Dependency Disorders: A New Challenge for Child Neurology," *Journal of the International Child Neurology Association* 17, no. 119 (2017), jicna.org/index.php/journal/article/view/67.

13. Terry E. Robinson and Kent C. Berridge, "The Incentive Sensitization Theory of Addiction: Some Current Issues," *Philosophical Transactions of the Royal Society B: Biological Sciences* 363, no. 1507 (2008): 3137–3146; Craig T. Werner et al., "Response of the Ubiquitin-Proteasome System to Memory Retrieval After Extended-Access Cocaine or Saline Self-Administration," *Neuropsychopharmacology* 40, no. 13 (2015): 3006–3014, https://doi.org/10.1038/npp.2015.156.

14. Juergen Vorges et al., "Deep Brain Stimulation Surgery for Alcohol Addiction," *World Neurosurgery* 80, no. 3–4 (September–October 2013): S28.e21–S28.e31, https://doi.org/10.1016/j.wneu.2012.07.011.

15. Kristine Rømer Thomsen et al., "Applying Incentive Sensitization Models to Behavioral Addiction," *Neuroscience and Biobehavioral Reviews* 45 (September 2014): 343–349, https://doi.org/10.1016/j.neubiorev.2014.07.009.

16. Specifically, the nucleus accumbens via a microinjection of amphetamine.

17. Berridge, "From Prediction Error to Incentive Salience."

18. Ellie Spanswick, "Children Should Have Access to iPads from Birth Says New Research," *Newsletter,* DayNurseries.co.uk, June 16, 2015, https://www.daynurseries.co.uk/news/article.cfm/id/1569466/children-should-have-access-to-ipads-from-birth.

19. Ari Brown, Donald L. Shifrin, and David L. Hill, "Beyond 'Turn It Off': How to Advise Families on Media Use," *AAP News*, October 2015; Donald L. Shifrin et al., *Growing Up Digital: Media Research Symposium* (Itasca, IL: American Academy of Pediatrics, 2015), https://web.archive.org/web/20210808173311/aap.org/en-us/Documents/digital_media_symposium_proceedings.pdf (site discontinued); Aric Sigman, "Erring on the Wrong Side of Precaution," commentary on "Prospective Associations between Television in the Preschool Bedroom and Later Bio-Psycho-Social Risks" by Linda S. Pagani, Marie Josée Harbec, and Tracie A. Barnett, *Pediatric Research* 85, no. 7: 925, https://doi.org/10.1038/s41390-019-0357-0.

20. Aric Sigman, "Invited Commentary on 'Prospective Associations between Television in the Preschool Bedroom and Later Bio-psycho-social Risks,'" *Pediatric Research* 85, no. 7 (June 2019): 925.

21. Marco Martuzzi and Joel A. Tickner, eds., *The Precautionary Principle: Protecting Public Health, the Environment, and the Future of Our Children* (Copenhagen: World Health Organization Regional Office for Europe, 2004).

22. Marisa Meier et al., "Advertising in Young Children's Apps: A Content Analysis," *Journal of Developmental and Behavioral Pediatrics* 40, no. 1 (January 2019): 32–39, https://doi.org/10.1097/dbp.0000000000000622.

23. Joe Pinsker, "Children Are Getting Great Practice at Being Sold to All the Time," *Atlantic*, November 1, 2018, theatlantic.com/family/archive/2018/11/kids-apps-ads-smartphones/574588.

24. Charlene Elliott, "Beauty and the Banana: It Is a Commercial Promotion, Not a Public Health Campaign," *Canadian Journal of Public Health* 109, no. 3 (2018): 436–438, https://doi.org/10.17269/s41997-018-0051-8.

25. Josh Golin and Melissa Campbell, "Reining in the Commercialization of Childhood," in *EarthEd: Rethinking Education on a Changing Planet*, ed. Worldwatch Institute (Washington, DC: Island Press, 2017); "We're Asking the FTC to Investigate Apps That Manipulate Kids," Fairplay, https://fairplayforkids.org/were-asking-ftc-investigate-apps-manipulate-kids.

26. See the website of Stanford University's Behavior Design Lab at https://behaviordesign.stanford.edu.

27. Farhad Manjoo, "We Have Reached Peak Screen Saturation. Now Revolution Is in the Air," *New York Times*, June 7, 2018, https://www.nytimes.com/2018/06/27/technology/peak-screen-revolution.html.

28. Adrian F. Ward et al., "Brain Drain: The Mere Presence of One's Own Smartphone Reduces Available Cognitive Capacity," *Journal of the Association for Consumer Research* 2, no. 2 (April 3, 2017): 140–154, https://doi.org/10.1086/691462.

29. Hal Hodson, "Speech Analyser Monitors Emotion for Call Centres," *New Scientist*, May 7, 2014, https://www.newscientist.com/article/mg22229683-800-speech-analyser-monitors-emotion-for-call-centres; John D. Williamson, Speech analyzer for analyzing frequency perturbations in a speech pattern to determine the emotional state of a person, U.S. Patent 4,142,067, filed April 11, 1978, and issued February 27, 1979; Björn W. Schuller, "Speech Emotion Recognition: Two Decades in a Nutshell, Benchmarks, and Ongoing Trends," *Communications of the ACM* 61, no. 5 (May 2018): 90–99, https://doi.org/10.1145/3129340.

30. Mario Beauregard, ed., *Consciousness, Emotional Self-Regulation and the Brain*, Advances in Consciousness Research 54 (Philadelphia: John Benjamins, 2004).

31. Morten L. Kringelbach and Kent C. Berridge, "Motivation and Pleasure in the Brain," in *The Psychology of Desire*, ed. Wilhelm Hofmann and Loran F. Nordgren (New York: Guilford, 2015).

32. Kent C. Berridge and Morten L. Kringelbach, "Pleasure Systems in the Brain," *Neuron* 86, no. 3 (2015): 646–664.

33. Alan P. Lightman, *Searching for Stars on an Island in Maine* (New York: Pantheon, 2018); Alan P. Lightman, *In Praise of Wasting Time* (New York: TED Books, 2018).

CHAPTER 13

1. Ellie Spanswick, "Children Should Have Access to iPads from Birth Says New Research," *Newsletter*, DayNurseries.co.uk, June 16, 2015, https://www.daynurseries.co.uk/news/article.cfm/id/1569466/children-should-have-access-to-ipads-from-birth; Rachael Bedford et al., "Toddlers' Fine Motor Milestone Achievement Is Associated with Early Touchscreen Scrolling," *Frontiers in Psychology* 7, no. 1108 (2016); "Inside the Psychologist's Studio: Annette Karmiloff-Smith," interview by BJ Casey, 26th APS Annual Convention, San Francisco, CA, May 23, 2014, https://youtu.be/qXmEe_m2Z9M; "Inside the Psychologist's Studio: Annette Karmiloff-Smith," *Observer*, December 2, 2015, http://www.psychologicalscience.org/video/karmiloff-smith-talks-future-of-developmental-science.html.

2. Richard House, ed., *Too Much, Too Soon? Early Learning and the Erosion of Childhood* (Stroud, UK: Hawthorn Press, 2011).

3. Annette Karmiloff-Smith and Richard House, "ICT—For and Against: Toddlers, TV and Touchscreens," *Nursery World*, September 21, 2015, https://www.nurseryworld.co.uk/news/article/ict-for-against-toddlers-tv-and-touchscreens.

4. Victoria Leong et al., "Speaker Gaze Increases Information Coupling between Infant and Adult Brains," *Proceedings of the National Academy of Sciences* 114, no. 50 (2017): 13290–13295, https://doi.org/10.1073/pnas.1702493114; Tae-Ho Lee, Yang Qu, and Eva H. Telzer, "Dyadic Neural Similarity During Stress in Mother-Child Dyads," *Journal of Research on Adolescence* 28, no. 1 (March 2018): 121–133, https://doi.org/10.1111/jora.12334.

5. John Bowlby, *A Secure Base: Parent-Child Attachment and Healthy Human Development* (New York: Basic Books, 2008); Klaus E. Grossman et al., eds., *Maternal Sensitivity: Mary Ainsworth's Enduring Influence on Attachment Theory, Research, and Clinical Applications* (London: Routledge, 2015); Mary D. Salter Ainsworth et al., *Patterns of Attachment: A Psychological Study of the Strange Situation* (New York: Psychology Press, 2015); Frans B. M. de Waal, "The 'Russian Doll' Model of Empathy and Imitation," in *On Being Moved: From Mirror Neurons to Empathy*, ed. Stein Bråten (Amsterdam: John Benjamins, 2007), 49–69.

6. "Parents See Impact of Screen Time on Babies, Toddlers," *Good Morning America*, ABC News, May 2, 2019, https://abcnews.go.com/GMA/News/video/parents-impact-screen-time-babies-toddlers-62774533; "Letting Your Phone Get between You and Your Child," *Nightline*, ABC News, May 3, 2019, https://abcnews.go.com/Nightline/video/letting-phone-child-62797697; Tracy Dennis-Tiwary, personal communication to the author, n.d.

7. Sarah Myruski et al., "Digital Disruption? Maternal Mobile Device Use Is Related to Infant Social-Emotional Functioning," *Developmental Science* 21, no. 4 (July 2018): e12610, https://doi.org/10.1111/desc.12610; Sarah Myruski, "Biological Signatures of Emotion Regulation in Children" (PhD diss., City University of New York, 2018), https://academicworks.cuny.edu/gc_etds/2533.

8. Jenny Radesky, "Digital Media and Symptoms of Attention-Deficit/Hyperactivity Disorder in Adolescents," *JAMA* 320, no. 3 (2018): 237–239, https://doi.org/10.1001/jama.2018.8932.

9. Craig Timberg and Rachel Siegel, "World Health Officials Take a Hard Line on Screen Time for Kids. Will Busy Parents Comply?," *Washington Post*, April 24, 2019, https://www.washingtonpost.com/business/2019/04/24/who-infants-under-year-old-shouldnt-be-exposed-any-electronic-screens; "YouTube Users Statistics 2023" (blog post), Global Media Insights, May 26, 2023, https://www.globalmediainsight.com/blog/youtube-users-statistics.

10. Linda K. Kaye et al., "The Conceptual and Methodological Mayhem of 'Screen Time,'" *International Journal of Environmental Research and Public Health* 17, no. 10 (2020): 3661.

11. Byron Reeves et al., "Screenomics: A Framework to Capture and Analyze Personal Life Experiences and the Ways That Technology Shapes Them," *Human–Computer Interaction* 36, no. 2 (2021): 150–201, https://doi.org/10.1080/07370024.2019.1578652; Byron Reeves, Thomas Robinson, and Nilam Ram, "Time for the Human Screenome Project," *Nature* 577, no. 7790 (2020): 314–317.

12. Philippe Verduyn, Nino Gugushvili, and Ethan Kross, "Do Social Networking Sites Influence Well-Being? The Extended Active-Passive Model," *Current Directions in Psychological Science* 31, no. 1 (2022): 62–68, https://doi.org/10.1177/09637214211053637.

13. Edward Miller and Joan Almon, *Crisis in the Kindergarten: Why Children Need to Play in School* (College Park, MD: Alliance for Childhood, 2009), https://web.archive.org/web/20190123212214/allianceforchildhood.org/sites/allianceforchildhood.org/files/file/kindergarten_report.pdf (site discontinued).

14. Vanessa Vietes, Shannon M. Pruden, and Bethany C. Reeb-Sutherland, "Childhood Wayfinding Experience Explains Sex and Individual Differences in Adult

Wayfinding Strategy and Anxiety," *Cognitive Research: Principles and Implications* 5, no. 12 (2020), https://doi.org/10.1186/s41235-020-00220-x.

15. Richard House, personal communication to author, January 2, 2018.

16. Joe Clement and Matt Miles, *Screen Schooled: Two Veteran Teachers Expose How Technology Overuse Is Making Our Kids Dumber* (Chicago: Chicago Review Press, 2017); Nellie Bowles, "A Dark Consensus About Screens and Kids Begins to Emerge in Silicon Valley," *New York Times*, October 26, 2018, https://www.nytimes.com/2018/10/26/style/phones-children-silicon-valley.html; Nellie Bowles, "Silicon Valley Nannies Are Phone Police for Kids," *New York Times*, October 26, 2018, nytimes.com/2018/10/26/style/silicon-valley-nannies.html.

17. "Wayback Machine," Internet Archive, https://web.archive.org.

18. *iPad in Education: Results*, Apple, March 2019, apple.com/education/docs/ipad-in-education-results.pdf. See also, for example, Clement and Miles, *Screen Schooled*, 194.

19. Cathy O'Neil, *Weapons of Math Destruction: How Big Data Increases Inequality and Threatens Democracy* (New York: Crown, 2016).

20. Arne Duncan, *Transforming American Education: Learning Powered by Technology*, prepared by the Office of Educational Technology, U.S. Department of Education (Washington, DC, 2010).

21. "Top Spenders," *Lobbying*, Open Secrets, last updated April 24, 2023, https://www.opensecrets.org/federal-lobbying/top-spenders.

22. Steven Pinker, "Can Students Have Too Much Tech?," *New York Times*, January 15, 2015, nytimes.com/2015/01/30/opinion/can-students-have-too-much-tech.html.

23. Kathy Hirsh-Pasek et al., "Putting Education in 'Educational' Apps: Lessons from the Science of Learning," *Psychological Science in the Public Interest* 16, no. 1 (2015): 3–34; Joe Clement, in-person interview with the author, June 20, 2019.

24. Clement, personal communication, June 20, 2019.

25. Erika Christakis, "The New Preschool Is Crushing Kids," *Atlantic*, January/February 2016, https://www.theatlantic.com/magazine/archive/2016/01/the-new-preschool-is-crushing-kids/419139; Erika Christakis, *The Importance of Being Little: What Young Children Really Need from Grownups* (New York: Viking, 2016).

26. Bedford et al., "Toddlers' Fine Motor Milestone Achievement."

27. Bedford et al., "Toddlers' Fine Motor Milestone Achievement"; Jin-Hwa Moon et al., "Smart Device Usage in Early Childhood Is Differentially Associated with Fine

Motor and Language Development," *Acta Paediatrica* 108, no. 5 (May 2019): 903–910, https://doi.org/10.1111/apa.14623.

28. Richard E. Cytowic, "Your Brain on Screens," *American Interest*, June 9, 2015, 53–61, the-american-interest.com/2015/06/09/your-brain-on-screens.

29. Personal communication to author, September 29, 2015.

30. Sean O'Neill, "Psychedelic Medicine: The Potential, the People, the Politics," *New Scientist*, August 18, 2017, https://www.newscientist.com/article/2144520-psychedelic-medicine-the-potential-the-people-the-politics.

31. Eckard H. Hess, "The Role of Pupil Size in Communication," *Scientific American* 233, no. 5 (1975): 110–119.

32. Richard E. Cytowic and David M. Eagleman, *Wednesday Is Indigo Blue: Discovering the Brain of Synesthesia* (Cambridge, MA: MIT Press, 2009).

CHAPTER 14

1. Nicholas Carr, *The Shallows: What the Internet Is Doing to Our Brains* (New York: W. W. Norton, 2011), 116; Nicolas Carr, "Is Google Making Us Stupid?," *Atlantic*, July/August 2008, theatlantic.com/magazine/archive/2008/07/is-google-making-us-stupid/306868.

2. A. G. A., "We Tested the Mommy Porn," *ELLE*, May 18, 2012, quoted in Maxime Derian, *Cognitive Prosthetics* (London: Elsevier, 2018).

3. Maxime Derian, "The Addictive Nature of Cognitive Prosthetics," in *Cognitive Prosthetics* (London: Elsevier, 2018), 125–145.

4. Susan Greenfield, *Mind Change: How Digital Technologies Are Leaving Their Mark on Our Brains* (New York: Random House, 2015); Edward Spector, "China Doesn't Have the Right Solution—but Excessive Video Gaming Is a Real Problem," *Washington Post*, September 6, 2021, https://www.washingtonpost.com/opinions/2021/09/06/excessive-video-gaming-real-problem-china-us.

5. *Web Junkie*, written and directed by Hilda Medalia and Shosh Shlam, aired July 13, 2015, on PBS, https://www.pbs.org/pov/films/webjunkie.

6. Javier C. Hernández and Albee Zhang, "90 Minutes a Day, Until 10 P.M.: China Sets Rules for Young Gamers," *New York Times*, November 6, 2019, nytimes.com/2019/11/06/business/china-video-game-ban-young.html; Keith Zhai, "China Limits Online Videogames to Three Hours a Week for Young People," *Wall Street Journal*, August 31, 2021, https://www.wsj.com/articles/china-sets-new-rules-for-youth-no-more-videogames-during-the-school-week-11630325781.

7. "China Mulls Limits for Children On Phones," *Wall Street Journal*, August 5, 2023, https://www.wsj.com/articles/china-set-to-impose-mobile-device-limits-for-minors-c547cac5.

8. Lindsay MacKenzie, "Lecture Halls for Learning, Not Watching Netflix," *Inside Higher Ed*, March 3, 2019, https://www.insidehighered.com/news/2019/03/04/purdue-university-extends-streaming-website-ban; Jenny Runevich, "Purdue to Ban Recreational Streaming in All Classrooms," WTHR, February 26, 2019, https://www.wthr.com/article/news/local/indiana/purdue-ban-recreational-streaming-all-classrooms/531-27e26815-1cc6-4f4d-8371-e6b45850a9a7.

9. Ian Bogost, "Why Is There a 'Gaming Disorder' but No 'Smartphone Disorder?,'" *Atlantic*, June 28, 2018, theatlantic.com/technology/archive/2018/06/whos-afraid-of-virginia-wolfenstein/563843.

10. See the website at netaddictionrecovery.com/the-problem/internet-addiction/22-dr-hilarie-cash-internet-addiction.html.

11. Emily Maloney, "The Dutch Have a Name for Doing Nothing. It's Called Niksen, and We Need More of It," *Washington Post*, December 12, 2019, washingtonpost.com/lifestyle/2019/12/12/dutch-have-name-doing-nothing-its-called-niksen-we-need-more-it.

12. Yalda T. Uhls et al., "Five Days at Outdoor Education Camp without Screens Improves Preteen Skills with Nonverbal Emotion Cues," *Computers in Human Behavior* 39 (2014): 387–392.

13. Jean M. Twenge, Gabrielle N. Martin, and Brian H. Spitzberg, "Trends in U.S. Adolescents' Media Use, 1976–2016: The Rise of Digital Media, the Decline of TV, and the (Near) Demise of Print," *Psychology of Popular Media Culture* 8, no. 4 (2019): 329–345, https://doi.org/10.1037/ppm0000203.

14. Joe Clement and Matt Miles, *Screen Schooled: Two Veteran Teachers Expose How Technology Overuse Is Making Our Kids Dumber* (Chicago: Chicago Review Press, 2017).

15. Aric Sigman, "Erring on the Wrong Side of Precaution," commentary on "Prospective Associations between Television in the Preschool Bedroom and Later Bio-Psycho-Social Risks" by Linda S. Pagani, Marie Josée Harbec, and Tracie A. Barnett, *Pediatric Research* 85, no. 7 (2019): 925, https://doi.org/10.1038/s41390-019-0357-0.

16. Joni Siani, NoAppForLife.com, https://www.noappforlife.com; Angelique Chrisafis, "French School Students to Be Banned from Using Mobile Phones," *Guardian* (international edition), https://www.theguardian.com/world/2018/jun/07/french-school-students-to-be-banned-from-using-mobile-phones.

17. Kep Kee Loh and Ryota Kanai, "How Has the Internet Reshaped Human Cognition?," *Neuroscientist* 22, no. 5 (2016): 506–520, https://doi.org/10.1177/1073858415595005.

18. Richard E. Cytowic, *Synesthesia* (Cambridge, MA: MIT Press, 2018).

19. Dan H. Sanes et al., eds., "Synapse Formation," in *Development of the Nervous System*, 4th ed. (London: Academic Press, 2019).

20. Richard E. Cytowic and David M. Eagleman, *Wednesday Is Indigo Blue: Discovering the Brain of Synesthesia* (Cambridge, MA: MIT Press, 2009).

21. Rich Parris, "Online T&Cs Longer than Shakespeare Plays—Who Reads Them?," *Which? Conversation*, March 23, 2012, https://web.archive.org/web/20220401224754/conversation.which.co.uk/technology/length-of-website-terms-and-conditions/ (site discontinued).

22. Myriam Balerna and Arko Ghosh, "The Details of Past Actions on a Smartphone Touchscreen Are Reflected by Intrinsic Sensorimotor Dynamics," *npj Digital Medicine* 1, no. 4 (2018), https://www.nature.com/articles/s41746-017-0011-3; Anne-Dominique Ghindrat et al., "Use-Dependent Cortical Processing from Fingertips in Touchscreen Phone Users," *Current Biology* 25, no. 1 (2015): 109–116, https://doi.org/10.1016/j.cub.2014.11.026; Arko Ghosh, "Linking Elementary Properties of the Human Brain to the Behaviour Captured on Touchscreen Smartphones," in *Internet Addiction: Neuroscientific Approaches and Therapeutical Implications Including Smartphone Addiction*, ed. Christian Montag and Martin Reuter, 2nd ed. (Cham, CH: Springer International, 2017).

23. Edward R. Tufte, *The Cognitive Style of PowerPoint: Pitching Out Corrupts Within*, 2nd ed. (Cheshire, CT: Graphics Press, 2016).

24. Donald L. Shifrin et al., *Growing Up Digital: Media Research Symposium* (Itasca, IL: American Academy of Pediatrics, 2015), https://web.archive.org/web/20210808173311/aap.org/en-us/Documents/digital_media_symposium_proceedings.pdf (site discontinued); Kathy Hirsh-Pasek et al., "Putting Education in 'Educational' Apps: Lessons from the Science of Learning," *Psychological Science in the Public Interest* 16, no. 1 (2015): 3–34; Erin Blakemore, "Those Educational Baby Toys Look Great, but They May Promise Too Much," *Washington Post*, June 15, 2018, washingtonpost.com/national/health-science/those-educational-baby-toys-look-great-but-they-may-promise-too-much/2018/06/15/764028ca-6e8a-11e8-bd50-b80389a4e569_story.html.

25. Jerry Mander, *Four Arguments for the Elimination of Television* (New York: Morrow, 1978).

26. Jeff Horwitz et al., "The Facebook Files," *Wall Street Journal*, September 13–December 29, 2021, https://www.wsj.com/articles/the-facebook-files-11631713039.

27. The actual number varies from seven ± two as an average.

28. Kyle Peterson, "The Conservative Wheelman," Pat Sajak interview, *Wall Street Journal*, September 25, 2015, wsj.com/articles/the-conservative-wheelman-1443218583.

Chapter 15

1. Johan Eklöf, *The Darkness Manifesto: On Light Pollution, Night Ecology, and the Ancient Rhythms That Sustain Life*, trans. Elizabeth DeNoma (New York: Scribner, 2023).

2. Christopher J. Preston, "The Evolutionary Power of Cities and Light," in *The Synthetic Age* (Cambridge, MA: MIT Press, 2020).

3. JCDecaux North America (website), accessed 2019, https://www.jcdecauxna.com.

4. Kathleen Elkins, "New York City Is Home to Nearly 1 Million Millionaires, More Than Any Other City in the World," Make It, CNBC, January 18, 2019, cnbc.com/2019/01/18/new-york-city-has-more-millionaires-than-any-other-city-in-the-world.html; *NYC & Company Annual Report: 2018–2019* (New York City Tourism + Conventions, 2019), https://indd.adobe.com/view/fcc4cd9f-7386-4b52-a39b-c401266a137f; Mike Follet, *Attention: The Common Currency for Media* (London: Lumen Research, 2018), https://www.jcdecaux.com/studies-documents/attention-common-currency-media-lumen-research-whitepaper.

5. Patrick Sisson, "Why Billboards and Outdoor Ads Are Booming in a Smartphone Age," *Curbed*, December 19, 2018, https://archive.curbed.com/2018/12/19/18148818/outdoor-digital-advertising-billboard.

6. Ballyhoo Media (website), ballyhooboats.com; Caroline Spivack, "Floating Digital Billboards Are Officially Banned from New York Waterways," *Curbed*, August 20, 2019, https://ny.curbed.com/2019/8/20/20813622/floating-digital-billboards-banned-new-york-waterways.

7. Worldwide Presence, JCDecaux, https://www.jcdecaux.com/brands/worldwide-presence.

8. Studies and Documents, JCDecaux, jcdecaux.com/studies-documents.

9. G. Brussel, telephone interview with JCDecaux North America and the author. December 18, 2018.

10. Follet, *Attention*.

11. Follet, *Attention*; Noel Scott et al., "A Review of Eye-Tracking Research in Tourism," *Current Issues and Tourism* 22, no. 1 (2017): 1244–1261.

12. Comments by Jean-Charles Decaux, Chairman of the Executive Board and Co-CEO of JCDecaux.

13. Fabian-Xosé Fernandez, "Current Insights into Optimal Lighting for Promoting Sleep and Circadian Health: Brighter Days and the Importance of Sunlight in the Built Environment," *Nature of Science and Sleep* 14 (2022): 25–39, https://doi.org/10.2147/NSS.S251712; Morgan P. Reid, Natalie D. Dautovich, and Joseph M. Dzierzewski, "Light and Sleep," in *Integrative Sleep Medicine*, ed. Valerie Cacho and Esther

Lum (New York: Oxford University Press, 2021), 101–112; Monica Motta, "American Medical Association Statement on Street Lighting," abstract, *Journal of the American Association of Variable Star Observers* 46, no. 2 (2018): 193.

14. Emily Payne, "Dark Skies Ordinance to Dim Pittsburgh's Light Pollution," *Carnegie Mellon University News*, September 27, 2021, cmu.edu/news/stories/archives/2021/september/light-pollution-ordinance.html.

15. An Trong Dinh Le et al., "Discomfort from Urban Scenes: Metabolic Consequences," *Landscape and Urban Planning* 160 (2017): 61–68.

16. Kristof Mollu et al., "Driving Simulator Study on the Influence of Digital Illuminated Billboards near Pedestrian Crossings," *Transportation Research Part F: Traffic Psychology and Behaviour*, 59 (2018): 45–56.

17. Malgorzata Zalesinka, "The Impact of the Luminance, Size and Location of Led Billboards on Drivers' Visual Performance-Laboratory Tests," *Accident Analysis and Prevention* 117 (August 2018): 439–448; Oscar Oviedo-Trespalacios et al., "The Impact of Road Advertising Signs on Driver Behaviour and Implications for Road Safety: A Critical Systematic Review," *Transportation Research Part A: Policy and Practice* 122 (April 2019): 85–98.

18. Jerry Wachtel, *Compendium of Recent Research Studies on Distraction from Commercial Electronic Variable Message Signs* (CEVMS) (Berkeley, CA: Veridian Group, February 2018), semanticscholar.org/paper/Compendium-of-Recent-Research-Studies-on-from-Signs-Wachtel/15d0c6b6091109aa6288c50cf8b25cf96625d2ee.

19. One candela radiating in all directions has a total flux of circa 12.57 lumens. However, 1 ANSI lumen is approximately equal to 3.426 regular lumens (American National Standards Institute). An ANSI lumen measures light reflected by a 1 candela source from a 1 m^2 area at a distance of 1 meter. It is used mostly to rate projectors. A laser pointer has a miniscule lumen value but a high candela rating since it doesn't give off much light but is visible from a great distance. And if you really want the definition of a lumen, it is 1 cd × steradians, which gets us into analytical geometry.

20. Raymond M. Soneira, "Watching Displays at Night," DisplayMate, last updated 2016, https://www.displaymate.com/Displays_At_Night_1.htm.

21. When comparing nits, screen size, and lumens, use the formula lumens = nits × screen area × π (3.1416). For screen area in square meters, multiply width by height measured in meters. The above-mentioned 500 nit 65-inch TV would have a 1.167 m^2 screen area with a lumen output of 1,833.

22. Anne-Marie Chang et al., "Evening Use of Light-Emitting eReaders Negatively Affects Sleep, Circadian Timing, and Next-Morning Alertness," *Proceedings of the National Academy of Sciences* 112, no. 4 (2014): 1232–1237, https://doi.org/10.1073/pnas.1418490112.

23. Raymond Soneira, "iPhone 7 Display Technology Shoot-Out: iPhone 6 and iPhone 7," DisplayMate, last updated 2016, https://www.displaymate.com/iPhone7_ShootOut_1.htm; Michael Herf and Lorna Herf, "Measuring the Brightest iPhone Ever: iPhone X," News, f.lux (website), last updated February 16, 2018, https://justgetflux.com/news/2018/02/16/OLED.html.

24. Soneira, "Watching Displays at Night."

25. Mariana G. Figueiro and David Pedler, "Red Light: A Novel, Non-pharmacological Intervention to Promote Alertness in Shift Workers," *Journal of Safety Research* 74 (2020): 169–177, https://www.sciencedirect.com/science/article/abs/pii/S0022437520300694?via%3Dihub.

26. Rohan Nagare et al., "Effect of White Light Devoid of 'Cyan' Spectrum Radiation on Nighttime Melatonin Suppression over a 1-h Exposure Duration," *Journal of Biological Rhythms* 34, no. 2 (2019): 195–204, https://doi.org/10.1177/0748730419830013; Mark Halper, "Worry about the Amount of Light at Night More Than the Blue Wavelengths," *LEDs Magazine*, April 15, 2019, ledsmagazine.com/manufacturing-services-testing/substrates-wafers/article/16699082/worry-about-the-amount-of-light-at-night-more-than-the-blue-wavelengths.

27. "Journi Mobile Task Light," Healthe, accessed September 28, 2020, https://web.archive.org/web/20200928155537/healthelighting.com/products/journi-mobile-task-light (site discontinued).

28. Gregory D. M. Potter, Janet E. Cade, and Laura J. Hardie, "Longer Sleep Is Associated with Lower BMI and Favorable Metabolic Profiles in UK Adults: Findings from the National Diet and Nutrition Survey," *PLoS One* 12, no. 7 (2017), e0182195, https://doi.org/10.1371/journal.pone.0182195.

29. Max Hirshkowitz et al., "National Sleep Foundation's Sleep Time Duration Recommendations: Methodology and Results Summary," *Sleep Health* 1, no. 1 (March 2015): 40–43.

30. Kathryn J. Reid et al., "Timing and Intensity of Light Correlate with Body Weight in Adults," *PLoS One* 9, no. 4 (2014), e92251, https://doi.org/10.1371/journal.pone.0092251; Edward L. Melanson, "Daytime Bright Light Exposure, Metabolism, and Individual Differences in Wake and Sleep Energy Expenditure during Circadian Entrainment and Misalignment," *Neurobiology of Sleep and Circadian Rhythms*, 4 (January 2018): 49–56, https://doi.org/10.1016/j.nbscr.2017.12.002.

31. Jiaxing Wang et al., "Progression of Myopia in School-Aged Children after COVID-19 Home Confinement," *JAMA Ophthalmology* 139, no. 3 (2021): 293–300, https://doi.org/10.1001/jamaophthalmol.2020.6239, https://www.wsj.com/tech/personal-tech/our-eyes-really-are-getting-worse-heres-how-to-save-your-kids-vision-de16d592.

32. Karla Zadnik et al., "Prediction of Juvenile-Onset Myopia," *JAMA Ophthalmology* 133, no. 6 (June 2015): 683–689; Karla Zadnik and Don Mutti, "Why Your Kids Might Be Able to See Better If They Play Outdoors More Often," *The Conversation*, September 25, 2017, https://theconversation.com/why-your-kids-might-be-able-to-see-better-if-they-play-outdoors-more-often-83693.

33. Sarah James et al., "Sleep Duration and Telomere Length in Children," *Journal of Pediatrics* 187 (August 2017): 247–252, https://doi.org/10.1016/j.jpeds.2017.05.014.

34. See, for example, "LOHAS LED Plug in Night Lights, Dusk to Dawn Sensor,Red, 0.3w," Lohas LED, https://www.lohas-led.com/red-night-lights-03w-p0187.html.

Chapter 16

1. Ortal Slobodin, Karen Frankel Heffler, and Michael Davidovitch, "Screen Media and Autism Spectrum Disorder: A Systematic Literature Review," *Journal of Developmental and Behavioral Pediatrics* 40, no. 4 (2019): 303–311; Donna Hermawati, "Early Electronic Screen Exposure and Autistic-Like Symptoms," *Intractable and Rare Diseases Research* 7, no. 1 (2018): 69–71; Himani Khanna and Puja Kapoor, "Is Excessive Electronic Screen Exposure One of the Culprits for Autism Spectrum Disorder," *The Lancet*, preprint, submitted September 12, 2018, https://papers.ssrn.com/sol3/papers.cfm?abstract_id=3218711; Ortal Slobodin, Karen Frankel Heffler, and Michael Davidovitch, "Screen Media and Autism Spectrum Disorder: A Systematic Literature Review," *Journal of Developmental & Behavioral Pediatrics* 40, no. 4 (2019): 303–311; Matthew Criscuola, "Critical Questions about Screens, AAC, and Kids aith Autism," *ASHA Leader*, August 2019, https://doi.org/10.1044/leader.FMP.24082019.8; Dimitri A. Christakis, "Early Media Exposure and Autism Spectrum Disorder: Heat and Light," *JAMA Pediatrics* 174, no. 7 (2020): 640–641, https://doi.org/10.1001/jamapediatrics.2020.0659.

2. Eric R. Kandel, *The Disordered Mind: What Unusual Brains Tell Us about Ourselves* (New York: Farrar, Strauss and Giroux, 2018).

3. Simon Baron-Cohen, *Zero Degrees of Empathy: A New Theory of Human Cruelty* (New York: Allen Lane, 2011); Florina Uzefovsky et al., "The Oxytocin Receptor Gene Predicts Brain Activity during an Emotion Recognition Task in Autism," *Molecular Autism* 10, no. 12 (2019).

4. McLeod Frampton Gwynette, Shawn S. Sidhu, and Tolga Atilla Ceranoglu, "Electronic Screen Media Use in Youth with Autism Spectrum Disorder," *Child and Adolescent Psychiatric Clinics of North America* 27, no. 2 (April 2018): 203–219, https://doi.org/10.1016/j.chc.2017.11.013.

5. Victoria L. Dunckley, "Autism and Screen Time: Special Brains, Special Risks" (blog post), Digital Media Treatment and Education Center, July 13, 2017, https://

web.archive.org/web/20210125102312/digitalmediatreatment.com/autism-screen-time-special-brains-special-risks (site discontinued).

6. Susan Greenfield, *Mind Change: How Digital Technologies Are Leaving Their Mark on Our Brains* (New York: Random House, 2015).

7. Common Sense Media, Social Media, Social Life, 2018, commonsensemedia.org/research/social-media-social-life-2018.

8. Christopher Mims, "Ask Not for Whom the Doorbell Tolls. They Won't Answer It," *Wall Street Journal*, August 28, 2017, wsj.com/articles/ask-not-for-whom-the-doorbell-tolls-they-wont-answer-it-1503864316.

9. Jenny S. Radesky, "Digital Media and Symptoms of Attention-Deficit/Hyperactivity Disorder in Adolescents," *JAMA Pediatrics* 320, no. 3 (2018): 237–239, https://doi.org/10.1001/jama.2018.8932; Chaelin K. Ra et al., "Association of Digital Media Use with Subsequent Symptoms of Attention-Deficit/Hyperactivity Disorder among Adolescents," *JAMA* 320, no. 3 (255–263), https://doi.org/10.1001/jama.2018.8931; Simon Baron-Cohen, Michael Lombardo, Helen Tager-Flusberg, *Understanding Other Minds: Perspectives from Developmental Social Neuroscience* (Oxford: Oxford University Press, 2013). See also Jenny S. Radesky, Jayna Schumacher, and Barry Zuckermann, "Mobile and Interactive Media Use by Young Children: The Good, the Bad, and the Unknown," *Pediatrics* 135, no. 1 (January 2015): 1–3, https://doi.org/10.1542/peds.2014-2251.

10. Leonard Oestreicher, written communication to author, January 24, 2016.

11. Gokul Ramaswami and Daniel H. Geschwind, "Genetics of Autism Spectrum Disorder," *Handbook of Clinical Neurology*, 147 (2018): 321–329, https://doi.org/10.1016/B978-0-444-63233-3.00021-X; Annio Posar and Paola Visconti, "To What Extent Do Environmental Factors Contribute to the Occurrence of Autism Spectrum Disorders?," *Journal of Pediatric Neuroscience* 9, no. 3 (September–December 2014): 297–298; Antonio Persico and Sara Merelli, "Environmental Factors and Autism Spectrum Disorder," in *Autism Spectrum Disorders: Phenotypes, Mechanisms and Treatments*, ed. Marion Leboyer and Pauline Chaste (Basel, Switzerland: Karger, 2015), 113–134.

12. Christakis, "Early Media Exposure and Autism Spectrum Disorder"; Karen Frankel Heffler, "Association of Early-Life Social and Digital Media Experiences with Development of Autism Spectrum Disorder–Like Symptoms," *JAMA Pediatrics* 174, no. 7 (2020): 690–696, https://doi.org/10.1001/jamapediatrics.2020.0230.

13. Hermawati, "Early Electronic Screen Exposure"; Lisa A. Ostrin, "Ocular and Systemic Melatonin and the Influence of Light Exposure," *Clinical and Experimental Optometry* 102, no. 2 (2019): 99–108.

14. Sean C. L. Deoni et al., "The COVID-19 Pandemic and Early Child Cognitive Development: A Comparison of Development in Children Born During the

Pandemic and Historical References," preprint, submitted August 16, 2022, https://doi.org/10.1101/2021.08.10.21261846; Melinda Wenner Moyer, "The COVID Generation: How Is the Pandemic Affecting Kids' Brains?," *Nature* 601, no. 7892 (2022): 180–183, https://doi.org/10.1038/d41586-022-00027-4.

15. Sudi Kate Gliebe, "The Effects of Video and Television on Young Children: Research and Reflection for Christian Educators," *Lutheran Education Journal* (August 12, 2011), https://web.archive.org/web/20210415074913/https://lej.cuchicago.edu/early-childhood-education/the-effects-of-video-and-television-on-young-children-research-and-reflection-for-christian-educators (site discontinued); Heather L. Kirkorian, "The Impact of Background Television on Parent–Child Interaction," *Child Development* 80, no. 5 (September–October 2009): 1350–1359. Elise Frank Masur, Valerie Flynn, and Janet Olson, "Infants' Background Television Exposure during Play: Negative Relations to the Quantity and Quality of Mothers' Speech and Infants' Vocabulary Acquisition," *First Language* 36, no. 2 (2016): 109–123.

16. Robert Pressman et al., "Examining the Interface of Family and Personal Traits, Media, and Academic Imperatives Using the Learning Habit Study," *American Journal of Family Therapy* 42, no. 5 (September 2014): 347–363.

17. Susan Linn, *Who's Raising the Kids?: Big Tech, Big Business, and the Lives of Children* (New York: New Press, 2022).

18. See the website for SaxeLab at https://saxelab.mit.edu.

19. Andrew N. Meltzoff and M. Keith Moore, "Newborn Infants Imitate Adult Facial Gestures," *Child Development* 54, no. 3 (June 1983): 702–709; Andrew N. Meltzoff and M. Keith Moore, "Explaining Facial Imitation: A Theoretical Model," in "Perceptual Development," special issue, *Infant and Child Development* 6, no. 3–4 (September–December 1997): 179–192; Andrew N. Meltzoff and Peter J. Marshall, "Human Infant Imitation as a Social Survival Circuit," *Current Opinion in Behavioral Sciences* 24 (December 2018): 130–136.

20. Michael Davidovich et al., "The Role of Cellular Phone Usage by Parents in the Increase in ASD Occurrence: A Hypothetical Framework," *Medical Hypotheses* 117 (August 2018): 33–36, https://doi.org/10.1016/j.mehy.2018.06.007.

21. Michael Davidovitch, Maayan Shrem, Nitzon Golovaty, Nurit Assaf, and Gideon Koren, "The Role of Cellular Phone Usage by Parents in the Increase in ASD Occurrence: A Hypothetical Framework," *Medical Hypotheses* 117 (2018): 33–36, https://doi.org/10.1016/j.mehy.2018.06.007.

22. Warren Jones and Ami Klin, "Attention to Eyes Is Present but in Decline in 2–6-Month-Old Infants Later Diagnosed with Autism," *Nature* 504, no. 7480 (2013): 427–431, https://doi.org/10.1038/nature12715.

23. Irva Hertz-Picciotto, Rebecca J. Schmidt, and Paula Krakowiak, "Understanding Environmental Contributions to Autism: Causal Concepts and the State of Science," *Autism Research* 11, no. 4 (April 2018): 554–586, https://doi.org/10.1002/aur.1938.

24. Sally J. Rogers, Geraldine Dawson, and Laurie A. Vismara, *An Early Start for Your Child with Autism: Using Everyday Activities to Help Kids Connect, Communicate, and Learn* (New York: Guilford, 2018).

25. Michael Rutter, "Developmental Catch-up, and Deficit, Following Adoption after Severe Global Early Privation," *Journal of Child Psychology and Psychiatry* 39, no. 4 (2003): 465–476; René Hoksbergen et al., "Post-Institutional Autistic Syndrome in Romanian Adoptees," *Journal of Autism and Developmental Disorders* 35, no. 5 (October 2005): 615–623; Charles A. Nelson, Charles H. Zeanah, and Nathan A. Fox, "How Early Experience Shapes Human Development: The Case of Psychosocial Deprivation," *Neural Plasticity* 2019, no. 1676285 (2019); Charles H. Zeanah et al., "Alternatives for Abandoned Children: Insights from the Bucharest Early Intervention Project," *Current Opinions in Psychology* 15 (June 2017): 128–188; Melissa Fay Greene, "30 Years Ago, Romania Deprived Thousands of Babies of Human Contact," *Atlantic*, July–August 2020, https://www.theatlantic.com/magazine/archive/2020/07/can-an-unloved-child-learn-to-love/612253.

26. Marius Teodor Zamfir, "The Consumption of Virtual Environment More than 4 Hours/Day, in Children between 0–3 Years Old, Can Cause a Syndrome Similar with the Autism Spectrum Disorder," *Journal of Romanian Literary Studies* 13 (2018): 956–969. See also Dimitri Christakis, "Understanding Media's Impact on Infants and Toddlers: A Pediatrician's Perspective" (presentation, Arthur M. Sackler Colloquium: "Digital Media and Developing Minds," Irvine, CA, October 13–16, 2015), youtu.be/HbUtvPfXt24.

27. Zamfir, "The Consumption of Virtual Environment."

28. Thomas R. Verny, *The Embodied Mind: Understanding the Mysteries of Cellular Memory, Consciousness, and Our Bodies* (New York: Pegasus, 2021).

29. Cris Rowan, "Unplug—Don't Drug: A Critical Look at the Influence of Technology on Child Behavior with an Alternative Way of Responding Other Than Evaluation and Drugging," *Ethical Human Psychology and Psychiatry* 12, no. 1 (April 2010): 60–68.

CHAPTER 17

1. Christine Soffing and Yvonne Anders (professor of early education, Free University, Berlin) in discussion with the author, Moscow, 2020.

2. Donna St. George, "Parents Investigated for Neglect after Letting Kids Walk Home Alone," *Washington Post*, January 14, 2015, https://washingtonpost.com/local/edu

cation/maryland-couple-want-free-range-kids-but-not-all-do/2015/01/14/d406c0be-9c0f-11e4-bcfb-059ec7a93ddc_story.html; Donna St. George, "'Free Range' Parents Cleared in Second Neglect Case after Kids Walked Alone," https://washingtonpost.com/local/education/free-range-parents-cleared-in-second-neglect-case-after-children-walked-alone/2015/06/22/82283c24-188c-11e5-bd7f-4611a60dd8e5_story.html.

3. Pasi Sahlberg and William Doyle, "To Really Learn, Our Children Need the Power of Play," *Wall Street Journal*, August 8, 2019, https://wsj.com/articles/to-really-learn-our-children-need-the-power-of-play-11565262002.

4. Vanessa Vietes, Shannon M. Pruden, and Bethany C. Reeb-Sutherland, "Childhood Wayfinding Experience Explains Sex and Individual Differences in Adult Wayfinding Strategy and Anxiety," *Cognitive Research: Principles and Implications* 5, no. 12 (2020), https://doi.org/10.1186/s41235-020-00220-x.

5. Francine Russo, "Loneliness Can Be Toxic," *Scientific American*, January 1, 2018, https://doi.org/10.1038/scientificamerican0118-64; Niobe Way et al., *The Crisis of Connection: Roots, Consequences, and Solutions* (New York: NYU Press, 2018).

6. William Davies, *Nervous States: How Feeling Took Over the World* (London: Jonathan Cape, 2018).

7. Jonathan Haidt and Tobias Rose-Stockwell, "The Dark Psychology of Social Networks," *Atlantic*, December 2019, https://theatlantic.com/magazine/archive/2019/12/social-media-democracy/600763; Frank Bruni, "The Real Campus Scourge," *New York Times*, September 2, 2017, https://www.nytimes.com/2017/09/02/opinion/sunday/college-freshman-mental-health.html; George F. Will, "We Have an Epidemic of Loneliness. How Can We Fix It?," *Washington Post*, October 12, 2018, https://www.washingtonpost.com/opinions/we-have-an-epidemic-of-loneliness-how-can-we-fix-it/2018/10/12/e8378a38-cd92-11e8-920f-dd52e1ae4570_story.html.

8. Anne Gulland, "Sixty Seconds on . . . Loneliness," *British Medical Journal* 360, no. 8138 (January 27, 2018): kt300, https://doi.org/10.1136/bmj.k300.

9. Ashley Fetters, "To Prevent Loneliness, Start in the Classroom," *Atlantic*, October 17, 2018, https://www.theatlantic.com/education/archive/2018/10/loneliness-prevention-schools/573280; Julianne Holt-Lundstad, "Friendship and Health," in *The Psychology of Friendship*, ed. Mahzad Hojjat and Anne Moyer (New York: Oxford University Press, 2017), 233–248; Julianne Holt-Lunstad, Theodore Robles, and David A. Sbarra, "Advancing Social Connection as a Public Health Priority in the United States," *American Psychologist* 72, no. 6 (2017): 517–530.

10. "Opioid Overdose Crisis," Opioids, National Institute on Drug Abuse, National Institutes of Health, https://web.archive.org/web/20220122212547/https://www.drugabuse.gov/drug-topics/opioids/opioid-overdose-crisis (page discontinued).

11. Rachel Wurzman, "How Isolation Fuels Opioid Addiction," TEDx video, 18:23, October 27, 2017, https://www.ted.com/talks/rachel_wurzman_how_isolation_fuels_opioid_addiction.

12. Markus Heilig et al., "Time to Connect: Bringing Social Context into Addiction Neuroscience," *Nature Reviews Neuroscience* 17, no. 9 (September 2016): 592–599, https://doi.org/10.1038/nrn.2016.67; Stephanie Cacioppo, John P. Capitano, and John T. Cacioppo, "Toward a Neurology of Loneliness," *Psychological Bulletin* 140, no. 6 (2014): 1464–1504, https://doi.org/10.1037/a0037618.

13. See, e.g., Washington Waldorf School (website), washingtonwaldorf.org.

14. Jennifer Page, interview with author, washingtonwaldorf.org.

15. Sarah P. Weeldreyer, "I Won't Buy My Teenagers Smartphones," *Atlantic*, September 11, 2019, https://www.theatlantic.com/family/archive/2019/09/i-wont-buy-my-teenagers-smartphones/597805.

16. Richard House, personal email communication to author, January 16, 2016.

17. Waldorf School of the Peninsula, n.d., waldorfpeninsula.org.

18. Nellie Bowles, "Silicon Valley Nannies Are Phone Police for Kids," *New York Times*, October 26, 2018, nytimes.com/2018/10/26/style/silicon-valley-nannies.html.

19. Anderson, quoted in Nellie Bowles, "A Dark Consensus about Screens and Kids Begins to Emerge in Silicon Valley," *New York Times*, October 26, 2018, https://www.nytimes.com/2018/10/26/style/phones-children-silicon-valley.html.

20. Nellie Bowles, "The Digital Gap between Rich and Poor Kids Is Not What We Expected," *New York Times*, October 26, 2018, https://www.nytimes.com/2018/10/26/style/digital-divide-screens-schools.html.

21. Center for Humane Technology (website), humanetech.com; Andrew Marantz, "Silicon Valley's Crisis of Conscience," *New Yorker*, August 19, 2019, https://www.newyorker.com/magazine/2019/08/26/silicon-valleys-crisis-of-conscience. See also Tristan Harris, Center for Human Technology, US Senate testimony on "Optimizing for Engagement: Understanding the Use of Persuasive Technology on Internet Platforms," June 25, 2019, https://www.youtube.com/watch?v=ZRrguMdzXBw.

22. Issie Lapowsky, "What Schools Must Learn from LA's iPad Debacle," *Wired*, May 8, 2015, https://www.wired.com/2015/05/los-angeles-edtech; Howard Blume, "U.S. Attorney Won't File Charges after Probe of $1.3-Billion iPads-for-All Project in L.A. Schools," *Los Angeles Times*, February 21, 2017, https://www.latimes.com/local/lanow/la-a-me-edu-no-charges-ipad-probe-20170221-story.html.

23. Nellie Bowles, "Silicon Valley Came to Kansas Schools. That Started a Rebellion," *New York Times*, April 21, 2019, https://www.nytimes.com/2019/04/21/technology/silicon-valley-kansas-schools.html.

24. Richard Freed (blog), https://richardfreed.com/blog. See also Screen Time Action Network, Fairplay, accessed 2019, http://screentimenetwork.org.

25. Admissions director at Pine Crest, interview with author.

26. Social worker at Anacostia, interview with author.

27. Alfredo Ardila, "Origins of Writing," in *Historical Development of Human Cognition, Perspectives in Cultural-Historical Research*, vol. 3 (Singapore: Springer Nature, 2018): 61–81; Anya Bartlett, "The Case for Cursive" (workshop, 2018 American Montessori Society Annual Conference, Denver, CO, March 24, 2018); Noella Maree Mackenzie and Rebecca Spokes, "The Why, Who, What, When and How of Handwriting Instruction," *Practical Literacy: The Early and Primary Years* 23, no. 1 (17–20).

28. Markham Heid, "Here's How Teachers Can Foil ChatGPT: Handwritten Essays," *Washington Post*, December 29, 2022, https://www.washingtonpost.com/opinions/2022/12/29/handwritten-essays-defeat-chatgpt.

29. Karen Heller, "From Punishing to Pleasurable, How Cursive Writing Is Looping Back into Our Hearts," *Washington Post*, September 2, 2018, https://www.washingtonpost.com/lifestyle/style/from-punishing-to-pleasurable-how-cursive-writing-is-looping-back-into-our-hearts/2018/08/31/aa180b9c-aa06-11e8-a8d7-0f63ab8b1370_story.html; Maria Konnikova, "What's Lost as Handwriting Fades," *New York Times*, June 2, 2014, https://www.nytimes.com/2014/06/03/science/whats-lost-as-handwriting-fades.html; Debra Bruno, "The National Archives Has Billions of Handwritten Documents. With Cursive Skills Declining, How Will We Read Them?," *Washington Post Magazine*, June 17, 2019, washingtonpost.com/lifestyle/magazine/the-national-archives-has-billions-of-handwritten-documents-with-cursive-skills-declining-how-will-we-read-them/2019/06/14/cbc44028-5c92-11e9-9625-01d48d50ef75_story.html.

30. Karin H. James and Laura Engelhardt, "The Effects of Handwriting Experience on Functional Brain Development in Pre-Literate Children," *Trends in Neuroscience Education* 1, no. 1 (December 2012): 32–42, https://doi.org/10.1016/j.tine.2012.08.001; Alyssa K. Jersey and Karin H. James, "Brain Activation Patterns Resulting from Learning Letter Forms through Active Self-Production and Passive Observation in Young Children," *Frontiers in Psychology* 4, no. 567 (2013), https://doi.org/10.3389/fpsyg.2013.00567.

31. Samuel Planton et al., "The 'Handwriting Brain': A Meta-Analysis of Neuroimaging Studies of Motor versus Orthographic Processes," *Cortex* 49, no. 10 (November–December 2013): 2772–2787, https://doi.org/10.1016/j.cortex.2013.05.011.

32. Pam A. Mueller and Daniel M. Oppenheimer, "The Pen Is Mightier Than the Keyboard: Advantages of Longhand over Laptop Note Taking," *Psychological Science* 25, no. 6: 1159–1168. https://doi.org/10.1177/0956797614524581.

33. Audrey L. H. van der Meer and F. Ruud van der Weel, "Only Three Fingers Write, but the Whole Brain Works: A High-Density EEG Study Showing Advantages of

Drawing over Typing for Learning," *Frontiers in Psychology* 8, no. 706 (2017), https://doi.org/10.3389/fpsyg.2017.00706.

34. Erika Christakis, *The Importance of Being Little: What Young Children Really Need from Grownups* (New York: Viking, 2016); Erika Christakis, "The New Preschool Is Crushing Kids," *Atlantic*, January/February 2016, https://www.theatlantic.com/magazine/archive/2016/01/the-new-preschool-is-crushing-kids/419139.

CHAPTER 18

1. Charles Duhigg, *The Power of Habit: Why We Do What We Do in Life and Business* (New York: Random House, 2014); Charles Duhigg, *Smarter Faster Better: The Transformative Power of Real Productivity* (New York: Random House, 2017); BJ Fogg, *Tiny Habits: The Small Changes That Change Everything* (Boston: Houghton Mifflin Harcourt, 2019); Josephine Tovey, "Before the Internet Broke My Attention Span I Read Books Compulsively. Now, It Takes Willpower," *Guardian* (US edition), October 2, 2019, https://www.theguardian.com/lifeandstyle/2019/oct/03/before-the-internet-broke-my-attention-span-i-read-books-compulsively-now-it-takes-willpower; Simon Doherty, "I Cured My Social Media Addiction by Reading Books," *Forge*, March 1, 2019, https://forge.medium.com/i-cured-my-social-media-addiction-by-reading-books-9376c3025079; Adam Alter, *Irresistible: The Rise of Addictive Technology and the Business of Keeping Us Hooked* (New York: Penguin, 2017); Cal Newport, *Digital Minimalism: Choosing a Focused Life in a Noisy World* (New York: Portfolio/Penguin, 2019); David T. Courtwright, *The Age of Addiction: How Bad Habits Became Big Business* (Cambridge, MA: Belknap Press of Harvard University Press, 2019); Brian Grazer, *Face to Face: The Art of Human Connection* (New York: Celadon Books, 2020); Kate Murphy, *You're Not Listening: What You're Missing and Why It Matters* (New York: Simon and Schuster, 2020).

2. Grazer, *Face to Face*.

3. Murphy, *You're Not Listening*.

4. If one fasts and restricts calories below the daily requirement, the body first burns muscle glycogen, fat, and finally muscle tissue itself to make up the difference. But the brain still consumes 20 percent of daily fuel no matter the source.

CHAPTER 19

1. Maxime Taquet et al., "6-Month Neurological and Psychiatric Outcomes in 236379 Survivors of COVID-19: A Retrospective Cohort Study Using Electronic Health Records," *Lancet Psychiatry* 8, no. 5 (May 2021): 416–427, https://doi.org/10.1016/S2215-0366(21)00084-5. The illnesses were intracranial hemorrhage; ischemic stroke; parkinsonism; Guillain-Barré syndrome; nerve, nerve root, and plexus

disorders; myoneural junction and muscle disease; encephalitis; dementia; psychotic, mood, and anxiety disorders; substance abuse; and insomnia.

2. Patrick Shanley, "Gaming Usage Up 75 Percent Amid Coronavirus Outbreak, Verizon Reports," *Hollywood Reporter*, March 17, 2020, https://www.hollywoodreporter.com/news/general-news/gaming-usage-up-75-percent-coronavirus-outbreak-verizon-reports-1285140.

3. Julie Jargon, "New Mothers Are Lonely during Coronavirus. Mom-Shaming on Social Media Is Making It Worse," *Wall Street Journal*, February 20, 2021, https://www.wsj.com/articles/new-mothers-are-lonely-during-coronavirus-mom-shaming-on-social-media-is-making-it-worse-11613829600; "Save Time Managing Your Group with Admin Assist," Facebook Community Leaders Blog, October 1, 2020, https://www.facebook.com/community/whats-new/admin-assist.

4. Cindy H. Liu, Carmina Erdei, and Leena Mittal, "Risk Factors for Depression, Anxiety, and PTSD Symptoms in Perinatal Women during the COVID-19 Pandemic," *Psychiatry Research* 295, no. 113552 (2021): https://doi.org/10.1016/j.psychres.2020.113552.

5. Richard Weissbourd et al., *Loneliness in America: How the Pandemic Has Deepened an Epidemic of Loneliness and What We Can Do About It* (Cambridge, MA: Harvard Graduate School of Education, 2021).

6. Abigail Shrier, "To Be Young and Pessimistic in America," *Wall Street Journal*, May 14, 2021, https://www.wsj.com/articles/to-be-young-and-pessimistic-in-america-11621019488; Jean M. Twenge et al., *Teens in Quarantine: Mental Health, Screen Time, and Family Connection* (Institute for Family Studies and Wheatley Institution, 2020).

7. Ryan M. Hill et al., "Suicide Ideation and Attempts in a Pediatric Emergency Department before and during COVID-19," *Pediatrics* 147, no. 3 (2021), e2020029280; Rebecca T. Leeb et al., "Mental Health–Related Emergency Department Visits among Children Aged <18 Years during the COVID-19 Pandemic—United States, January 1–October 17, 2020," *Morbidity and Mortality Weekly Report* 69, no. 45 (November 13, 2020): 1675–1680.

8. Robyn A. Cree et al., "Characteristics Associated with Presence of Pediatric Mental Health Care Policies in Emergency Departments," *Pediatric Emergency Care* 37, no. 12 (December 2021): e1116–e1121; Benedict Carey, "For Some Teens, It's Been a Year of Anxiety and Trips to the E.R.," *New York Times*, February 23, 2021, https://www.nytimes.com/2021/02/23/health/coronavirus-mental-health-teens.html.

9. Heather Kelly, "Growing Up on Screens: How a Year Lived Online Has Changed Our Children," *Washington Post*, March 5, 2021, https://www.washingtonpost.com/technology/2021/03/05/screen-time-one-year-kids.

10. Andrea Petersen, "Loneliness, Anxiety and Loss: The Covid Pandemic's Terrible Toll on Kids," *Wall Street Journal*, April 9, 2021, https://www.wsj.com/articles/pandemic-toll-children-mental-health-covid-school-11617969003.

11. Maya L. Rosen et al., "Promoting Youth Mental Health during COVID-19: A Longitudinal Study Spanning Pre- and Post-pandemic," preprint, submitted June 28, 2021, https://doi.org/10.31234/osf.io/n5h8t.

12. Marcela Lopez et al., "The Social Ecology of Childhood and Early Life Adversity," *Pediatric Research* 89, no. 2 (2021): 353–367, https://doi.org/10.1038/s41390-020-01264-x.

13. Carl F. Weems et al., "Translating the Neuroscience of Adverse Childhood Experiences to Inform Policy and Foster Population-Level Resilience," *American Psychologist* 76, no. 2 (February–March 2021): 188–202, https://doi.org/10.1037/amp0000780.

14. Ceren Hayran and Laila Anik, "Well-Being and Fear of Missing Out (FOMO) on Digital Content in the Time of COVID-19: A Correlational Analysis among University Students," in "Problematic Internet Use: A Biosocial Model," special issue, *International Journal of Environmental Research and Public Health* 18, no. 4 (2021): 1974; Erica N. Madero et al., "Environmental Distractions during Unsupervised Remote Digital Cognitive Assessment," *Journal of Prevention of Alzheimer's Disease* 8, no. 3 (July 2021): 263–266, https://doi.org/10.14283/jpad.2021.9; Dimitra Kostaki and Irene Karayianni, "Houston, We Have a Pandemic: Technical Difficulties, Distractions and Online Student Engagement," *Student Engagement in Higher Education Journal* 4, no. 2 (2022): 105–127.

15. Anjel Vahratian et al., "Symptoms of Anxiety or Depressive Disorder and Use of Mental Health Care among Adults during the COVID-19 Pandemic—United States, August 2020–February 2021," *Morbidity and Mortality Weekly Report* 70, no. 13 (April 2, 2019): 490–494, https://doi.org/10.15585/mmwr.mm7013e2.

16. Heather Kelly, "From Podcasts to Clubhouse, Audio Is Filling More of Our Time. For Some Families, That's a Problem," *Washington Post*, March 15, 2021, https://www.washingtonpost.com/technology/2021/03/15/audio-time-clubhouse-podcasts.

17. "Clubhouse May Fade. Group Voice Chat Is Here to Stay," *Economist*, April 10, 2021, 61.

18. Christian Stirling Haig, Katherine Schmidt, and Samuel Brannen, *The Age of Mass Protests: Understanding an Escalating Global Trend* (Washington, DC: Center for Strategic and International Studies, 2020), https://www.csis.org/analysis/age-mass-protests-understanding-escalating-global-trend.

19. Kwame Anthony Appiah, "YouTube Videos Brainwashed My Father. Can I Reprogram His Feed?," *New York Times Magazine*, April 20, 2021, https://www.nytimes.com/2021/04/20/magazine/youtube-radicalization.html.

20. Dan Ariely, *Misbelief: What Makes Rational People Believe Irrational Things* (New York: HarperCollins, 2023).

21. Mikołaj Kamiński, Cyntia Szymańska, and Jan Krzysztof Nowak, "Whose Tweets on COVID-19 Gain the Most Attention: Celebrities, Political, or Scientific Authorities?," *Cyberpsychology, Behavior, and Social Networking* 24, no. 2 (February 2021): 123–128, https://doi.org/10.1089/cyber.2020.0336.

22. Richard E. Cytowic, "Escape from Zoom Fatigue, and What to Do about It," *The Fallible Mind* (blog), *Psychology Today*, May 23, 2020, https://www.psychologytoday.com/us/blog/the-fallible-mind/202005/escape-zoom-fatigue-and-what-do-about-it.

23. My Verizon fiber-optic connection shoots data into my home at 8,589,934,592 bits per second, about 71,600,000 times the rate my gray matter can handle. At most, a nerve cell can discharge electrical spikes 1,000 times per second down one nerve axon to another cell. The fastest a signal can cross a synaptic gap is about 1 millisecond. Considering both numbers, the brain can therefore carry out a maximum of 1,000 operations per second, 10 million times slower than an old laptop.

24. Geraldine Fauville et al., "Nonverbal Mechanisms Predict Zoom Fatigue and Explain Why Women Experience Higher Levels Than Men," preprint, submitted April 14, 2021, https://doi.org/10.2139/ssrn.3820035.

25. Anne Branigin, "Has Zoom Changed How You See Yourself? 6 Women Share Their Stories," *Lily*, March 4, 2021, https://www.thelily.com/has-zoom-changed-how-you-see-yourself-5-women-share-their-stories.

26. See the website for Stanford University's Virtual Human Interaction Lab at https://stanfordvr.com. Vignesh Ramachandran, "Stanford Researchers Identify Four Causes for 'Zoom Fatigue' and Their Simple Fixes," *Stanford News*, February 23, 2021, https://news.stanford.edu/2021/02/23/four-causes-zoom-fatigue-solutions; Shauna M. Rice et al., "Zooming into Cosmetic Procedures during the COVID-19 Pandemic: The Provider's Perspective," *International Journal of Women's Dermatology* 7, no. 2 (March 2021): 213–216, https://doi.org/10.1016/j.ijwd.2021.01.012.

27. Mario Luis Small, *Someone to Talk To* (New York: Oxford University Press, 2017).

28. Susan Orlean (@susanorlean), "Good morning to everyone but especially to the sentence I just rewrote for the tenth time.," Twitter, March 23, 2021, 7:22 p.m., https://twitter.com/susanorlean/status/1374441410569535495.

29. Sarah Lyall, "We Have All Hit a Wall," *New York Times*, April 3, 2021, https://www.nytimes.com/2021/04/03/business/pandemic-burnout-productivity.html.

30. Paulina Firozi, "Yo-Yo Ma Played a Surprise Concert for a Clinic during His Post-Vaccination Waiting Period," *Washington Post*, March 14, 2021, https://www.washingtonpost.com/lifestyle/2021/03/14/yo-yo-ma-vaccine-concert.

31. Pete Buttigieg, "Pete Buttigieg on the Pandemic Year: How Little We Communicate with Words," *Wall Street Journal*, March 20, 2021, https://www.wsj.com/articles/pete-buttigieg-on-the-pandemic-year-how-little-we-communicate-with-words-11616212861.

32. Parmy Olson and Tripp Mickle, "Google to Invest $7 Billion in Bet on Post-Pandemic Office," *Wall Street Journal*, March 18, 2021, https://www.wsj.com/articles/google-invests-7-billion-across-u-s-as-it-bets-on-the-post-pandemic-office-11616069081.

Appendix

1. Ira Progoff, *At a Journal Workshop: Writing to Access the Power of the Unconscious and Evoke Creative Ability* (New York: Jeremy P. Tarcher/Putnam, 1992).

Index

Note: Page numbers appearing in italics indicate illustrations.

Aaker, Jennifer, 120
Academy Awards (the Oscars), 18–19, 45, 189
Accretion, principle of, 42
Addiction, 54–56
   and dopamine, 155, 157
Addictum, xviii, 154
Advanced Baby Imaging Lab (Brown University Medical School), 99
Affectiva, 163
Affective Computing (Picard), 92
Age of Surveillance Capitalism, The (Zuboff), 10
Ali, Muhammed and Parkinsonism, 211
Alien hand syndrome, 25
Alliance for Childhood, 172–173
All-Party Parliamentary Group on a Fit and Healthy Childhood (UK), 159–160
Alzheimer's disease (AD), 76, 127–129
Amazon, 160
Amazon Prime, 11
American Academy of Child and Adolescent Psychiatry, 214
American Academy of Facial Plastic and Reconstructive Surgery (AAFPRS), 21
American Academy of Pediatrics (AAP), 3–4, 120, 130, 159, 170–171, 188, 222
American Bandstand (TV show), 214
American Interest (Garfinkle), 8, 177
American Medical Association, 199–200
American Psychiatric Association, 184
American Psychological Association (APA), 88–89
American Society of Plastic Surgeons, 254
American University, 87
Anacostia High School (Washington, D.C.), 228
Anderson, Chris, 225–226
Andromeda Strain, The (Crichton), 8
Angry Birds (game), 181–182
Antarctica, 137
Apollo program, 41
Apple, 11, 12, 162, 170, 174, 227. *See also* iPad, iPhone
   Night Shift setting, 133, 203–204
   Screen Time, 162
Apps, and behaviors, 162
App Store (Apple), 133, 188
Apptivity Seat (Fisher-Price), 100, 112, 113
Ariely, Dan, 251
Aristotle, 118, 142
Arnold-Foster, Mary, 260

Arpanet, 186
Artificial intelligence (AI), and data about users, 9–11
Asperger disorder, 219
Association of Nature and Forest Therapy, 71, 72
Atlantic, The (magazine), 15, 79, 160, 176, 237, 238
Attendere, 119–120
Attention and attention spans, 1, 2, 9, 13, 28, 75, 240
  alternating, 29
  biological limits of (capacity), 16, 148, 197
  and burnout, 254
  in children and infants, 160, 161, 170, 177, 181, 187–188, 217, 220
  and context, 82
  and distractions, 31, 46, 95, 97, 100, 121, 250
  Dutch *niksen*, 184
  focused, 248–252
  and "forced viewing," 200
  fragmenting of, 12
  and Googlized mind, 189
  joint, 217
  and learning, 59
  and lighting, 194
  and metacognition, 106
  of parents, 169
  rapid shifts of, 32, 96–97, 190
  reading and, 62, 190
  and reinforcement, 145
  and relationships, 34
  searchlight hypothesis of, 29–30
  seeking, 17
  selective, 29, 50
  and shallow brain, 186
  sustained, 29
  switching, 84, 105, 289
  visual, 178–179
  and working memory, 19
Attention deficit disorder, 213

Autism spectrum disorder (ASD), 32, 211–218
  virtual autism, xvi, 218–220
Autism Spectrum Foundation, 212
Automobiles, and distractions, 23, 61
Avatar (film), 240
Awake and Alert bulbs (Phillips), 134
Awarē, 68

Bailenson, Jeremy, 254
Ballyhoo Media, 197
Bankman-Fried, Sam, 63
Barnaby Rudge (Dickens), 125
Baron-Cohen, Simon, 212
Beatty, Warren, 18
"Beauty and the Banana" (Disney-Dole), 161
Bernard (fictional biographer), 73
Berns, Gregory, 151
Berridge, Kent, 156–158, 163–164
Beyond Freedom and Dignity (Skinner), 15
Biden, Joe, 250
Bing, 10
Binge watching, 11
Binocular rivalry, 101
Biophilia (Wilson), 72
Birkbeck Centre for Brain and Cognitive Development, 177
Blackberry phone, 229
Blackboard, 174
Black Lives Matter demonstrations, 251
Blade Runner 2049 (film), 195, 199
Bluemink, Matt, 119
B-19 (patient), 144
Body dysmorphic disorder, 21
Bolshoi Theatre, 193–194
Books, 7, 11, 63, 70
Borg, the, xiv, 66
Bosker, Bianca, 79–80
Boston Public Garden, 76
Boston University School of Medicine, 21
Bowles, Nellie, 225–226

"Boy Detective, The" (Rosenblatt), 73
Braasch, Jonas, 74
Bradbury, Ray, 7
Braille, 187
Brain
    energy use by, xv–xvi, 48–50
    evolution of, 39–44
    and neural networks, 48, 140, 144, 147, 170, 217
    Stone Age, xv
Brain and Cognition (journal), 83
Brain and Language (journal), 83
Brand, Matthias, 55
Brave New World (Huxley), 71
Brenner, Andrew, 34–35
Brickman, Philip, 141
Brin, David, 106
Broca's (brain) area, *230*
Brown, Ramsay, 87
Bruni, Frank, 223
Brussel, Gabrielle, 197–198, 200
Bucharest Early Intervention Project (BEIP), 118–119
Buddha, the, 150
Buddhism, 71
Buffett, Warren, 151
Bunn, Elizabeth, 229
Burgess, Harry, 77
Burnout, 254
Buttigieg, Pete, 255
Buxton, Rachel, 79

Cain, Susan, 151
Cambridge University, 134
Campaign for a Commercial-Free Childhood, 112, 162
Campbell, Donald T., 141
Candy Crush (game), 14
Canterbury Tales, The (Chaucer), 125
Cardoso, Tiago, 77
Carnegie Mellon University, 87
Carr, Nicholas, 181, 190
Cash, Hilarie, 184

"Celebrate 90 Years of Mickey Mouse with Dole," 160–161
Celling Your Soul (film), 185–186
Cellphones
    and distracted walking and driving, 22–24
    and selfie deaths, 17–22
Center for Humane Technology, 162, 226
Center for Internet and Technology Addiction, 32–33
Center for Strategic and International Studies, 250
Centers for Disease Control (CDC), 248, 250
Central Park (New York), 76
Change blindness, 46
Charlton, Bruce, 106–107
ChatGPT, 10, 91
Chaucer, Geoffrey, 125
Chelsea Flower Show, 62
Chernobyl, 125
Child Protective Services (CPS), 222
Children, and smart screens, 3–6
Children's Digital Media Center (UCLA), 185
Children's Screen Time Action Network, 228
China and the Chinese Communist Party, 182–183
Christakis, Erika, 176–177, 232–233
City lights, Earth's, *196*
Clayton, Russell, 102
Clement, Joe, 114–115, 176
Clinton, Hillary, 72
Clubhouse (app), 250
Cognition and Emotion Lab (Florida State University), 102
Cohen, Richard, 66
Colorado State University, Department of Fish, Wildlife, and Conservation Biology, 79
Common Sense Media, 171, 213, 226

Comparison, of self to others, 20
Computer Power and Human Reason (Weizenbaum), 91
Context, 32–33, 82–83
    and screen time, 171–172
    social, 47
CoolSculpting, 21–22
COVID-19 pandemic, 89, 207, 209, 233
    lessons from the lockdown years, 245–255
Crawford, Matthew, 60
Crichton, Michael, 8–9
Critical Institute, The, 167
Critical Thinking (Haber), 59
Crowne Plaza hotels, 136–137
Cruise, Tom, 100, 199
Csikszentmihalyi, Mihaly, 60
Cullinan, Brian, 18–19, 45
Cursive writing, 228–233

Danbury Museum and Historical Society, 229
Daniels, Mitch, 30
Darkness Manifesto, The (Eklov), 195
Dark sky ordinance, 200
Darpanet, 186
Darwin, Charles, 108
Davidovich, Michael, 215–216, 218
Davies, William, 223
Davis, Bette, 153
Daxing boot camp, 182–183
De Brevitate Vitae (Seneca), 70
Decaux, Jean-Claude, 197
Declaration of Independence, 141
Defense Meteorological Satellite Program, 196
Dell, 174
Dennis-Tiwari, Tracy, 169
Der Zauberlehrling (Goethe), 15
Detroit, Michigan, 222
Development illness, and myelination, 53

Diagnostic and Statistical Manual of Mental Disorders (DSM-5), 32, 140
DiCaprio, Leonardo, 18
Dickens, Charles, 125
Digital dementia, 58
"Digital Life and Youth Well-Being, Social Connectedness, Empathy, and Narcissism" (AAP), 120
Disney, 160–161
DisplayMate Technologies, 202
Distraction-industrial complex, 60
Distractions, 45–46. *See also* Attention and attention spans
"Doctor Kids" (app) 160
Dopamine, xii–xiii, 54, 94, 140, 144, 148–149, 156, 241
    and addiction, 155, 157
    and Parkinson's disease, 211
    and sunlight, 207
Dopamine-addiction feedback loop, 2–3, 22
Dopamine fasting, 30
Dopaminergic network, 147
Dopamine, Inc., 87
Dole Foods, 160–161
Down syndrome, 167
Dream diary, 259–263
Drexel Autism Institute, 214
Drexel University, 87
Dr. Strangelove (film), 25
Duchenne, Guillaume, 109, *110*
Duchenne smile, 108, *109*
Duke University, 66
Dunaway, Faye, 18–19
Duncan, Arne, 175
Dunckley, Victoria, 212
Dupont Circle (Washington, D.C.), 76

Eagleman, David, 59, 154
Earl (a millennial New Yorker), 197, 199
Early Start (Denver), 218
Economist (magazine), 238

Edison, Thomas A., 123
Ed Sullivan (TV show), 51
Educational Technologists Limited, 174
Eight Sleep cooling pad, 208
Ekirch, Roger, 125
Eklof, Johan, 195
ELIZA (demonstration program), 91–92
ELLE (magazine), 181
Elliot, Charlene, 161
Emergency departments (EDs), 247
Emotional attachments, 114, 168, 183, 241
Emotion and emotions, xvii, 20, 33, 89, 97–102
  and change, 84
  development, 212
  and faces, 111
  feedback, 218
  immaturity, 215
  intelligence and temperament, 117, 119, 173, 177
  and intentions, 113
  valence, 91
ENIAC (Electronic Numerical Integrator and Computer), 40–41
Evans, Arthur, Jr., 88–89
Expression of the Emotions in Man and Animals, The (Darwin), 108
Exxon Valdez oil spill, 125
Eyes, and vision, 95–101

Face, human, *107*, 108–111
Facebook, 9, 11, 88, 142, 227, 246
"Facebook Files, The" (*Wall Street Journal*), 12, 88
FaceTime, 51
FaceTune (app), 21
Facial Plastic Surgery (journal), 19–20
Fahrenheit 451 (Bradbury), 7
Fahrney's Pens, 229
"Fallible Mind, The" (magazine column), 2
Fan, Larry, 21

Fear of missing out. *See* FOMO
Fiebelkorn, Ian, 30
Fifty Shades of Grey (James), 181
Figueiro, Mariana, 136, 137, 204–205
Finland, 222
Fischer, Adam, 63
Fisher-Price, 100, 112
Flinders University, 133
Flow, 60
Floyd, George, 251
f.lux (software), 133
FOMO (fear of missing out), 33, 35, 100, 101, 102, 103, 142, 223, 240
"Forced viewing," 194–200
Forest Therapy Society, 71
Fox, Jesse, 22
France, 186, 222
Francis of Assisi, 150
Franklin, Benjamin, 208
Freed, Richard, 88, 228
Fremont Street Experience (Las Vegas), 195
Frobel, Friedrich, 220, 224
Future Shock (Toffler), 27

Gamification, 87
Garfinkle, Adam, 8
GeekDad.com, 226
Genie (American girl), 47–48, 112
Gen-Z, 250
Geography of Bliss, The (Weiner), 141
Georgetown University, 66
George Washington University, 11, 46, 92
German Aerospace Center, 203
Gessell Institute for Human Development, 232
Ghosh, Arko, 31–32, 144, 155, 189
Glitter Gulch (Las Vegas), 195
Goethe, Johann Wolfgang von, 15, 72
Golden Gate Park, 76
Golin, Josh, 162
Good Night light bulbs, 134

Google, 9, 11, 12, 63, 101, 116, 159, 162, 174, 181, 190, 242, 255
  and Time Well Spent (Center for Humane Technology), 226
Google Glass, 101
Google Play, 160
Googlized mind, 181–182, 189
Gosling, Ryan, 195, 199
Governors Highway Safety Association, 23
Grand Canyon National Park, 79
Grazer, Brian, 238
Greenfield, Susan, 13, 51, 182, 212
Guinness World Records, 130
Guy Montag (fictional character), 7

Haber, Jonathan, 59
Hamrah, Daria, 21
Hạnh, Thích Nhất, 70–71
Happiness, 149
Happy to Chat benches, 76, 78
Harlow, Harry, 113–116, 182–183
Harris, Tristan, 162, 226
Harvard Business Review, 58
Harvard Medical School, 203
Harvard University, 247
Hastings, Reed, 11
Haugen, Frances, 88
HBO, 184
HBO Max, 11
Healthe bedside lamps, 137
Heath, Robert, 144
Hedonic hotspots, 164
Hedonic treadmill and hedonic desires, 141, 142, 149, 151, 241
Heffler, Karen, 214, 218
Hein, A., 93
Held, R., 93
Henreid, Paul, 153
Her (film), 92
Hill, David, 4
Hobson, Allan, 262

Holberton, Betty Snyder, 40
Holt-Lunstad, Julianne, 223–224
Homeostasis, 49–50, 105
Hong Kong, 195
House, Richard, 167, 172, 174, 225
Household Pulse Survey (CDC), 250
"How the New Preschool is Crushing Kids" (Christakis), 176
Hubel, David, 117–118
Hulu, 11
"Human Downgrading" (Harris), 226
Huxley, Aldous, 71

IBM, 22, 30
IdeaLight, 77
"iGen" and iGeners, 247, 250
Imagination, undeveloped, 218–220
Importance of Being Little, The (Christakis), 176, 232–233
Inattentional blindness, 46
Infinite scroll and scrolling, 67, 87, 139, 213, 232
In Praise of Wasting Time (Lightman), 7, 164–165
Instagram, 246
Institute for Family Studies, 247
Institute of Neuroinformatics, 31
Intensive Journal (method), 263
InterContinental Hotels Group, 137
International Classification of Diseases (ICD-11), 54–55, 184
International Space Station, 205
Interrupt function, 84–85
Inuit people, 126–127
iPad, 6, 100
  in nursery, 167–179
iPhone, 6, 41, 53, 74, 131, 167, 248–249
iPotty, 100, 113
"Is Google Making Us Stupid?" (Carr), 181
ISIS, beheading viewed by second grader, 57

INDEX 327

JAMA Pediatrics (journal), 118–119, 214
Jargon, Julie, 246–247
JCDecaux, 195, 198, 199
Jefferson, Thomas, 141
Jeopardy! (game show), 30
Jobs, Steve, 19
Johnny Mnemonic (film character), 46
Joi (fictional character), 195, 197
Journal of the American Medical Association (JAMA), 153, 213
JOURNI circadian lamp, 206
Jung, Carl, 259
Jurassic Park (Crichton), 8

Kael, Pauline, 260
Kagge, Erling, 80
Kaiser Family Foundation, 214–215
Kanai, Ryota, 186–187, 189–190
Kardashian, Kim, 21
Karl Marx Monument, 194
Karmiloff-Smith, Annette, 159, 167, 177
Kawabata, Yasunari, 69
Kekule, August, 190
Keller, Helen, 220
Kindergarten, 221
Kindertagesstätte, 221
Kindle, 203
Kitafahrten, 222, 233
Kittinger-Sereinig, Rupert, 77
Klass, Perri, 135
Kleitman, Nathaniel, 261–262
Kline, Greg, 184
Kluge: The Haphazard Evolution of the Human Mind (Marcus), 82
Koffka, Kurt, 67
Koren, Leonard, 68
Krishnamurti, Jiddu, 143
Krizan, Zlatan, 18
Kuhl, Patricia, 93

La chasse au bonheur ("the hunt for happiness"), 141

La La Land (film), 18–19
Lamb Chop (puppet), 99
Lancet Psychiatry, (journal), 245–246
Lanier, Jaron, 10, 155
Lassie (TV show), 51
Las Vegas, 195
Learning, 59–63
 and cursive writing, 228–233
 passive, 92–93
 social, 221–224
LED screens, 74–76
Le Guin, Ursula, 72
Leonardo da Vinci, 259
Levitin, Daniel, 2
Lewis, Shari, 99
Life in the Dark, A (Kael), 260
Lighting Research Center (LRC), 204–205
Lighting Science, 133–134
Lightman, Alan, 7, 164–165
Lights and lighting, 194
 blue light and sleep, 130–137
 in cities, 195–196
 in Moscow, 193–194
Limbic system, 89–90
LinkNYC, 198
Loh, Kep Kee, 186–187, 189–190
Loneliness, 69, 70, 223, 224, 241
Loot boxes, 148
Los Angeles Unified School District, 227
Lubyanka Square (Moscow), 194
Lucy and Limitless (film characters), 47
"Lusting While Loathing" (experiment), 155–156

Ma, Yo-Yo, 255
Maly Theatre (Moscow), 194
Manjoo, Farhad, 162
Marcus, Gary, 82
Marcus Aurelius, 101

Marcus Autism Center (Emory University), 217–218
Mark, Gloria, 28–29
Marriott Royal Aurora (Moscow), 193
Marshall, Eliot, 119
Matrix (films), 59, 232
Matrix Resurrections, The (film), xv
Max (author's grandnephew), 97
McGurk illusion, 179
McLuhan, Marshall, 61
Medicare beneficiaries, 76
Meitiv, Alex, 222–223
Meitiv, Danielle, 222–223
Meltzer, Marlyn Wescoff, *41*
Merton, Thomas, 150
Meta, 12, 88, 159
Miami, 196–197
Michigan, Upper Peninsula of, 222
Microsoft, 11, 174
Microsoft Research Canada, 28–29
Miles, Matt, 114–115, 176, 182–183
Miller, Benjamin, 229
Miller, David, 229
Mims, Christopher, 60
Mind Change (Greenfield), 212
Minority Report (film), 100–101, 199
Misbelief (Ariely), 251
Mister Rogers' Neighborhood (TV show), 170
MIT, 87
Mittimatalik (Inuit settlement), 126
Moleskine notebooks, 69
Molt, Emil, 224
Montaigne, Michael, 73, 101–102, 103
Moonlight (film), 19
Moore, Alan, 92
Moore's law, 41
Moscow, 193–194
Mounjaro, 35, 54
Multitasking, 44–45. *See also* Attention and attention spans
Musk, Elon, 170

Myelin and myelination, 52–53, 241
Myruski, Sarah, 168

Naclario, Mike, 198
NASA, 41
Nass, Clifford, 44, 51, 119
National Bureau of Economic Research, 89
National Education Technology Plan, 175
National Geographic, 23
National Institutes of Health (NIH), 4, 211–212
National Merit finalists, 175
National Park Service, 22, 195
National Safety Council, 22–23
National Science Foundation, 137
National Sleep Foundation, 206
National Toxicology Program, 137
National Transportation and Safety Board (NTSB), 34
Naturliebe, 72, 221
Na'vi (fictional beings), 240
NBC, 34
Nelson, Charles, 219
Neotony, 105–106, 108
Nervous States: How Feeling Took Over the World (Davies), 223
Nest, 213
Netflix, 9, 11, 74, 184
Newport, Cal, 44, 66
New York City, 195–196
  Metro Transportation Authority, 198
New Yorker (magazine), 79, 238
New York Times (newspaper), 237, 254
New York University, 87
Nietzsche, Friedrich, 73, 229
Nikolskaya Ulitsa (Moscow), 194
Niksen, 184
No Child Left Behind, 233
Noise and noise pollution, 79–80
Nook, 203

INDEX 329

Novelty, xii, xiv, 53, 57, 82–83, 139–140, 158, 200, 252
Now, Voyager (film), 153
Nursery World (British publication), 167, 225

Obama, Barack, 175–176
Odysseus, 139
Oestreicher, Leonard, 213, 214, 218
O'Hara, Frank, 73
Ohio State University, 22
Olds, James, 144
Olmsted, Frederick Law, 76
O'Neill, Cathy, 175
O'Neill, John, 134
Orefice, Lauren, 32
Orienting reflex, 81–82
Orlean, Susan, 254
Orphans, Romanian, 118–119, 218–219
Ostrin, Lisa, 133
Outdoor Advertising Association of America, 197
OUTFRONT, 198
"Out-of-home" (OOH) advertising, 197, 199
Owens, Judith, 135
Oxford English Dictionary, 17, 149
Ozempic, 54

Page, Jennifer, 225
Pali Institute, 185
Pandora, 184
Parker (30-month-old with issues looking at screens), 4, 5
Parkinson, James, 211
Parkinson's disease, 76, 211
Pascal, Blaise, 164
Pattern context, 83
Pavlov, Ivan, 147
Pavlov's dogs, 145–147, 164. *See also* Reinforcement
Pearson, 227
Pediatric Research (journal), 58

Perseveration, 84
Persuasive Tech Lab (Stanford University), 87
Petrovka Ulitsa (Moscow), 193
Pew Research Center, 57
Philadelphia Flower Show, 62
Phillips, 133–134
Picard, Rosalind, 92
Piedmont Park (Atlanta), 76
Pied Pipers of Autism, The (Oestreicher), 213
Pine Crest (Fort Lauderdale), 228
Pinker, Steven, 81–82, 111
Pinsker, Joe, 160
Pinterest, 161
Pittsburgh, Pennsylvania, 200
"Place memory," 232
Plasticity, 111–112
Pleasure, 140
Pocket Universe (app), 209
Poincaré, Henri, 190
Pokémon GO, 70, 181–182
Positive intermittent reinforcement, 14, 93–94, 146, 147–148, 151, 155, 186, 213
Present context, 83
PricewaterhouseCoopers, 18
Progoff, Ira, 259, 263
ProPublica, 22
Psychology Today (magazine), 1, 2, 51, 150
Purdue University, 184

Quiet (Cain), 151

Radesky, Jenny, 213
Rapid eye movement (REM) sleep, 261–262
Rationality (Pinker), 81–82
Reader Come Home (Wolf), 2
Reading, 62–63
by candlelight, 208
from tablets, 201–203

Reagle, Joseph, 103
Red Square (Moscow), 194
Reinforcement, 93, 146, 147, 194.
    See also Positive intermittent reinforcement
Rensselaer Polytechnic University, 87
Resonance Consortium, 99
ReSTART, 184
Re-Timer (device), 133, 209
Revenge of Analog: Real Things and Why They Matter, The (Sax), 69
Reynard, Alex, 112
Right to a Quiet Society, The, 79–80
Ring, 213
Romania, Socialist Republic of, 118–119
Rosen, Larry, 102–103, 145
Rosenblatt, Roger, 73
Rosenstein, Justin, 19
Ruiz, Martha, 18
Rutter, Michael, 219

Sabbath (digital), 184
Sabi, 67–68
Sajak, Pat, 190–191
Salience, 33, 82, 85, 89, 90
Sasse, Ben, 2
Sax, David, 69
Saxe, Rebecca, 216
Sbarra, Joseph, 33–34
Schmidt, Eric, 63
Science (journal), 119
Screen Schooled (Miles and Clement), 114, 176
Searching for Stars on an Island in Maine (Lightman), 164–165
Searchlight hypothesis, of attention, 29–30
Seattle, "summer of love," 251
Second Life (game), 61
Seib, Gerald, 250
Selfie deaths and injuries, 17–22
Sellers, Peter, 25
Selye, Hans, 27

Seneca, 59, 70
Sequoia Capital, 63
Serene Reflection School of Buddhism, 67
Sesame Street (TV show), 51, 170
Shallow learning, 190, 242
Shallow mind, 181, 186, 190
Shallows: What the Internet Is Doing to Our Brains, The (Carr), 181, 190
Shibui, 68
Shirky, Clay, 57–58, 90, 162
Siddhārtha Gautama (prince), 150
Sigman, Aric, 157, 159
Silence, 65–69
    forest bathing, 71–73
    and loneliness, 69–71
    and walking, 73
Silence: The Power of Quiet in a World Full of Noise (Hanh), 70–71
Silence in the Age of Noise (Kagge), 80
Silicon Valley, 226
Sisyphus, 141
"Sketch of the Past, A" (Woolf), 73
Skinner, B. F., 15
Skype, 51
Sleep, 123–124
    and blue light, 130–137
    in children and infants, 129–130
    normal, 124–129
Sleep pressure, 127
Sleepy Baby light bulbs, 134
Small, Mario, 254
Snapchat, 162
Snow Country (Kawabata), 69
Soma (fictional drug), 71
Someone to Talk To (Small), 254
Soneira, Raymond, 202, 204
"Sorcerer's Apprentice, The" (Goethe), 15
South Korea, 183
Spence, Charles, 70
Spirited Away (film), 68–69
Sputnik, 186

"Squire's Tale, The" (Chaucer), 125
SSRIs, 154
Stand Together and Rethink Technology (START), 227
Stanford University, 19, 87, 139, 155–156, 253
  Virtual Human Interaction Lab, 254
Starlink satellites, 170
Star Trek (TV series), 66
Star Wars (film series), 69
Steiner, Rudolph, 224
Stern, Isaac, 67
Steyer, Jim, 162
Stone, Emma, 18–19
Stone Age brain, xv
St. Regis Nikolskaya, 193, 194
Strip, the (Las Vegas), 195
Studies in Dreams (Arnold-Foster), 260
Styron, William, 62–63
Sullivan, Annie, 220
Summit Learning, 227
Synesthesia, 2–3
Syracuse University, 66

Tantalus, 164
Tao Ran, 182–183
Teatralny Proyezd, 194
Technoference, 17, 33–34, 58, 34
Technology, unintended effects of, 8–9
TED talks, 51
Teitelbaum, Ruth Lichterman, *41*
Tempe (Arizona) Police Department, 34
Ten Arguments for Deleting Your Social Media Accounts Right Now (Lanier), 155
Thoreau, Henry David, 72, 73
3D Robotics, 226
Three Poles Challenge, 80
TikTok, 7, 21, 183, 250
Times Square (NYC), 196
Time Well Spent. *See* Center for Humane Technology
Toffler, Alvin, 27

Tokyo Story (film), 68–69
Too Much, Too Soon? Early Learning and the Erosion of Childhood (House, ed.), 167
Torey, Zoltan, 105–106
To the Lighthouse (Woolf), 73
TracFones, 53
True North (successor to Watson), 30
Tufte, Edward, 188
Tulane University, Department of Neurology, 144
TV, nighttime, 202–203
Twenge, Jean, 247, 250
*Twilight of the Idols (*Nietzsche), 73
Twitter, 250

Uber, 10, 34–35
Umwelt, 86–87
Understanding Media (McLuhan), 61
United Nations, World Happiness Report, 141
United States
  Department of Education, 175
  Department of Health and Human Services, 137
  Department of Transportation, 17
  Environmental Protection Agency, 80
  and gaming addiction, 183–184
  Navy, 137
  Securities and Exchange Commission, 88
  Surgeon General, 153
University of California, 113
University College London, 59
University of Houston, 133
University of Michigan Medical School, Department of Pediatrics, 160
University of Pennsylvania, 40
University of Utah, 87
"Using Eye Tracking to Understand the Reality of Attention to Advertising Across Media," 198

Utah, 227–228
Utah Valley University, 23

Vasquez, Rafaela, 34–35
Ventura, Ana, 77
Verizon, 43, 246
V for Vendetta (Moore), 92
Victor of Aveyron, 46
Visual overload, 27–28
Voice, The (TV show), 34

Wabi, 68
Wabi-sabi, 68–69
Walden (Thoreau), 72
Waldorf Education, 224–225, 232
Waldorf School of the Peninsula, 225, 233
"Walking" (Thoreau), 73
Walkman, 238
Wallace, David Foster, 120
Wall-E (film), 185
Wall Street Journal (newspaper), 12, 88, 237, 250
Walt Disney Company, 160–161
Wanting, 140–141
Ward, Adrian, 162
WarGames (film), 235
Washington Post (newspaper), 237
Watson (supercomputer), 30
Waves, The (Woolf), 73
Wayback Machine, 174
Weapons of Math Destruction (O'Neill), 175
Web Junkie (documentary), 182–183
Wegovy, 35
Wehr, Thomas, 125
Weiner, Eric, 141, 149
Weizenbaum, Joseph, 91–92
Wernicke's (brain) area, *230*
West, Kanye, 63
Westworld (Crichton), 8
Whitaker, Harry, 83
Whitman, Walt, 73

"Why Everything Is Getting Louder" (Bosker), 79–80
Wiesel, Torstein, 117–118
Wilbrecht, Linda, 120, 121
Wildlife World Zoo, 22
Wilkins, Arnold, 74, 75–76
Wilson, E. O., 72
Wilson, Timothy, 70
Wired Child: Reclaiming Childhood in a Digital Age (Freed), 88, 228
Wolf, Maryanne, 2
Wood, Evelyn, 59
Woolf, Virginia, 73
Wordsworth, William, 73
World beyond Your Head, The (Crawford), 60
World Health Organization (WHO), 4, 160, 184
World without Email, A (Newport), 66
Wurzman, Rachel, 224

Ye (Kanye West), 63
Yellowstone National Park, 22
"Your Brain on Screens" (Cytowic), 177
YouTube, 7, 9, 11, 46, 179, 249–250, 251
Yugen, 68

Zamfir, Marius, 218–219
Zarr, Robert, 76
Zeitgeber, 126, 134
Zen, 150
  Sōtō, 67
Zicherman, Gabe, 87
Zoom and Exhaustion Fatigue Scale, 253
Zoom and Zoom sessions, 223, 246, 252–253
Zuboff, Shoshana, 10
Zuckerberg, Mark, 12, 63, 66, 88